Straight Motion of Road Vehicles

Straight Motion of Road Vehicles

ALESSANDRO LATTUADA
GIAMPIERO MASTINU
GIUSEPPE MATRASCIA

SAE INTERNATIONAL®

Warrendale, Pennsylvania, USA

400 Commonwealth Drive
Warrendale, PA 15096-0001 USA
E-mail: CustomerService@sae.org
Phone: 877-606-7323 (inside USA and Canada)
724-776-4970 (outside USA)
Fax 724-776-0790

Library of Congress Catalog Number 2020930090
http://dx.doi.org/10.4271/9781468601312

ISBN-Print 978-1-4686-0129-9

To purchase bulk quantities, please contact: SAE Customer Service

E-mail: CustomerService@sae.org
Phone: 877-606-7323 (inside USA and Canada)
724-776-4970 (outside USA)
Fax: 724-776-0790

Visit the SAE International Bookstore at books.sae.org

Chief Product Officer
Frank Menchaca

Publisher
Sherry Dickinson Nigam

Director of Content Management
Kelli Zilko

Production Associate
Erin Mendicino

Manufacturing Associate
Adam Goebel

dedication

To my family and everyone who has supported me along this (straight) road.
Especially Agnese, who ended up knowing most of the pages of this book by heart, I'm afraid.

—Alessandro

To Nettie and Hans Pacejka, remembering the nice times spent together during Simposia around the world and in Italy.

—Giampiero

To Paolo Bandel, for his teaching.

—Giuseppe

contents

preface

The book deals with the straight motion of road vehicles, with particular focus on the role played by tires, vehicle suspensions, and road cross slope. The aim is to describe in a mathematical, analytical way the phenomenon of straight ahead running of vehicles and the driver's reactions to achieve this. Since the straight motion is relatively simple to be modeled, even considering rather complex vehicle models, the aim of the book is to provide accurate analytical formulae for hopefully a ultimate engineering contribution on this topic.

The aim is to enable proper engineering solutions for vehicles, which naturally run straight ahead, saving fuel and minimizing tire wear. Since the very beginning of automotive engineering, obtaining the straight ahead running of road vehicles has been a known issue. The book aims at instructing engineers on how to make a vehicle run straight, making the driver feel comfortable and safe.

The aim of the book is to encompass issues pertaining both to current production vehicles and to future automated vehicles. Actually, a full analytical knowledge of the straight ahead running phenomenon is a primary need for the development of any road vehicle.

In the design process of a vehicle, straight running is often obtained after tuning of tires and suspension parameters. The feel of tire pull or weak stiffness at the steering wheel by the driver strongly affects the perceived level of the quality of the vehicle. Presently, a number of complex nonlinear vehicle models are available that can be used effectively to tune relevant parameters affecting straight ahead running of vehicles. However, no analytical formulae are available for the same purpose. This book offers such formulae that highlight the trends and the relationships among parameters, enabling engineers to have an insight into and a sound understanding of the technical solutions they adopt. More specifically, engineers may get in-depth knowledge of the basic interactions between vehicle and tires. Such understanding and knowledge cannot be obtained by numerical simulations.

Authors do believe that this book is a preliminary contribution, able to stimulate future studies for safer, smarter, and more comfortable vehicles.

disclaimer

The book is not lo be used as a manual, that is, as a book containing practical instruction on how to build a mechanical system. The publisher and the authors agree that the book has the only aim of explaining to readers both the relevant physical phenomena and how the solution of technical problems can be obtained by means of a scientific approach. Technical applications have then to be realized by experienced engineers only on the basis of national or international standards and/or practical rules that are omitted in the book.

To the fullest extent of the law, neither the publisher nor the authors assume any liability for any injury and/or damage to persons or property as a matter of products liability, negligence, or otherwise, or from any use, operation of any methods, products, instructions, or ideas contained in the material herein.

This book contains information obtained from authentic and highly regarded sources. Reasonable efforts have been made to publish reliable data and information, but the authors and publisher cannot assume responsibility for the validity of all materials or the consequences of their use. The authors and publisher have attempted to trace the copyright holders of all material reproduced in this publication and apologize to copyright holders if permission to publish in this form has not been obtained. If any copyright material has not been acknowledged, please write and let us know so we may rectify in any future reprint.

Trademark Notice: Product or corporate names may be trademarks or registered trademarks, and are used only for identification and explanation without intent to infringe.

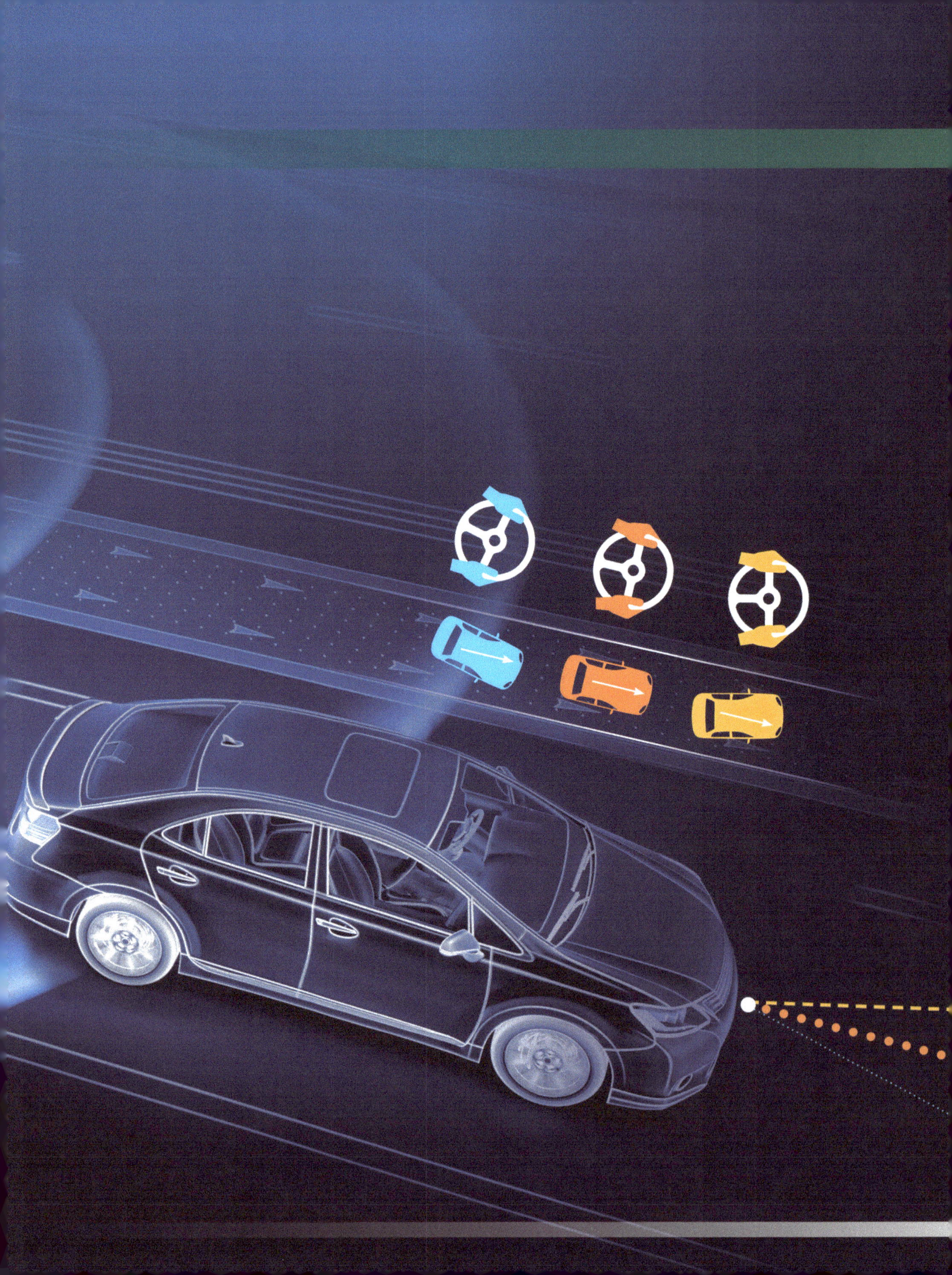

Introduction

You have to make a car go straight before you make it turn.

—Unknown Vehicle Dynamics Engineer

Vehicle pull arises whenever the driver has to exert a discernible steering torque (called *pull*) in order for the vehicle to move straight ahead, which can become a tiresome task in case of long-distance driving along straight roads (e.g., when traveling on highways). A lateral deviation (called *drift*) occurs when the vehicle, despite being not steered, does not maintain the intended straight path. These two expressions of the same issue represent a critical aspect for vehicle straightability performance [1]. Actually, these handling features are often ascribed among the vehicle (or component) *basic needs*, or *Must-Be Qualities*, to use the terminology of Kano model of product development [2]: customers expect and take for granted that the vehicle tracks straight and free of pull in the absence of steering inputs and external disturbances. Non-fulfillment of this requisite has deleterious consequences on customers' opinion and satisfaction, impacting on the perception of both comfort and safety. Therefore, vehicle pull represents a key problem for automotive Original Equipment Manufacturers (OEMs), and it is often examined as a crucial factor during approval tests with tire manufacturers.

The factors producing vehicle pull can be divided into

- *External factors*, such as crosswind or road cross slope, that are part of the boundary conditions or environment in which the vehicle system operates.
- *Internal factors* or vehicle system related parameters. Among them, a key role is played by specific tire properties and vehicle unbalances (e.g. non-symmetric weight distribution, improper suspension alignment, dimensional tolerances).

"Nuisance" is the term that Pottinger juxtaposes to the vehicle pull issue [3]. *Nuisance* can be related to the continuous driver's action to keep a straight motion. Understanding such a phenomenon is of crucial importance not only for the comfort of the driver, but also for

- reducing energy losses
- designing future autonomous vehicles

State of the Art

Among the different causes of vehicle pull, tire characteristics are of particular interest. In his comprehensive reference [3], Pottinger provides a definition for the *pull forces*, describing them as the offsets in tire side force F_y and aligning moment M_z responsible for vehicle pull. These offsets, which give rise to non-negligible forces and moments when the tire is not steered nor cambered, are inherent characteristics common to all tire constructions. They have become increasingly relevant with the establishment of the radial ply tire as the prevalent tire construction type, supplanting the bias tire [3–6]. Side force and aligning moment offsets are included also in the semi-empirical *Magic Formula Tire models* developed by Pacejka [7], which are widely used as accurate portrayals of steady-state tire force and moment characteristics, for vehicle dynamics analyses and simulations.

Furthermore, the aforementioned tire force and moment offsets are commonly divided into two components, called *ply-steer* and *conicity*, to discern the two characteristically different types of sources generating them [3, 4, 8]

- *Ply-steer* is defined as the lateral force offset contribution whose direction relative to the tire face depends on the direction of rotation. In the axis system attached to the tire, the sign of ply-steer lateral force is independent of the direction of rotation. It is an intrinsic characteristic of the tire structural design, and its effect can be considered analogous to that of a *pseudo-side slip angle*.
- *Conicity* is defined as the contribution of the lateral force offset whose direction relative to tire face is independent of the direction of rotation. In the axis system attached to the tire, the sign of conicity lateral force changes with direction of rotation. It is mainly related to finite manufacturing precision, and its effect is analogous to that of a *pseudo-camber angle*.

Extensive research has been conducted with regard to tire design characteristics producing these two components, in particular for what concerns ply-steer. Among the others, the works performed by Pottinger [4] and Bert [6] are worth mentioning: they made use of the laminated composite theory to describe how the belt layers contribute to the generation of ply-steer side force and moment, with experimental measure validations. However, the subsistence of a considerable influence on ply-steer from tire tread was pointed out experimentally by Matyja [5], and then corroborated with FEM studies performed by Mundl et al. [9] and *Toyo Tire & Rubber Co., Ltd.* [10, 11].

In the last decades, the relationships between tire ply-steer and conicity features and vehicle behavior in straight-ahead motion have been explored. On the one hand, Lindenmuth [12] is indicated as one of the first authors to have evidenced experimentally the existence of a relationship between vehicle pull and tire conicity and ply-steer [3, 13]. On the other hand, Topping [14] is claimed to be the first author offering an easy grasp of the origin of steady-state tire-induced vehicle pull [3, 4]. Topping's theory, reported by Pottinger [3], considers front and rear axles as separated entities, and provides a basic explanation of how vehicle pull occurs, looking at tire side force F_y and aligning moment M_z characteristics as functions of side slip angle α. The theory is based on the assumption that no lateral deviation occurs if no lateral force acts on the axles of the vehicle [3, 14]. Two kinds of driving conditions are generally considered in literature in which the different expressions of vehicle pull arise

- *Fixed control condition*: in this condition, the driver holds the steering wheel to make the vehicle run along a straight path. Front and rear axles arrange themselves with respective side slip angles α_1 and α_2 in order for the sum of tire side forces to become null on both axles (front $F_{y,1} = 0$, rear $F_{y,2} = 0$).

  On the one hand, the front axle side slip angle α_1 – nullifying $F_{y,1}$ – determines a residual aligning moment $M_{z,1}$ on the front axle, since in general either the side force F_y or the aligning moment M_z do not become null at the same slip angle. This has a direct effect on the average steering torque (steering *pull*) that the driver has to exert during vehicle straight motion, sometimes called *Vehicle Residual Aligning Torque* (*VRAT*) in literature [15, 16].

  On the other hand, the rear axle side slip angle α_2 determines vehicle attitude β, causing the vehicle to exhibit the so-called *dog-tracking* straight ahead motion.

  Finally, the difference between front and rear axle side slip angles causes a non-null steering wheel angle (or *misalignment*).
- *Free control condition*: in this condition, the driver releases the steering wheel. The front axle arranges itself with a side slip angle α_1 for which the aligning moment $M_{z,1}$ becomes null, since it is no more counterbalanced by a steering torque.

  The sum of tire side forces is not equal to zero for the front and rear axles, and a lateral deviation (steering *drift*) from the straight path occurs.

Thereafter, the analyses focus on the connection between vehicle straight motion dynamics and specific tire parameters. Particular relevance was attributed to the tire Residual Aligning Torque (RAT), remaining when side force is null. As an alternative, other related parameters were considered, such as the one introduced by Pottinger [3, 4]: he defined the Aligning Torque Shift Phase (ATSP) as the difference between the slip angle at which side force is null and the slip angle at which the aligning moment is null. Pottinger's theory, despite being straightforward and sufficiently accurate, does not explain completely the underlying interactions between vehicle and tires in the straight motion [17]. For this reason, an analysis was conducted by Lee [17], who expanded upon Pottinger's and Topping's contributions with a rather more complete approach, including tire moments in the equilibrium equations and performing dynamic simulations both with a simple single-track model and with a full-vehicle model. Actually, Lee pointed out that little had been examined up to then on the tire-induced vehicle pull pertaining to dynamic simulations: the same *drift* experienced in real-life driving situations is obtained when running numerical simulations, in case tire properties from experimental data are used.

Other studies considered experimentally the effect of suspension geometry on vehicle pull, e.g., by examining cross camber, cross caster, and other suspension alignment tolerances [13, 15, 18, 19].

In addition, the effect of road cross slope on vehicle pull has been commonly recognized as non-negligible [3, 16, 19]: original equipment manufacturers (OEMs) specify a limit for the acceptable conicity lateral force for a pair of tires to be installed on a given vehicle axle. Moreover, OEMs specify a target range for tire ply-steer residual aligning torque (PRAT), whose magnitude depends on vehicle weight, which is aimed at counteracting road cross slope according to the destination market (i.e. left- or right-hand drive countries). Nevertheless, in the abovementioned studies, the road cross slope effect was either not analytically discussed or just included with simplifying considerations. As a consequence, the outcomes turn out to be unsuitable to provide an accurate description of the actual impact of road cross slope on vehicle pull.

It is important to underline that vehicle pull issue is still currently an object of studies and is taken into consideration with particular care in the automotive field. This is confirmed by researches on robust design optimization aimed at steering pull reduction, such as the one conducted by Park et al. [20]. Moreover, in the last few decades, the establishment of the Electric Power-Assisted Steering (EPS) has opened possibilities in the active reduction of vehicle pull. This option to overcome passive control limitations, proposed, just to name a few, by *Nissan Motor Co., LTD.* [21] and *General Motors Co.* [2], has proved to be quite effective in the reduction of steering torque (or *pull*) issue. Nevertheless, as for steering wheel *misalignment* and vehicle attitude (or *dog-tracking*) problems, EPS constitutes a solution just to a limited or no extent.

This book is a full rework (revision, correction, and substantial extension with data and full mathematical proofs) of a Master of Science dissertation, produced and discussed at the *Politecnico di Milano* [22]. A similar dissertation has been discussed publicly at the *KTH Royal Institute of Technology* in Stockholm [23].

Book Approach and Objectives

The focus of this book is on the influence of steady-state tire characteristics on vehicle pull and straight motion behavior. In particular, the tire-associated pull problem is considered in the free rolling condition, isolating it from the torque steer and the influence of longitudinal tire forces that arise when the vehicle is driven or brakes applied, respectively [3].

The main objective of the book is the study and accurate description of the tire-associated vehicle pull problem, with reference to the effects on vehicle straight motion and on-center handling, based on a thorough theoretical study by using available tire data.

Analytical equations and solutions are derived for the addressed problem. They are obtained by means of the study of a rather complete single-track vehicle model, able to accommodate different left and right tire characteristics and load transfer and static roll induced by road cross slope. The presence of road cross slope effect is extensively considered by pointing out the different effects that descend from it. Moreover, proper (symmetric) suspension alignment contribution is analyzed in presence of different left and right tire characteristics, which can be due to tire manufacturing tolerances or induced by road cross slope.

The interaction between the tires and vehicle, namely the effect of elastokinematics and compliances in suspension and steering systems, is taken into account by means of the effective axle characteristics. The formulation of effective axle characteristics, described by Pacejka [7], is specifically extended in this book in order to account for the tire force and moment offsets (*pull forces*), together with the different left and right tire characteristics. The purpose is to draw an analytical representation and explanation for problems still open in literature.

Furthermore, the handling diagram theory [7, 24] is suitably extended and adapted in this book to accommodate the aforementioned tire and effective axle characteristics. The adapted handling diagram provides the possibility of conveniently and simultaneously linking the axle's side slip angles, steering angle, and steering torque by expressing all of them as functions of vehicle lateral acceleration. For small lateral accelerations, which are the range of interest in this study, the adapted linear handling diagram allows accurate and effective evaluation of all the vehicle pull related parameters, namely steering torque, steering angle and vehicle attitude.

Analytical expressions and a single-track model are developed in *MATLAB*®. Such expressions are validated by means of proper simulations performed with a full-vehicle model in the multibody dynamics simulation software *MSC Adams*™. The effect of road irregularity is studied by comparative analyses of the computed variances and the results of full-vehicle simulations.

Finally, the existence of a relationship between the tire offsets and the on-center handling of the vehicle is explored. Based on the company's experience and available subjective feedback of a test driver, the tire sets with the worst performances with regards to the vehicle pull are also the ones producing the worst results during on-center handling tests. The simulation of an adapted weave test is employed to analyze the correlation between simulated steering torque objective metrics and analytically computed steering torque offsets. The observed results may provide an explanation to the subjective feedback and demonstrate a relationship between tire characteristics offsets and steering feel performance.

Structure of the Book

The book is divided into two parts.

The first part is dedicated to the thorough analytical discussion of the various aspects related to tire-induced vehicle pull issue and straight motion

Chapter 1 (at page 9): the ply-steer and conicity components of tire side force and aligning moment are presented, together with the pertaining formulae and description of the test methods used to measure them.

Chapter 2 (at page 25): the effective axle cornering stiffness formulation is extended in order to include the tire force offsets. The single-track vehicle model is introduced.

Chapter 3 (at page 41): the effect of tire characteristics on vehicle straight motion is analyzed. The straight-driving expressions are described and linked to the extended handling diagram theory.

Chapter 4 (at page 53): an expression for the steering torque as a function of tire characteristics is obtained, and combined with the extended handling diagram theory.

Chapter 5 (at page 69): different test methods for the analysis of vehicle pull are described. A section is dedicated to the on-center handling evaluation.

Chapter 6 (at page 77): the effects of road cross slope are thoroughly described.

Chapter 7 (at page 87): contributions of wheel alignment and tire characteristics to straight-driving steering angle, steering torque, and vehicle attitude are examined. Analytical solutions are given.

Chapter 8 (at page 103): road irregularity (or unevenness) effect is investigated, with single-and double-track excitation, to have a formulation for the variance of tires' normal load, straight-driving slip angles, and vehicle's body lateral displacement.

The second part is dedicated to the numerical simulations, with discussion of the results and comparison between analytical expressions, namely the *MATLAB®* single-track model and *MSC Adams™* full-vehicle model

Chapter 9 (at page 137): the available data, referring to four different sets of tires, are presented together with test driver's subjective feedback.

Chapter 10 (at page 145): the analytical expressions and *MATLAB®* single-track model are validated by means of null steer test simulations performed with *MSC Adams™* full-vehicle model.

Chapter 11 (at page 157): the analytical expressions for straight-driving slip angles, vehicle attitude and steering torque are validated by comparison with fixed control test simulations performed with *MSC Adams™* full-vehicle model. Furthermore, the influence of road irregularity is examined.

Chapter 12 (at page 169): an adapted weave test is considered to study on-center handling, and simulations are performed both with *MATLAB®* single-track model and *MSC Adams™* full-vehicle model. The results are combined with the analytically computed steering torque offset, showing a relationship between tire characteristics and vehicle on-center handling.

Chapter 13 (at page 185): the contributions of wheel alignment, highlighted in Chapter 7, are discussed numerically. Moreover, the results provided by analytical expressions, *MATLAB®* single-track model, and *MSC Adams™* full-vehicle model are compared to discuss reliability and limitations.

Finally, in the Appendix (at page 207) further details are provided on how solutions and expressions presented in the analytical part are obtained, on sign conventions adopted and on vehicle and tire parameters.

Part I

Derivation of Analytical Formulae

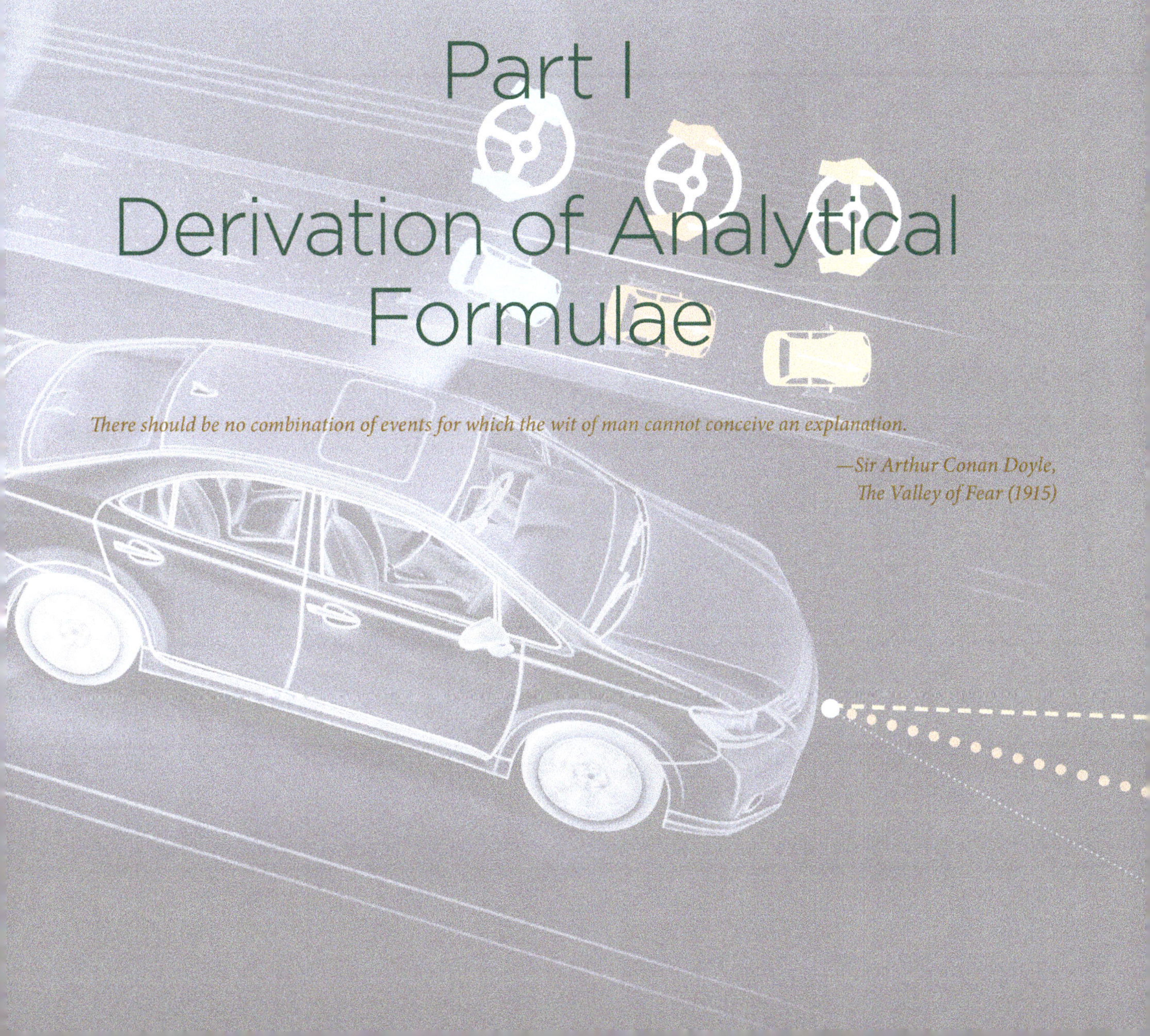

There should be no combination of events for which the wit of man cannot conceive an explanation.

—Sir Arthur Conan Doyle,
The Valley of Fear (1915)

CHAPTER 1

Ply-Steer and Conicity

"*Pull forces*," as they are called by Pottinger [3], are tire side force F_y and aligning moment M_z, exhibited when the tire is not steered nor cambered. In other words, a tire will *always* exhibit a side force F_y and a self-aligning moment M_z when the lateral slip α and the camber γ are null. This force and moment *shift* the $F_y - \alpha$ and $M_z - \alpha$ characteristics from the origin, i.e. tire characteristics do have offsets from the origin (see Figure 1.10). Shift terms such as S_{Hy}, S_{Vy}, and M_{zr0} – appearing in *Magic Formulae* (Pacejka [7]) – are the quantitative parameters that represent pull forces.

These offsets may cause a tire-associated vehicle pull issue when the driver needs to apply a discernible correcting effort (steering torque) in order for the vehicle to run straight ahead [3]. The tire-generated free rolling pull became particularly relevant when radial ply tire was adopted as the predominant type of tire construction [3–6].

Ply-steer and conicity are the two different sources of side force and aligning moment offsets. They can be related to tire nonuniformities or anisotropies. This chapter is committed to the description of ply-steer and conicity and to the introduction of the test methods and formulae aimed at their description and assessment.

The sign conventions adopted in this book for tire's lateral force F_y, aligning moment M_z, and lateral slip angle α are reported in Appendix A.2 and are consistent with the *Pacejka's modified SAE* [7], with *z-axis* directed downward.

1.1 Pure Ply-Steer

A tire without conicity but only ply-steer exhibits the same behavior (same direction of side force) when it is assembled flipped (or reversed) on the rim, while an opposite side force is obtained when the rolling direction is inverted (**Figure 1.1**).

The physical interpretation or explanation of this characteristic is beyond the intent of this book. Nevertheless, it is briefly mentioned here. In literature, the sources of ply-steer are mainly the tire composite structure and tire tread.

- Tire composite structure: Its contribution to ply-steer derives from the construction of the belt layers, which are placed between the carcass and the tread (see **Figure 1.2**). Belt layers are composed of rubber coating with alternate stacking of non-anchored steel cords, at opposite angles with respect to the tire meridian plane, in order to resist side forces [25].

Pottinger [3, 4] described the effect of the belt layers using the classical theory for laminated composites. The same approach was adopted in [6], comparing theoretical predictions with experimental measurements. Moreover, the results of an experimental analysis on the effects of belt-related parameters [5] were in line with the laminated composite theory.

According to this theory, the tire structure can be described as a laminate with Eq. 1.1 [27]

$$\begin{Bmatrix} N_X \\ N_Y \\ N_{XY} \\ M_X \\ M_Y \\ M_{XY} \end{Bmatrix} = \begin{bmatrix} A_{11} & A_{12} & A_{16} & B_{11} & B_{12} & B_{16} \\ A_{21} & A_{22} & A_{26} & B_{21} & B_{22} & B_{26} \\ A_{61} & A_{62} & A_{66} & B_{61} & B_{62} & B_{66} \\ B_{11} & B_{12} & B_{16} & D_{11} & D_{12} & D_{16} \\ B_{21} & B_{22} & B_{26} & D_{21} & D_{22} & D_{26} \\ B_{61} & B_{62} & B_{66} & D_{61} & D_{62} & D_{66} \end{bmatrix} \begin{Bmatrix} \varepsilon_{0XX} \\ \varepsilon_{0YY} \\ \gamma_{0XY} \\ \kappa_X \\ \kappa_Y \\ \kappa_{XY} \end{Bmatrix} \tag{1.1}$$

FIGURE 1.1 Pure ply-steer: effects of rim assembly and rolling direction (adapted from Pacejka [7]). *Flipped* means that the tire is flipped on the rim with respect to the *original* assembly.

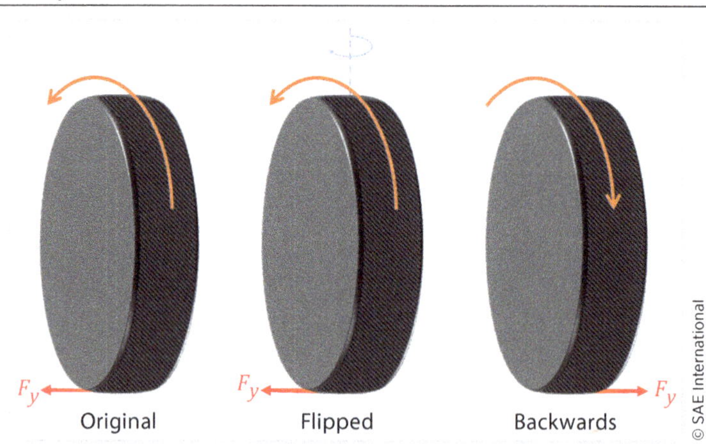

FIGURE 1.2 Radial ply pneumatic tire structure (adapted from [25] and [26]).

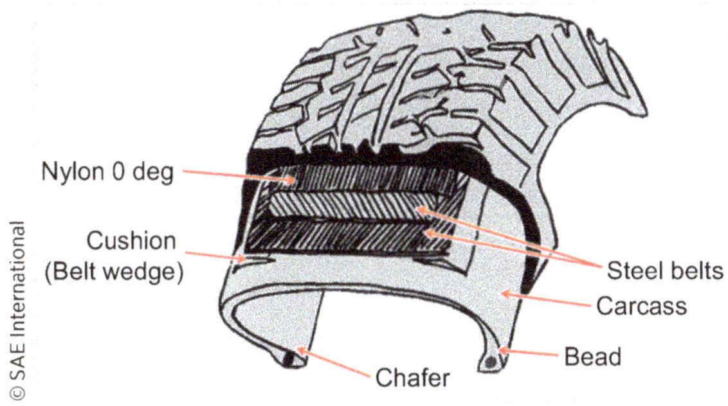

where (see Figure 1.3)

- $N_X = \int_{-t/2}^{t/2} \sigma_{XX} dz$ In-plane force per unit length [N/m]
- $N_Y = \int_{-t/2}^{t/2} \sigma_{YY} dz$ In-plane force per unit length [N/m]
- $N_{XY} = \int_{-t/2}^{t/2} \tau_{XY} dz$ In-plane force per unit length [N/m]
- $M_X = \int_{-t/2}^{t/2} \sigma_{XX} z dz$ Out-of-plane bending moment per unit length [Nm/m]
- $M_Y = \int_{-t/2}^{t/2} \sigma_{YY} z \, dz$ Out-of-plane bending moment per unit length [Nm/m]
- $M_{XY} = \int_{-t/2}^{t/2} \tau_{XY} z \, dz$ Torsional moment per unit length [Nm/m]
- $\varepsilon_{0XX}, \varepsilon_{0YY}, \gamma_{0XY}$ Mid-plane strains
- $\kappa_X, \kappa_Y, \kappa_{XY}$ are Curvatures
- A_{ij} is the membrane (*in-plane*) stiffness submatrix
- B_{ij} is the membrane-bending (*in-plane/out-of-plane*) coupling submatrix
- D_{ij} is the bending (*out-of-plane*) stiffness submatrix

The tire structure is flattened under load during rotation. If the coupling stiffness terms are nonnull ($B_{16}, B_{26} \neq 0$), a circumferential curvature change κ_X (assumed uniform under the hypothesis of pure cylindrical bending) causes a deformation

FIGURE 1.3 Laminate theory nomenclature (adapted from [27]).

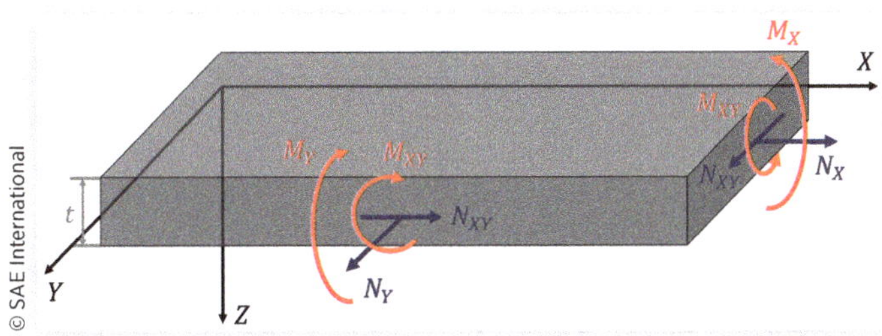

FIGURE 1.4 Pottinger's physical interpretation of ply-steer induced by composite structure (adapted from [4]).

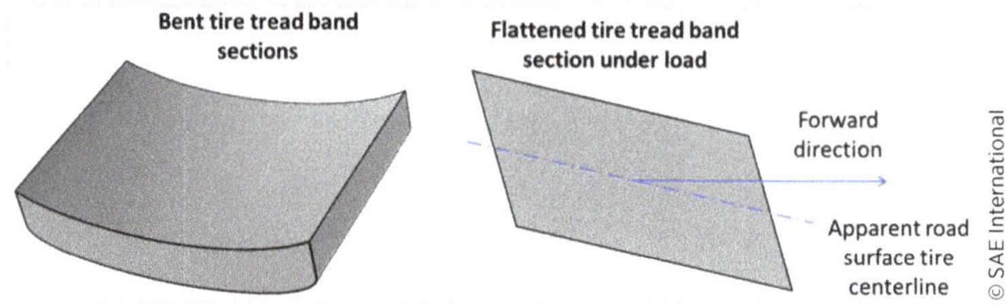

equivalent to a small steer angle (ply-steer) about the *z-axis* [6], as in **Figure 1.4**. This is estimated in [6] under the assumption of pure cylindrical bending during rolling under load (with approximately null N_{XY}, κ_Y, and κ_{XY}), according to

$$\alpha_{ply,belt} = \frac{1}{2}\left(\frac{\partial v}{\partial x} - \frac{\partial u}{\partial y}\right) \approx \frac{1}{2}\left(-\frac{\partial u}{\partial y}\right) = -\frac{\gamma_{0XY}}{2} = \frac{B_{16}}{A_{66}}\frac{\kappa_X}{2} = \frac{B_{16}}{A_{66}}\frac{1}{2 \cdot R_r}$$

where R_r is the rolling radius.

Moreover, Ply-steer Residual Aligning Torque (PRAT, i.e., the residual aligning moment M_z when ply-steer side force is null) is caused by nonnull in-plane stiffness coupling terms, between tension and shear (namely A_{16}, $A_{26} \neq 0$) [3, 4].

Pottinger presented a patent of a radial pneumatic tire with reduced belt-layers-induced ply-steer, by using an equal number of layers with the same thickness and stiffness with opposite cord angles, symmetric with respect to the laminate mid-plane [28]. This construction requires a minimum of four layers with stacking order $\theta/-\theta/-\theta/\theta$. This configuration, known as *paired-ply construction* [6], has $[B_{ij}] = 0$ due to its symmetry with respect to the laminate mid-plane. Moreover, for each layer with orientation θ, there is another layer with opposite orientation $-\theta$. As a consequence of this arrangement, the in-plane shear-normal stiffness terms are null (namely $A_{16} = A_{61} = 0$ and $A_{26} = A_{62} = 0$).

- Tire tread: A considerable contribution to ply-steer is provided by the tread pattern and design. This aspect was not included in [4, 6, 28], but an experimental analysis performed by Matyja [5] pointed out the effect of tread pattern on ply-steer characteristics, confirmed in the studies conducted by *Toyo Tire & Rubber Co., Ltd.* [10, 11]. Moreover, an analysis using finite element method (FEM) [9] was conducted on the effects of the tread pattern alone: it confirmed that the shape and the location of tread blocks have a strong influence on the Ply-steer Residual Aligning Torque (PRAT). Actually, the asymmetric shape of tread blocks in the longitudinal and lateral directions (with respect to the tire plane) causes them to have coupling stiffness. This aspect, combined with lateral and longitudinal shear stress in the contact patch, generates coupled reaction forces, which can provide a mechanical explanation to the tread-induced PRAT.

Thus, ply-steer is considered a deterministic feature, related to the inherent structural design of a specific tire.

The ply-steer is also called *pseudo slip angle* since the effect is similar to a tire rolling with a small slip angle α_{ply}.

It provides a contribution to the aligning moment with the same sign of the one given by side slip.

The corresponding peculiar footprint and force diagrams are reported respectively in **Figure 1.5** and in **Figure 1.6**.

FIGURE 1.5 Pure ply-steer: skewed footprint on a frictionless surface (adapted from [7]).

1.2 Pure Conicity

A tire without ply-steer but only conicity exhibits the same behavior (same direction of side force) when it is rolled in the opposite direction, while an opposite side force is obtained when it is assembled flipped on the rim (**Figure 1.7**).

Conicity is mainly due to finite manufacturing process precision, producing differential belt constraint, making the belt band conical toward the more constrained side, and thus deviating from nominal design [3]. It can therefore be considered a random characteristic among a specific design tire. However, a certain conicity contribution can be a design controlled parameter, e.g., by means of the nylon belt [25] (see Figure 1.2) or with intentional differential belt constraint as in the *Michelin XAS* tire [29], with the aim of compensating for suspension alignment effects [4].

The conicity is also called *pseudo camber* since the effect is similar to a cambered wheel with camber angle γ_{con}. It provides a contribution to the aligning moment with opposite sign with respect to the one given by side slip.

The corresponding peculiar footprint and force diagrams are reported, respectively, in **Figure 1.8** and in **Figure 1.9**.

FIGURE 1.6 Pure ply-steer: side force and aligning moment diagrams (adapted from Pacejka [7]).

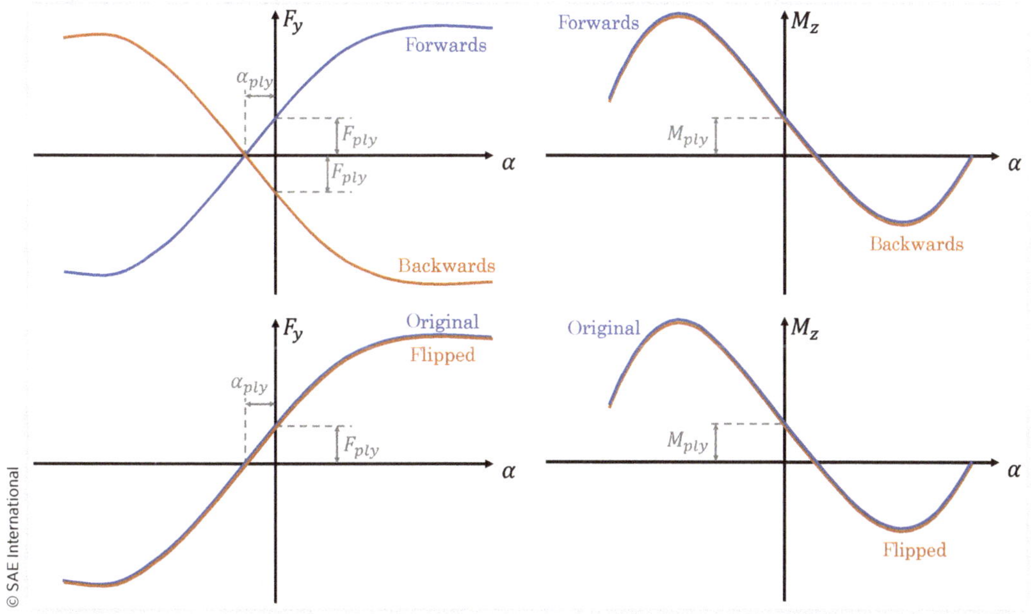

FIGURE 1.7 Pure conicity: effects of rim assembly and rolling direction (adapted from Pacejka [7]).

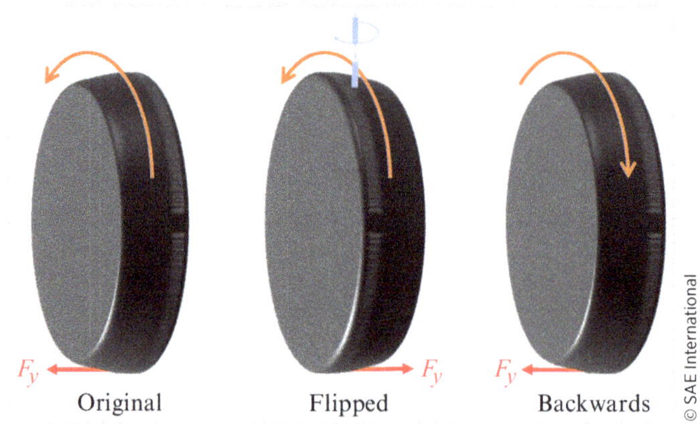

1.3 Combined Ply-Steer and Conicity

A tire with both conicity and ply-steer presents a characteristic diagram (side force-side slip) shifted according to **Figure 1.10**. The footprint is a combination of skewed and curved footprints, respectively, of pure ply-steer and pure conicity (**Figure 1.11**).

A qualitative example of the statistical distribution of ply-steer and conicity components is reported in **Figure 1.12**. It is referred to side force F_y contributions in a country following right-hand traffic; in a country following left-hand traffic, the belt layup and tread band are reversed, thus the ply-steer contribution has the opposite sign. One can observe that ply-steer is a more deterministic component, while conicity can be considered a random characteristic within a tire specification. Therefore, tires are usually sorted by conicity value [5].

1.4 Ply-steer and Conicity as Equivalent Side Slip and Camber Angles

An approach implemented by Pacejka [7], assumes that ply-steer can be considered as an equivalent side-slip α_{ply} and conicity as an equivalent camber γ_{con} (**Figure 1.13**). In this way, the effect of ply-steer and conicity can be introduced into the computation of vehicle motion in a compact and effective fashion.

FIGURE 1.8 Pure conicity: curved footprint on a frictionless surface (adapted from [7]).

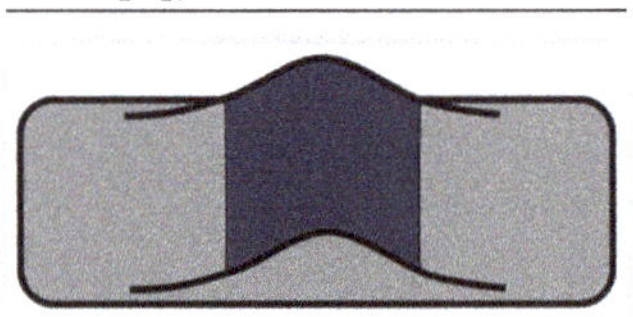

From Figure 1.13, the equivalent camber angle γ_{con} and side slip angle α_{ply} can be obtained by means of geometrical considerations

$$\alpha_{FM0} = \frac{C_{M\gamma} \cdot \gamma_{con}}{C_{M\alpha}} + \Delta\alpha_\gamma = \frac{C_{M\gamma} \cdot \gamma_{con}}{C_{M\alpha}} + \frac{C_{F\gamma} \cdot \gamma_{con}}{C_{F\alpha}} \Rightarrow \gamma_{con} = \frac{\alpha_{FM0}}{\dfrac{C_{M\gamma}}{C_{M\alpha}} + \dfrac{C_{F\gamma}}{C_{F\alpha}}} \tag{1.2}$$

$$\alpha_{ply} = \Delta\alpha_0 - \Delta\alpha_\gamma = \Delta\alpha_0 - \frac{C_{F\gamma}}{C_{F\alpha}}\gamma_{con}$$

FIGURE 1.9 Pure conicity: side force and aligning moment diagrams (adapted from Pacejka [7]).

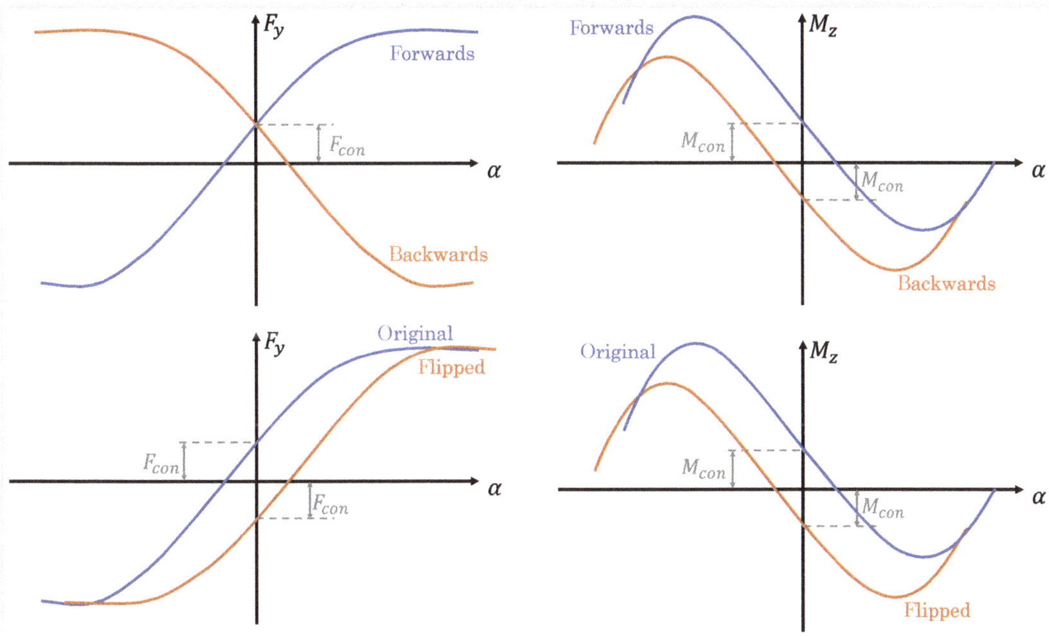

The parameters appearing in the formulae and in Figure 1.13 can be expressed in terms of Magic Formula shifts [7]

$$
\begin{aligned}
\Delta\alpha_0 &= \frac{F_{y0}}{C_{F\alpha}} = S_{Hf} \,|\left(\varphi = 0, \gamma = 0\right) \\
\alpha_{FM0} &= \frac{F_{y0}}{C_{F\alpha}} + \frac{M_{z0}}{C_{M\alpha}} = \frac{M_{zr0}}{C_{M\alpha}}
\end{aligned}
\tag{1.3}
$$

FIGURE 1.10 Combined ply-steer and conicity: side force diagram (adapted from Pacejka [7]).

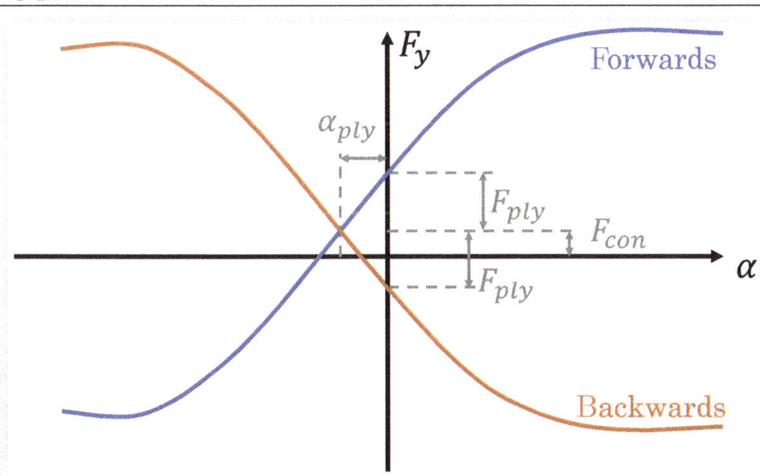

FIGURE 1.11 Combined ply-steer and conicity: footprint on a frictionless surface (adapted from [7]).

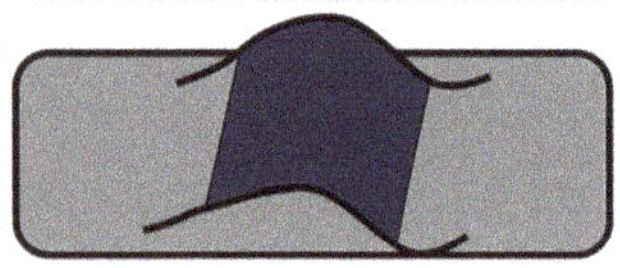

where

$$M_{zr0} = D_{r0} = F_z R_0 \left(q_{Dz6} + q_{Dz7}\ df_z\right) \cdot \lambda_{Mr}$$
$$C_{M\alpha} = K_{z\alpha 0} = D_{t0} K_{y\alpha}\Big|_{\gamma=0} \quad (1.4)$$

Camber coefficients for force ($C_{F\gamma}$) and aligning moment ($C_{M\gamma}$) are not provided as test results, but have to be estimated. According to Pacejka [7], it is possible to define them as

$$C_{F\gamma} = c_5 \cdot F_z$$
$$C_{M\gamma} = c_6 \cdot \frac{\left(w_{tire}/2\right)^2}{r_{eff}} \cdot C_{F\kappa} = c_6 \cdot \frac{\left(w_{tire}/2\right)^2}{r_{eff}} \cdot \left(c_8 \cdot F_z\right) \quad (1.5)$$

where

- c_5, c_6, and c_8 are coefficients
- w_{tire} is the width of tire contact patch (assumed constant)
- $r_{eff} = \frac{V_x}{\Omega}$ is the effective rolling radius
- $C_{F\kappa}$ is the longitudinal slip stiffness

For small camber angles γ (as expected in this study), side force camber coefficient $C_{F\gamma}$ can be assumed approximately equal to normal force F_z, so $c_5 \approx 1$.

Examples of values for the remaining coefficients can be found in literature [7]. More details pertaining to these estimations are reported in Appendix A.8.

1.5 Tire Pull

When driving in a straight line, different cases on the effect of ply-steer and conicity are available to be examined by considering the lateral forces and aligning moments they are causing. In the following cases, for simplicity, the characteristics of left and right tire of the same axle are considered equal. In addition, ply-steer and conicity are taken into account only for front tires, while they are vanishing for rear tires.

- **Ideal tires**. When driving straight, neither side forces nor aligning moments are generated at $\alpha = 0$. This case is represented in **Figure 1.14a**.

FIGURE 1.12 Qualitative example of statistical distribution of ply-steer and conicity side force contributions, referred to a right-hand traffic country (adapted from [3]).

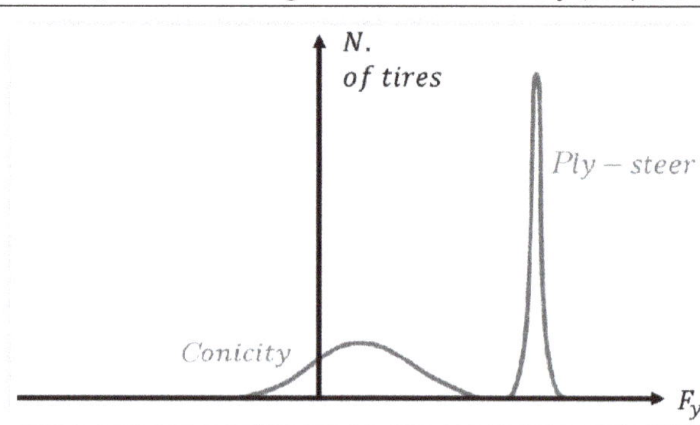

FIGURE 1.13 Conicity and ply-steer equivalent angles (adapted from [7]).

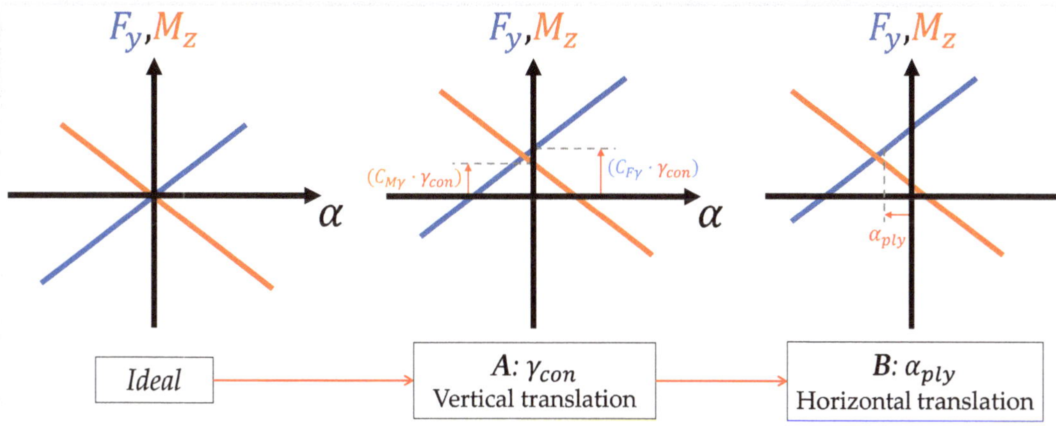

(a) *Plots sequential construction.*

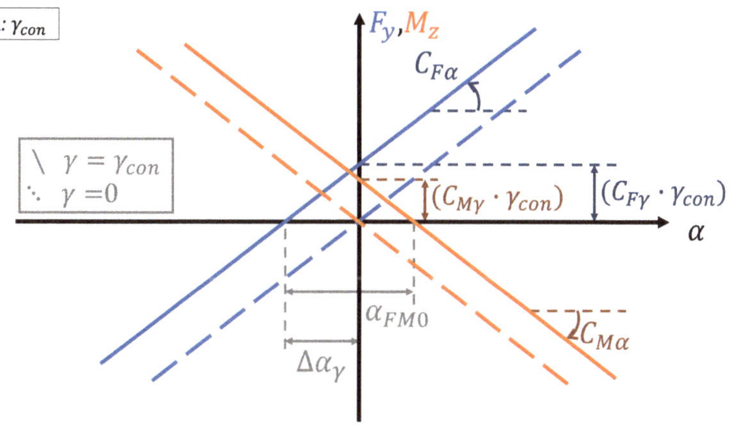

(b) *Plot A. Conicity equivalent camber angle* γ_{con}.

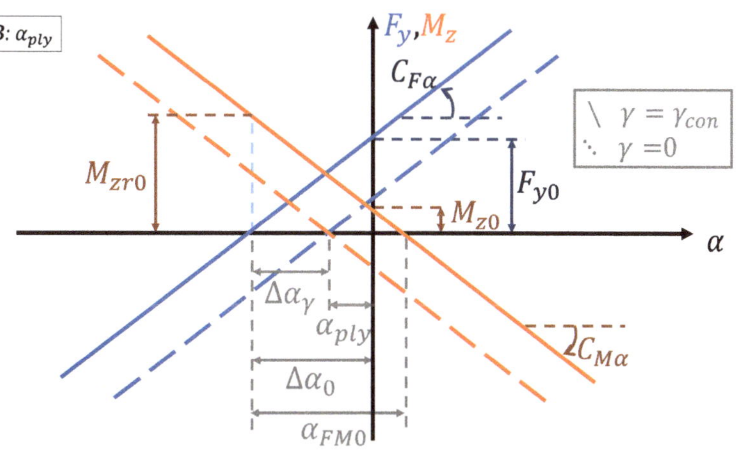

(c) *Plot B. Ply-steer equivalent side slip angle* α_{ply}.

FIGURE 1.14 Different tire pull cases. Solid lines: right tire, dashed lines: left tire.

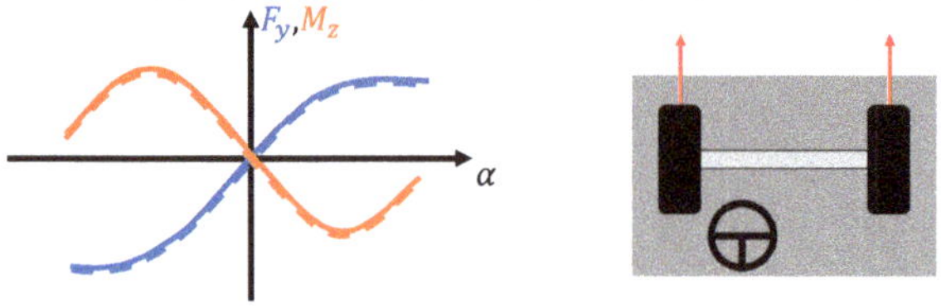

(a) *Ideal front tires.*

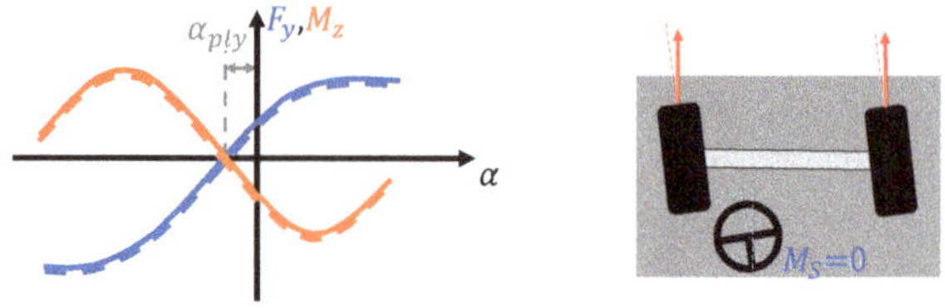

(b) *Front tires with ply-steer.*

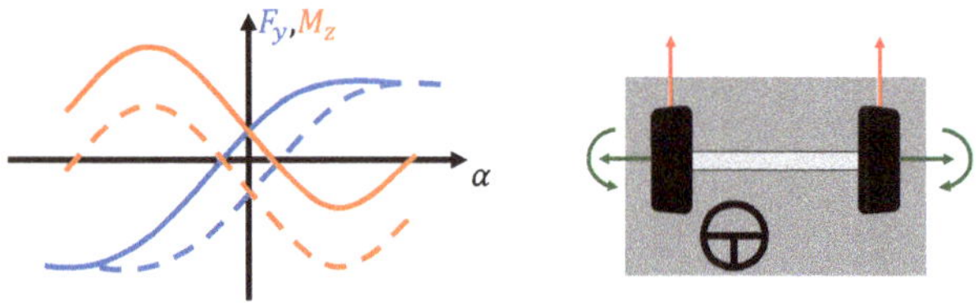

(c) *Front tires with opposite conicity.*

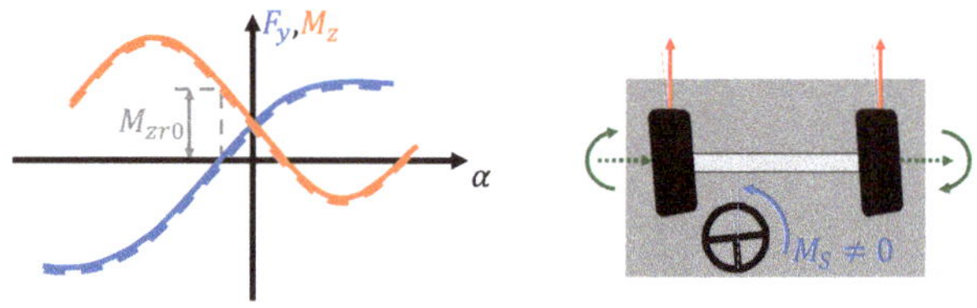

(d) *Front tires with same-side conicity.*

- **Ply-steer**. The equally directed side forces are compensated by side slip angle of the front axle (provided by steering). This is the situation where both F_y and M_z are null for the same slip angle, thus no residual aligning torque is left. This case is represented in **Figure 1.14b**.
- **Opposite conicity**. Side force and aligning moment are opposite and thus compensating at $\alpha = 0$. This case is represented in **Figure 1.14c**.
- **Same-side conicity**. For $\Sigma F_y = 0$, the residual torque M_{zr0} has to be compensated by an opposite steering torque to maintain a straight path (tire pull). This M_{zr0} is the Residual Aligning Torque often called *RAT* in literature [3, 5, 17]. This case is represented in **Figure 1.14d**.

Another simplified case can be considered when the rear axle is also equipped with tires having the same ply-steer characteristics as of the front tires. The whole vehicle will have a nonnull attitude, equal to the side slip angle where the lateral forces and aligning moments are null, as represented in **Figure 1.15**.

FIGURE 1.15 Front and rear tires with same ply-steer case. Solid lines: front tires, dashed lines: rear tires.

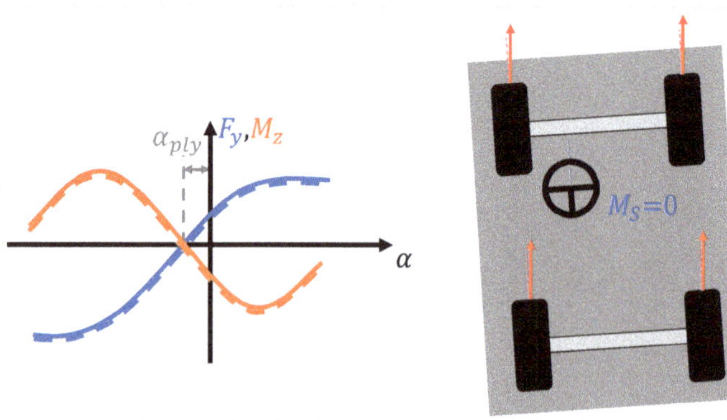

FIGURE 1.16 MTS Flat-Trac® Tire Test Systems (adapted from [30]).

1.6 Ply-steer and Conicity: Tire Test Methods

There are two methods to test the tires with the aim of measuring the ply-steer and conicity components [3]. A continuous flat belt tire test system (as the one represented in Figure 1.16) is often used.

- **Flipped assembly on rim**. The tire is tested mounted alternatively on the two sides. Thus, rotation direction is not changed, while the assembly on rim is reversed (flipped tire on rim).

  The tire "*right*" side is conventionally identified with the one for which the conicity force F_{con} is defined positive when directed as the *y-axis* in Figure 1.17, with tire face outboard.

  Ply-steer and conicity components of forces and moments can therefore be computed according to Figure 1.18 as

$$F_{ply} = \frac{F_{y0,right} + F_{y0,left}}{2}$$
$$F_{con} = \frac{F_{y0,right} - F_{y0,left}}{2} \tag{1.6}$$

$$M_{ply} = \frac{M_{z0,right} - M_{z0,left}}{2}$$
$$M_{con} = \frac{M_{z0,right} - M_{z0,left}}{2} \tag{1.7}$$

FIGURE 1.17 Flipped assembly on rim test (adapted from [3]).

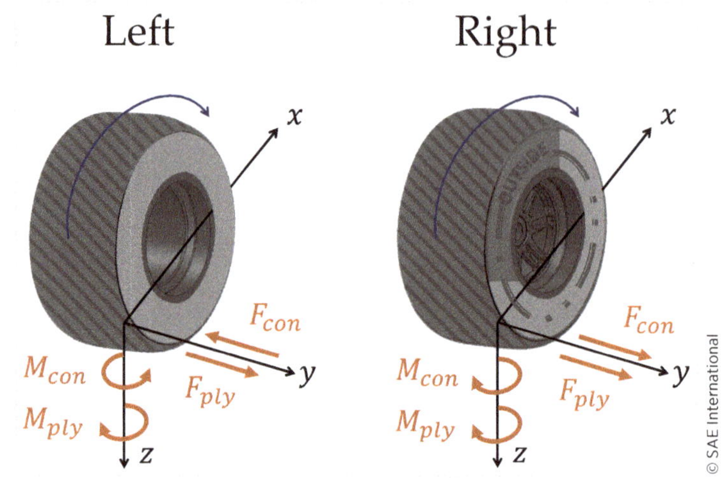

FIGURE 1.18 Side force and aligning moment: ply-steer and conicity components (flipped assembly on rim test).

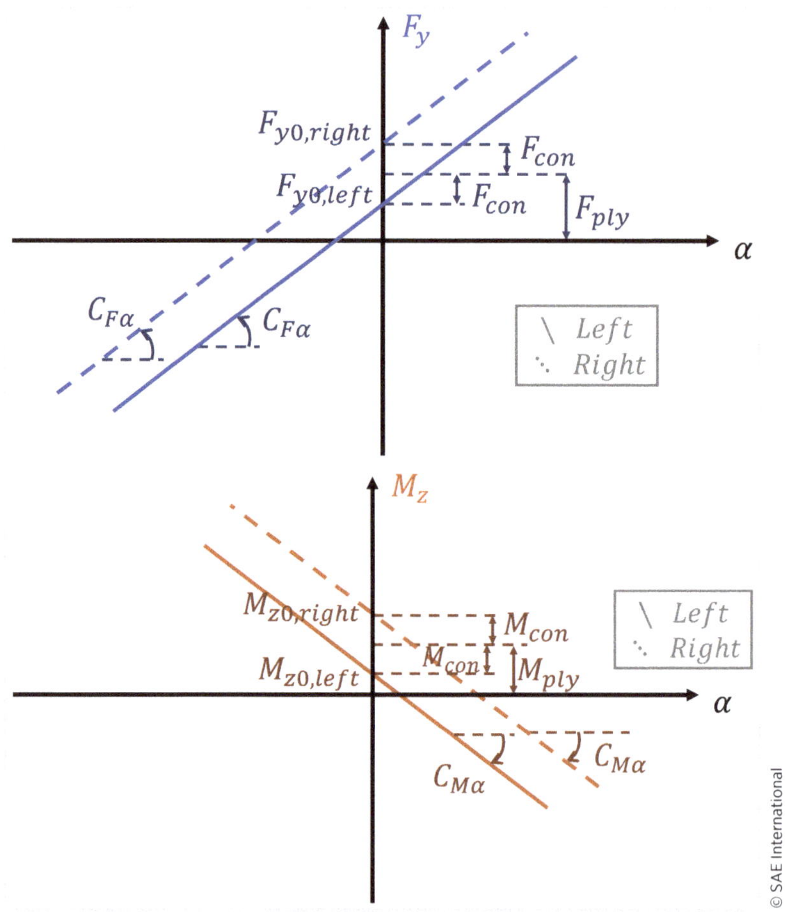

FIGURE 1.19 Reverse direction of rotation test (adapted from [3]).

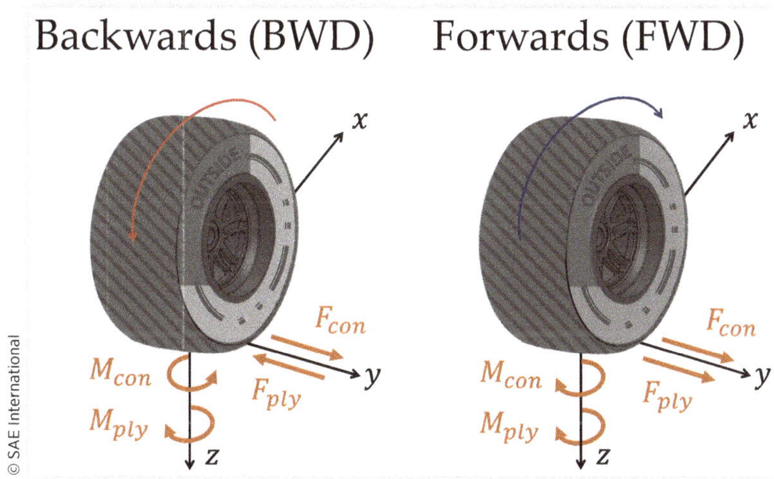

- **Reverse direction of rotation**. This method is not suitable to test directional tires for which the flipped assembly on rim test should be used. The tire is tested rotating alternatively in clockwise and counterclockwise direction. Thus, the assembly on rim is not changed, while rotation direction is reversed, as portrayed in **Figure 1.19**. This is the easiest testing method of the two, since it does not require removal of the tire from the rim or from the testing machine, but just inversion of the direction of motion of the belt (see Figure 1.16).

  Ply-steer and conicity components of forces and moments can therefore be computed according to **Figure 1.20** as

$$F_{ply} = \frac{F_{y0,FWD} - F_{y0,BWD}}{2} \qquad F_{con} = \frac{F_{y0,FWD} + F_{y0,BWD}}{2} \tag{1.8}$$

$$M_{ply} = \frac{M_{z0,FWD} + M_{z0,BWD}}{2} \qquad M_{con} = \frac{M_{z0,FWD} - M_{z0,BWD}}{2} \tag{1.9}$$

  It is to be underlined that the previous equations apply in the absolute axis system represented in Figure 1.19, not linked with tire rotation direction.

It is important to note that the two test methods provide equivalent results, as stated in [3], unless the tire has directional tread. As previously discussed, tread design and geometry have substantial effect on tire force shifts. Therefore, directional tires should be tested with the flipped assembly on rim test.

FIGURE 1.20 Side force and aligning moment: ply-steer and conicity components (reverse direction of rotation test).

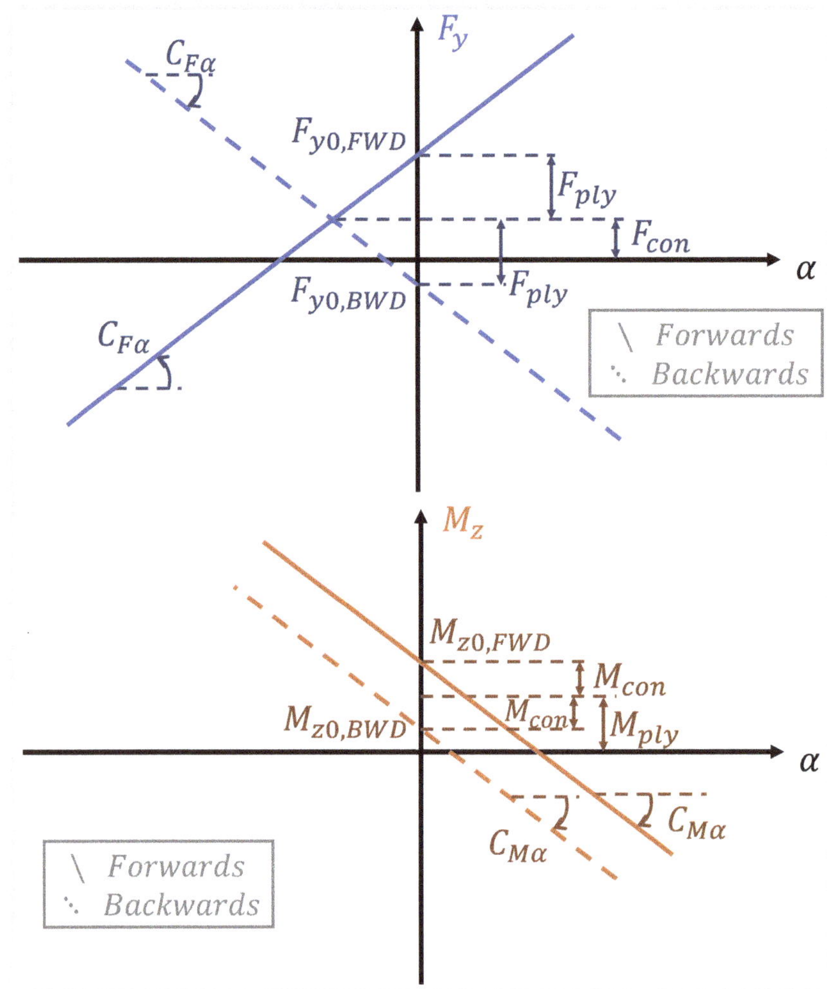

1.7 PRAT and CRAT: Residual Aligning Torques

The Ply-steer Residual Aligning Torque (PRAT) and Conicity Residual Aligning Torque (CRAT) can be computed according to the following expressions, as represented in **Figure 1.21**

$$
\begin{aligned}
PRAT &= M_{ply} + C_{M\alpha} \cdot \frac{F_{ply}}{C_{F\alpha}} \\
CRAT &= M_{con} + C_{M\alpha} \cdot \frac{F_{con}}{C_{F\alpha}}
\end{aligned}
\tag{1.10}
$$

PRAT can be modified by specific design of tire tread.

FIGURE 1.21 Ply-steer and Conicity Residual Aligning Torques (namely *PRAT* and *CRAT*). PRAT graph can be obtained from curves of Figure 1.18 as half of the sum of *Left* and *Right* curves. CRAT graph can be obtained from curves of Figure 1.18 by subtracting ply-steer components (namely F_{ply} and M_{ply}) to *Right* side force and aligning moment curves.

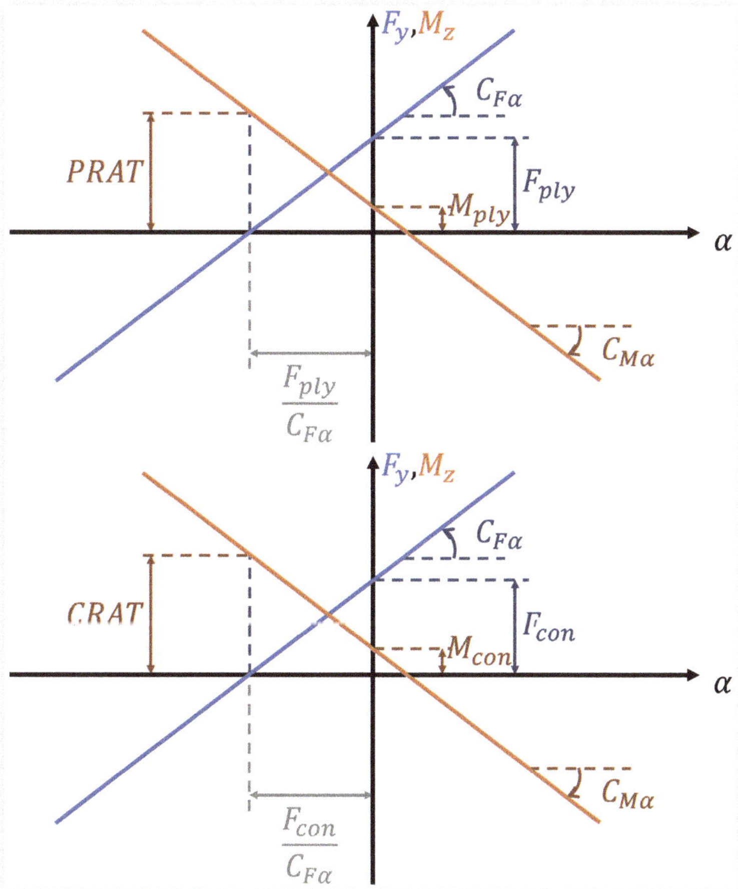

CHAPTER 2

Effective Axle Generalized Forces with Ply-Steer and Conicity

Suspension kinematics and elastokinematics, together with compliances from steering system or vehicle roll angle contributions, are parameters affecting the so-called *effective axle cornering stiffness*. Actually, the cornering stiffness of an axle does not depend only on the cornering stiffnesses of the two tires. The description and computations from Mastinu [24] and Pacejka [7] are integrated here with the ply-steer and conicity equivalent angles described in Chapter 1.

Henceforth, the study will be referring to a single-track vehicle model, as the one represented in Figure 2.6.

It is important to note here that as slip angles are small during straight motion, a linearized vehicle model can be used. In this particular case, the linearized model is exact, despite its simplicity. Starting from the linearized (complex) vehicle model, we can derive a simple linearized model, which provides solutions close to the exact ones.

2.1 Approximated Generalized Forces

In this analysis, small slip angles are expected, which are referred to the effect of tire ply-steer and conicity (pull forces) on straight driving.

The side force F_y can be considered linearly dependent on two variables, namely slip angle α and camber γ, and normal load parameter F_z (a hint on this is reported in Appendix A.3).

$$F_y = C_{F\alpha}\left(F_z\right)\cdot\alpha + C_{F\gamma}\left(F_z\right)\cdot\gamma \tag{2.1}$$

With first-order Taylor expansion, it is possible to linearize the coefficients' dependence on load transfer ΔF_z in the neighborhood of the tire nominal load F_{z0} ($\xi_\gamma = 1\frac{1}{\text{rad}}$ can be considered)

$$
\begin{aligned}
C_{F\alpha} &= C_{F\alpha 0} + \left(\left. \frac{\partial C_{F\alpha}}{\partial F_z} \right|_{F_{z0}} \right) \cdot \Delta F_z = C_{F\alpha 0} + \xi_\alpha \cdot \Delta F_z \\
C_{F\gamma} &= C_{F\gamma 0} + \left(\left. \frac{\partial C_{F\gamma}}{\partial F_z} \right|_{F_{z0}} \right) \cdot \Delta F_z = C_{F\gamma 0} + \xi_\gamma \cdot \Delta F_z
\end{aligned}
\tag{2.2}
$$

Analogously, the aligning moment M_z can be considered to be linearly dependent on two variables, namely slip angle α and camber γ, and normal load parameter F_z

$$
M_z = -C_{M\alpha}\left(F_z\right) \cdot \alpha + C_{M\gamma}\left(F_z\right) \cdot \gamma \tag{2.3}
$$

With first-order Taylor expansion, it is possible to linearize the coefficients' dependence on load transfer ΔF_z in the neighborhood of the tire nominal load F_{z0}

$$
\begin{aligned}
C_{M\alpha} &= C_{M\alpha 0} + \left(\left. \frac{\partial C_{M\alpha}}{\partial F_z} \right|_{F_{z0}} \right) \cdot \Delta F_z = C_{M\alpha 0} \pm \xi_{M\alpha} \cdot \Delta F_z \\
C_{M\gamma} &= C_{M\gamma 0} + \left(\left. \frac{\partial C_{M\gamma}}{\partial F_z} \right|_{F_{z0}} \right) \cdot \Delta F_z = C_{M\gamma 0} \pm \xi_{M\gamma} \cdot \Delta F_z
\end{aligned}
\tag{2.4}
$$

2.2 Steady-State Cornering

Henceforth, the subscript i is used to distinguish between the front axle (i = 1) and the rear axle (i = 2). The steady-state cornering condition, with centripetal acceleration $a_y = v_x^2/R$, is considered to introduce some effects that are described as follows. In this way, it is possible to obtain algebraic expressions for roll and load transfer as functions of lateral acceleration (or side forces).

2.2.1 Roll Angle

Let us consider the moment equilibrium of the car body around roll axis given as (for the meaning of symbols, please refer to Figures 2.1, 2.2, 2.6)

$$
\left(k_{\varphi 1} + k_{\varphi 2}\right) \cdot \varphi - mgh' \cdot sin\varphi + ma_y h' \cdot cos\varphi = 0
$$

By introducing the small angle approximation, the roll angle can be written as

$$
\varphi = \frac{-h'}{k_{\varphi 1} + k_{\varphi 2} - mgh'} \cdot ma_y = \frac{-h'}{k_{\varphi 1} + k_{\varphi 2} - mgh'} \cdot \frac{l}{l - a_i} F_{y,i} \tag{2.5}
$$

FIGURE 2.1 Center of gravity elevation with respect to roll axis.

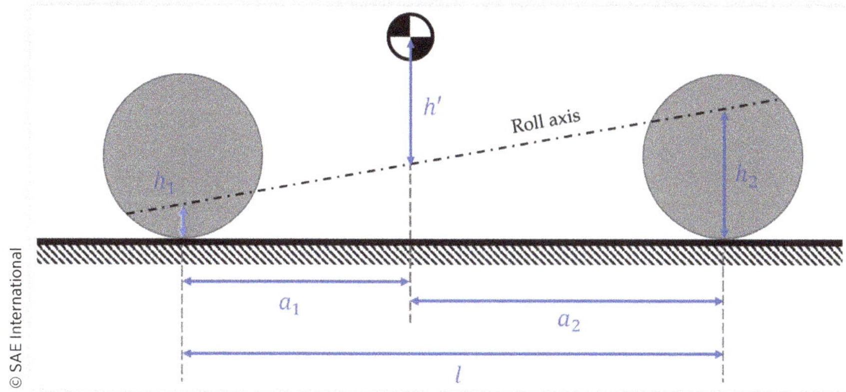

The center of gravity elevation h' with respect to the roll axis can be computed from geometrical considerations in Figure 2.1, with h_G representing the center of gravity height with respect to the ground

$$h' = h_G - \frac{a_1 h_2 + a_2 h_1}{l} \tag{2.6}$$

2.2.2 Load Transfer

Let us consider the moment equilibrium of the *i-th* axle around a wheel's contact point in the rolling vehicle scheme of Figure 2.2. Since

$$F_{y,i} = m_i \cdot a_y = \left(\frac{l - a_i}{l} m \right) \cdot a_y$$

we can write

$$-mh_i \frac{l - a_i}{l} \cdot a_y + k_{\varphi i} \cdot \varphi + \Delta F_{zi} \cdot 2s_i = 0$$

$$\Rightarrow \Delta F_{zi} = \frac{1}{2s_i} \left(-k_{\varphi i} \cdot \varphi + \frac{l - a_i}{l} h_i \cdot m\ a_y \right)$$

Substituting the expression of roll angle conveyed in Equation 2.5, the load transfer on each axle can be computed

$$\Delta F_{zi} = m a_y \sigma_i = F_{y,i} \frac{l}{l - a_i} \sigma_i$$
$$\sigma_i = \frac{1}{2s_i} \left(\frac{k_{\varphi i}}{k_{\varphi 1} + k_{\varphi 2} - mgh'} h' + \frac{l - a_i}{l} h_i \right) \tag{2.7}$$

FIGURE 2.2 Load transfer on single axle.

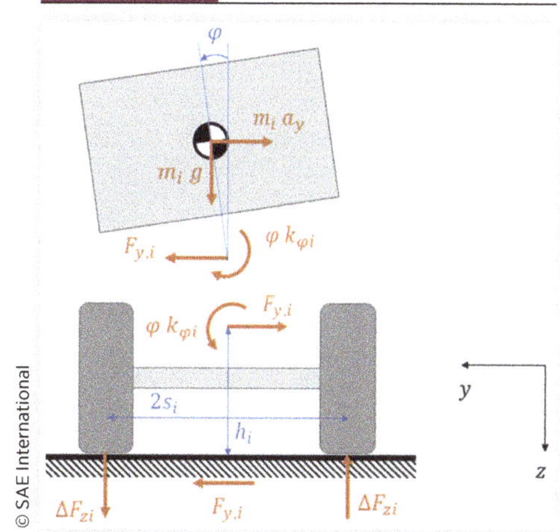

2.2.3 Static Camber and Toe Angles

Initial angles for toe and camber are reported with their positive conventions, respectively, in Figure 2.3 and 2.4.

FIGURE 2.3 Static toe angle sign convention: positive when the forward portion of wheel points farther away from the vehicle centerline (toe-out).

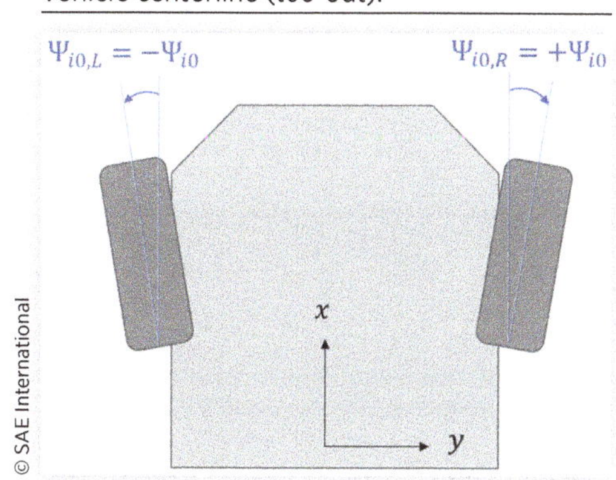

FIGURE 2.4 Static camber angle sign convention: positive when wheel leaning outward at the top. Camber angle pertains to the orientation of the wheel plane relative to the vehicle, whereas inclination angle pertains to the orientation of the wheel plane relative to the road plane.

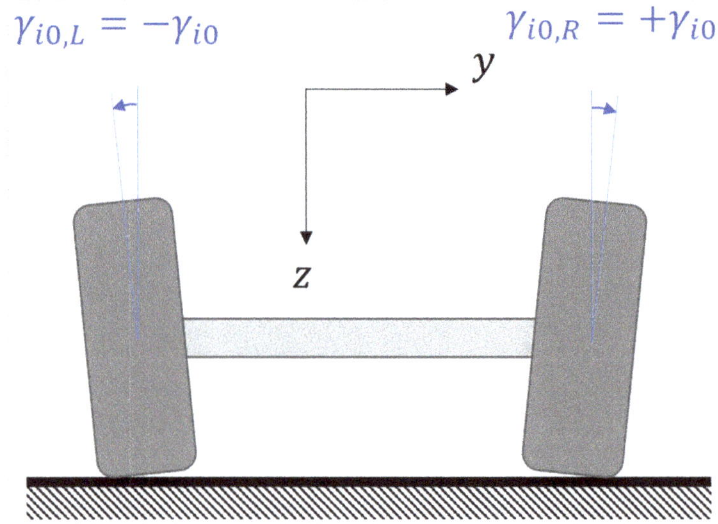

Static toe angle is positive when the forward portion of wheel is pointing outward (toe-out). It is underlined that this sign convention is in line with Pacejka's [7] and opposite to SAE J670 [8] sign convention. Nevertheless, the term "*static toe angle*" is retained to maintain the terminology adopted by Pacejka [7].

Static camber angle is positive when the top of the wheel is leaning outward.

2.3 Elastokinematics Contributions to Cornering Stiffness

Several contributions that affect the actual slip and camber angles are highlighted in [7, 24].

- **Imposed steering angle**. Wheels are rotated around king-pins. The average rotation angle δ is produced by means of a rotation of the steering wheel δ_{sw}. The two angles are related by the average steering ratio n_{st}

$$\delta = \frac{\delta_{sw}}{n_{st}} \tag{2.8}$$

- **Roll-induced steering and camber angles**. Roll angle φ can contribute through suspension kinematics to steering angle and camber with the assumed linear relationships

$$\begin{aligned} \Psi_{r,i} &= \varepsilon_i \cdot \varphi \\ \gamma_{r,i} &= \tau_i \cdot \varphi \end{aligned} \tag{2.9}$$

- **Suspension compliance.** Elastic elements in the suspension (e.g. bushings) can cause an additional steering angle under the action of the lateral force

$$\Psi_{sf,i} = C_{sf,i} \cdot F_{y,i} \tag{2.10}$$

- **Steer compliance.** Side force exerts an external torque about the king-pin axis, with a moment arm given by caster length n and pneumatic trail t_0. This can give rise to an additional steering angle by means of the total steering stiffness around the king-pins $c_{\Psi 1}$

$$\Psi_{c,1} = -\frac{F_{y,1}\left(n - t_0\right)}{c_{\Psi 1}} \tag{2.11}$$

The pneumatic trail for the axle can be obtained as the mean between left and right wheels' pneumatic trails that are defined as

$$t_0 = -\frac{C_{M\alpha}}{C_{F\alpha}} \tag{2.12}$$

- **Ply-steer and conicity.** As already mentioned in Chapter 1, ply-steer and conicity can be expressed respectively as an equivalent side slip angle α_{ply} and an equivalent camber angle γ_{con}. The related definitions are reported in Equation 1.2.

They can be seen as an alternative way of expressing the tire's side force and aligning moment shifts (appearing in Magic Formulae [7]). The introduction of these two terms (α_{ply} and γ_{con}) makes it easier to include tire characteristics' shifts into the expression of effective axle side force.

2.4 Effective Axle Side Force

2.4.1 Effective Axle Side Force with Same Left and Right Characteristics

The full derivation of the effective axle cornering stiffness is reported in Appendix A.4.

It is recalled that the analysis refers to a single-track vehicle model (see Figure 2.6), which is properly manipulated in order to consider distinct contributions of the left and right tires.

Let us introduce

- α_i the axle slip angle due to tire and steer due to suspension kinematics, suspension compliances, and steer compliances previously introduced
- α_{ai} the difference between axle slip angle due to tires and the steer angle due to suspension kinematics, suspension compliances and steering compliances, i.e., α_{ai} is the axle slip angle due to tires. Pacejka [7] calls this angle *virtual slip angle*, defining it as the angle between the direction of motion of the axle i for low speed (thus $F_{y,i} \to 0$) and the actual direction of motion (refer to **Figure 2.5**)

$$\alpha_{ai} = \alpha_i - \Psi_{r,i} - \Psi_{sf,i} - \Psi_{c,i} \tag{2.13}$$

FIGURE 2.5 Virtual slip angle α_{ai}

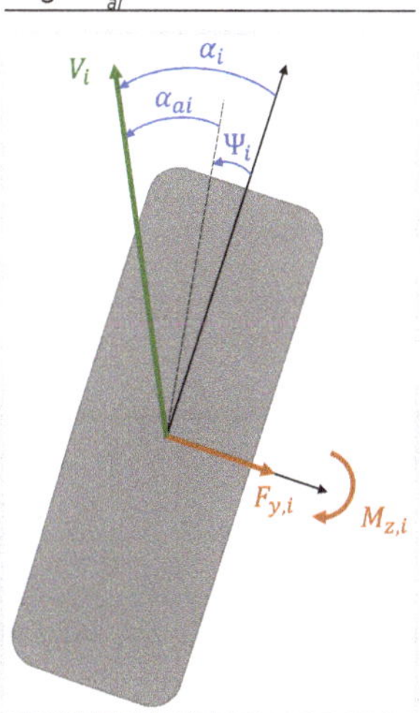

The hypothesis is that the vehicle system is linear, so the superposition of effects can be assumed.

The effective camber is just influenced by the roll-induced contribution

$$\gamma_i = \gamma_{r,i} \tag{2.14}$$

From Equation 2.1, by including the equivalent angles representative of ply-steer and conicity, the lateral forces for left and right wheel of each axle are

$$\begin{aligned} F_{y,iL} &= \left(C_{F\alpha,i0} + \xi_{\alpha,i}\Delta F_{z,i}\right)\left(\alpha_i - \Psi_{i0} + \alpha_{ply,iL}\right) + \left(C_{F\gamma,i0} + \xi_{\gamma,i}\Delta F_{z,i}\right)\left(\gamma_i - \gamma_{i0} + \gamma_{con,iL}\right) \\ F_{y,iR} &= \left(C_{F\alpha,i0} - \xi_{\alpha,i}\Delta F_{z,i}\right)\left(\alpha_i + \Psi_{i0} + \alpha_{ply,iR}\right) + \left(C_{F\gamma,i0} - \xi_{\gamma,i}\Delta F_{z,i}\right)\left(\gamma_i + \gamma_{i0} + \gamma_{con,iR}\right) \end{aligned} \tag{2.15}$$

where the sign of static toe angle Ψ_{i0} and camber angle γ_{i0} for left and right tires depends on the sign convention represented in Figures 2.3, 2.4.

Thus, the axle lateral force is

$$\begin{aligned} F_{y,i} &= F_{y,iL} + F_{y,iR} = \\ &= 2C_{F\alpha,i0}\cdot\alpha_i + 2CF_{\gamma,i0}\cdot\gamma_i - 2\xi_{\alpha,i}\Delta F_{z,i}\cdot\Psi_{i0} - 2\xi_{\gamma,i}\Delta F_{z,i}\cdot\gamma_{i0} + \\ &+ C_{F\alpha,i0}\cdot\left(\alpha_{ply,iL} + \alpha_{ply,iR}\right) + \xi_{\alpha,i}\Delta F_{z,i}\cdot\left(\alpha_{ply,iL} - \alpha_{ply,iR}\right) + \\ &+ C_{F\gamma,i0}\cdot\left(\gamma_{con,iL} + \gamma_{con,iR}\right) + \xi_{\gamma,i}\Delta F_{z,i}\cdot\left(\gamma_{con,iL} - \gamma_{con,iR}\right) \end{aligned} \tag{2.16}$$

The effective axle cornering stiffness is defined as the slope of the lateral force-virtual slip angle diagram, slightly different from Pacejka's definition [7], since the curve is not passing through the origin

$$C_{eff,i} = \frac{dF_{y,i}}{d\alpha_{ai}} \tag{2.17}$$

The derivative is a total derivative since the only variable left is α_{ai}: the effects of camber and load transfer have been included in the effective cornering stiffness. By substituting Equations 2.5-2.13 into Equation 2.16, the effective axle cornering stiffness, accounting also for ply-steer and conicity effect, is (complete computation in Appendix A.4).

$$C^*_{eff,i} = \frac{2C_{F\alpha,i0}}{1 + C_{roll,i} - C_{susp,i} + C_{steer,i} + C^*_{\Delta F_{z,i}}} \tag{2.18}$$

The superscript * highlights the terms which are modified with respect to the dissertations in [7, 24] in order to take into account the role played by ply-steer and conicity. The terms of Equation 2.18 are defined as follows

$$C_{roll,i} = \frac{2l}{l - a_i}\cdot\frac{h'}{k_{\varphi1} + k_{\varphi2} - mgh'}\left(\varepsilon_i C_{F\alpha,i0} + \tau_i C_{F\gamma,i0}\right) \tag{2.19}$$

$$C_{susp,i} = 2C_{F\alpha,i0}\cdot c_{sf,i} \tag{2.20}$$

$$C_{steer,i} = 2C_{F\alpha,i0}\cdot\frac{n_i - t_{0,i}}{c_{\Psi,i}} \tag{2.21}$$

$$C^*_{\Delta F_{z,i}} = \sigma_i \frac{l}{l - a_i} [\xi_{\alpha,i} \left(2\Psi_{i0} - \alpha_{ply,iL} + \alpha_{ply,iR}\right) + \\ + \xi_{\gamma,i} \left(2\gamma_{i0} - \gamma_{con,iL} + \gamma_{con,iR}\right)] \tag{2.22}$$

The effective axle side force can thus be expressed as

$$F_{y,i} = C^*_{eff,i} \cdot \left[\alpha_{ai} + \left(\frac{\alpha_{ply,iL} + \alpha_{ply,iR}}{2} \right) + \left(\frac{\gamma_{con,iL} + \gamma_{con,iR}}{2} \right) \cdot \frac{C_{F\gamma,i0}}{C_{F\alpha,i0}} \right] \tag{2.23}$$

with the ply-steer equivalent angles $\alpha_{ply,iL/R}$ defined from Equation 1.2 as

$$\alpha_{ply,iL/R} = \Delta\alpha_{0,iL/R} - \frac{C_{F\gamma,i0L/R}}{C_{F\alpha,i0L/R}} \cdot \gamma_{con,iL/R} \tag{2.24}$$

We can write the terms referring to the ply-steer and conicity as the axle side force shifts, i.e., the tires' side force horizontal shifts $\Delta\alpha_{0,iL/R}$ (from the Magic Formula or from the experimental data)

$$F_{y,i} = C^*_{eff,i} \cdot \left(\alpha_{ai} + \frac{\Delta\alpha_{0,iL} + \Delta\alpha_{0,iR}}{2} \right) \\ F_{y,i} = C^*_{eff,i} \cdot \alpha_{ai} + F_{y0,i} \tag{2.25}$$

where the side force vertical shift is

$$F_{y0,i} = C^*_{eff,i} \cdot \left(\frac{\Delta\alpha_{0,iL} + \Delta\alpha_{0,iR}}{2} \right) \tag{2.26}$$

2.4.2 Effective Axle Side Force with Different Left and Right Tire Characteristics

From Equation 2.1, by including the equivalent angles representative of ply-steer and conicity, the lateral forces for left and right wheel of each axle are written, considering different cornering stiffnesses and camber coefficients for the left and the right wheel

$$F_{y,iL} = \left(C_{F\alpha,i0L} + \xi_{\alpha,i}\Delta F_{z,i}\right)\left(\alpha_i - \Psi_{i0} + \alpha_{ply,iL}\right) + \left(C_{F\gamma,i0L} + \xi_{\gamma,i}\Delta F_{z,i}\right)\left(\gamma_i - \gamma_{i0} + \gamma_{con,iL}\right) \\ F_{y,iR} = \left(C_{F\alpha,i0R} - \xi_{\alpha,i}\Delta F_{z,i}\right)\left(\alpha_i + \Psi_{i0} + \alpha_{ply,iR}\right) + \left(C_{F\gamma,i0R} - \xi_{\gamma,i}\Delta F_{z,i}\right)\left(\gamma_i + \gamma_{i0} + \gamma_{con,iR}\right) \tag{2.27}$$

It is possible to introduce the mean value between left and right tire's cornering stiffness and camber coefficient for each *i-axle* as

$$C_{F\alpha,i0} = \frac{C_{F\alpha,i0L} + C_{F\alpha,i0R}}{2} \\ C_{F\gamma,i0} = \frac{C_{F\gamma,i0L} + C_{F\gamma,i0R}}{2} \tag{2.28}$$

Thus, the axle lateral force is

$$\begin{aligned} F_{y,i} &= F_{y,iL} + F_{y,iR} = \\ &= 2C_{F\alpha,i0} \cdot \alpha_i + 2C_{F\gamma,i0} \cdot \gamma_i - 2\xi_{\alpha,i} \Delta F_{z,i} \cdot \Psi_{i0} - 2\xi_{\gamma,i} \Delta F_{z,i} \cdot \gamma_{i0} + \\ &+ \xi_{\alpha,i} \Delta F_{z,i} \cdot (\alpha_{ply,iL} - \alpha_{ply,iR}) + \xi_{\gamma,i} \Delta F_{z,i} \cdot (\gamma_{con,iL} - \gamma_{con,iR}) + \\ &+ [C_{F\alpha,i0L} \cdot (\alpha_{ply,iL} - \Psi_{i0}) + C_{F\alpha,i0R} \cdot (\alpha_{ply,iR} + \Psi_{i0})] + \\ &+ [C_{F\gamma,i0L} \cdot (\gamma_{con,iL} - \gamma_{i0}) + C_{F\gamma,i0R} \cdot (\gamma_{con,iR} + \gamma_{i0})] \end{aligned} \tag{2.29}$$

The effective axle cornering stiffness is defined as in Equation 2.17

$$C_{eff,i} = \frac{dF_{y,i}}{d\alpha_{ai}}$$

By substituting Equations 2.5-2.13 into Equation 2.29, the effective axle cornering stiffness accounting also for ply-steer and conicity effect is

$$C^*_{eff,i} = \frac{2C_{F\alpha,i0}}{1 + C_{roll,i} - C_{susp,i} + C_{steer,i} + C^*_{\Delta F_{z,i}}} \tag{2.30}$$

The terms of Equation 2.30 are defined below and are equal to the ones in Equations 2.19-2.22, obtained when the left and right tire characteristics are the same. This is a direct consequence of the terms defined in Equation 2.28, which are thus common in the two cases.

$$C_{roll,i} = \frac{2l}{l - a_i} \cdot \frac{h'}{k_{\varphi 1} + k_{\varphi 2} - mgh'} \left(\varepsilon_i C_{F\alpha,i0} + \tau_i C_{F\gamma,i0} \right) \tag{2.31}$$

$$C_{susp,i} = 2C_{F\alpha,i0} \cdot c_{sf,i} \tag{2.32}$$

$$C_{steer,i} = 2C_{F\alpha,i0} \cdot \frac{n_i - t_{0,i}}{c_{\Psi,i}} \tag{2.33}$$

$$C^*_{\Delta F_z,i} = \sigma_i \frac{l}{l - a_i} \left[\xi_{\alpha,i} \left(2\Psi_{i0} - \alpha_{ply,iL} + \alpha_{ply,iR} \right) + \xi_{\gamma,i} \left(2\gamma_{i0} - \gamma_{con,iL} + \gamma_{con,iR} \right) \right] \tag{2.34}$$

The effective axle side force can thus be expressed as

$$F_{y,i} = C^*_{eff,i} \cdot \alpha_{ai} + F_{y0,i} \tag{2.35}$$

where the force vertical shift is different with respect to the case with equal left and right characteristics (see Equation 2.26)

$$F_{y0,i} = C^*_{eff,i} \cdot \left[\Psi_{i0} \cdot \left(\frac{C_{F\alpha,i0R} - C_{F\alpha,i0L}}{C_{F\alpha,i0R} + C_{F\alpha,i0L}} \right) + \gamma_{i0} \cdot \left(\frac{C_{F\gamma,i0R} - C_{F\gamma,i0L}}{C_{F\alpha,i0R} + C_{F\alpha,i0L}} \right) + \left(\frac{F_{y0,iR} + F_{y0,iL}}{C_{F\alpha,i0R} + C_{F\alpha,i0L}} \right) \right] \tag{2.36}$$

It is important to underline that the last term of Equation 2.36 can be associated with the term expressed in Equation 2.26.

2.4.2.1 Simplified Cases

- **Case 1.**
  - *Hp. 1:* no camber coefficient difference: $C_{Fy,i0L} = C_{Fy,i0R}$

    The axle side force shift expression becomes

$$F_{y0,i} = C^{*}_{eff,i} \cdot \left[\Psi_{i0} \cdot \left(\frac{C_{F\alpha,i0R} + C_{F\alpha,i0L}}{C_{F\alpha,i0R} + C_{F\alpha,i0L}} \right) + \left(\frac{F_{y0,iR} + F_{y0,iL}}{C_{F\alpha,i0R} + C_{F\alpha,i0L}} \right) \right] \tag{2.37}$$

- **Case 2.**
  - *Hp. 1:* elastokinematic effects are not considered (thus, the axle cornering stiffness is $C^{*}_{eff,i} = 2C_{F\alpha,i0}$)
  - *Hp. 2:* no camber coefficient difference: $C_{Fy,i0L} = C_{Fy,i0R}$

    The axle side force shift expression becomes

$$F_{y0,i} = F_{y0,iL} + F_{y0,iR} + \Psi_{i0} \cdot \left(C_{F\alpha,i0R} - C_{F\alpha,i0L} \right) \tag{2.38}$$

This result is meaningful. Actually, it gives evidence of the fact that the toe angle does influence the lateral force at null lateral slip angle α, in case left and right tires have different cornering stiffness, as it is expressed in [18].

2.4.3 Effective Axle Side Force with Cross-Toe and Cross-Camber

It is possible to introduce also wheel alignment tolerances, or misalignments, in the formulation of the effective axle side force. Notably, cross-toe and cross-camber are considered, with different static toe angle and static camber angle for left and right wheels, respectively, being $\Psi_{i0,L} \neq \Psi_{i0,R}$ and $\gamma_{i0,L} \neq \gamma_{i0,R}$.

The effect of wheel alignment tolerance (with difference between left and right side toe and camber) has been studied in [13]. In this book, a more comprehensive description is pursued, with analytical expressions considering

- the difference between left and right tire characteristics
- suspension elastokinematics by means of the effective axle characteristics.

In order to effectively adjust the axle side force expression, here the mean values between left and right tire's static toe and static camber angles are introduced as

$$\Psi_{i0} = \frac{\Psi_{i0,R} + \Psi_{i0,L}}{2} \tag{2.39}$$

$$\gamma_{i0} = \frac{\gamma_{i0,R} + \gamma_{i0,L}}{2} \tag{2.40}$$

Please notice that $\Psi_{i0,R}$, $\Psi_{i0,L}$, $\gamma_{i0,R}$, and $\gamma_{i0,L}$ are intended to be the absolute values of the angles. The correct sign is introduced in the following expressions according to the conventions specified in Figures 2.3, 2.4.

Equation 2.27 turn out to be slightly modified when accommodating different left and right wheel alignment angles

$$\begin{aligned} F_{y,iL} &= \left(C_{F\alpha,i0L} + \xi_{\alpha,i}\Delta F_{z,i}\right)\left(\alpha_i - \Psi_{i0,L} + \alpha_{ply,iL}\right) + \\ &+ \left(C_{F\gamma,i0L} + \xi_{\gamma,i}\Delta F_{z,i}\right)\left(\gamma_i - \gamma_{i0,L} + \gamma_{con,iL}\right) \\ F_{y,iR} &= \left(C_{F\alpha,i0R} - \xi_{\alpha,i}\Delta F_{z,i}\right)\left(\alpha_i + \Psi_{i0,R} + \alpha_{ply,iR}\right) + \\ &+ \left(C_{F\gamma,i0R} - \xi_{\gamma,i}\Delta F_{z,i}\right)\left(\gamma_i + \gamma_{i0,R} + \gamma_{con,iR}\right) \end{aligned} \tag{2.41}$$

The axle lateral force is thus

$$\begin{aligned} F_{y,i} &= F_{y,iL} + F_{y,iR} = \\ &= 2C_{F\alpha,i0}\cdot\alpha_i + 2C_{F\gamma,i0}\cdot\gamma_i - 2\xi_{\alpha,i}\Delta F_{z,i}\cdot\Psi_{i0} - 2\xi_{\gamma,i}\Delta F_{z,i}\cdot\gamma_{i0} + \\ &+ \xi_{\alpha,i}\Delta F_{z,i}\cdot(\alpha_{ply,iL} - \alpha_{ply,iR}) + \xi_{\gamma,i}\Delta F_{z,i}\cdot(\gamma_{con,iL} - \gamma_{con,iR}) + \\ &+ [C_{F\alpha,i0L}\cdot(\alpha_{ply,iL} - \Psi_{i0,L}) + C_{F\alpha,i0R}\cdot(\alpha_{ply,iR} + \Psi_{i0,R})] + \\ &+ [C_{F\gamma,i0L}\cdot(\gamma_{con,iL} - \gamma_{i0,L}) + C_{F\gamma,i0R}\cdot(\gamma_{con,iR} + \gamma_{i0,R})] \end{aligned} \tag{2.42}$$

It is highlighted that this Equation 2.42 is substantially equal to Equation 2.29;: thanks to the mean values of static toe and camber angles Ψ_{i0} and γ_{i0} previously introduced, only the last two terms in square brackets are different when cross-toe and cross-camber are introduced.

As a consequence, the only difference with respect to Equation 2.35 consists in the formula of the force vertical shift $F_{y0,i}$

$$\begin{aligned} F_{y0,i} = C^*_{eff,i}\cdot &\left[\left(\frac{\Psi_{i0,R}\cdot C_{F\alpha,i0R} - \Psi_{i0,L}\cdot C_{F\alpha,i0L}}{C_{F\alpha,i0R} + C_{F\alpha,i0L}}\right) + \left(\frac{\gamma_{i0,R}\cdot C_{F\gamma,i0R} - \gamma_{i0,L}\cdot C_{F\gamma,i0L}}{C_{F\alpha,i0R} + C_{F\alpha,i0L}}\right)\right. \\ &\left. + \left(\frac{F_{yi0,R} + F_{y0,iL}}{C_{F\alpha,i0R} + C_{F\alpha,i0L}}\right)\right] \end{aligned} \tag{2.43}$$

Further details pertaining to the derivation of the above formulae are reported in Appendix A.4.

2.5 Effective Axle Aligning Moment

The same approach adopted for obtaining the effective axle side force $F_{y,i}$ can be considered to find an expression for the effective axle aligning moment $M_{z,i}$. The full derivation is reported in Appendix A.5.

From Equation 2.3, by including the equivalent angles representative of ply-steer and conicity, the aligning moments for left and right wheel of each axle are written, considering different aligning stiffnesses and aligning camber coefficients for the left and the right wheel

$$M_{z,iL} = \left(-C_{M\alpha,i0L} - \xi_{M\alpha,i}\Delta F_{z,i}\right)\left(\alpha_i - \Psi_{i0} + \alpha_{ply,iL}\right) + \left(C_{M\gamma,i0L} + \xi_{M\gamma,i}\Delta F_{z,i}\right)\left(\gamma_i - \gamma_{i0} + \gamma_{con,iL}\right)$$
$$M_{z,iR} = \left(-C_{M\alpha,i0R} + \xi_{M\alpha,i}\Delta F_{z,i}\right)\left(\alpha_i + \Psi_{i0} + \alpha_{ply,iR}\right) + \left(C_{M\gamma,i0R} - \xi_{M\gamma,i}\Delta F_{z,i}\right)\left(\gamma_i + \gamma_{i0} + \gamma_{con,iR}\right) \tag{2.44}$$

It is possible to introduce the mean value between left and right tire's aligning stiffness and aligning camber coefficient for each *i-axle* as

$$C_{M\alpha,i0} = \frac{C_{M\alpha,i0L} + C_{M\alpha,i0R}}{2}$$
$$C_{M\gamma,i0} = \frac{C_{M\gamma,i0L} + C_{M\gamma,i0R}}{2} \tag{2.45}$$

Thus, the axle aligning moment is

$$\begin{aligned} M_{z,i} &= M_{z,iL} + M_{z,iR} = \\ &= -2C_{M\alpha,i0}\cdot\alpha_i + 2C_{M\gamma,i0}\cdot\gamma_i + 2\xi_{M\alpha,i}\Delta F_{z,i}\cdot\Psi_{i0} - 2\xi_{M\gamma,i}\Delta F_{z,i}\cdot\gamma_{i0} - \\ &- \xi_{M\alpha,i}\Delta F_{z,i}\cdot(\alpha_{ply,iL} - \alpha_{ply,iR}) + \xi_{M\gamma,i}\Delta F_{z,i}\cdot(\gamma_{con,iL} - \gamma_{con,iR}) + \\ &+ [-C_{M\alpha,i0L}\cdot(\alpha_{ply,iL} - \Psi_{i0}) - C_{M\alpha,i0R}\cdot(\alpha_{ply,iR} + \Psi_{i0})] + \\ &+ [C_{M\gamma,i0L}\cdot(\gamma_{con,iL} - \gamma_{i0}) + C_{M\gamma,i0R}\cdot(\gamma_{con,iR} + \gamma_{i0})] \end{aligned} \tag{2.46}$$

The effective axle aligning stiffness is defined as

$$C_{M\alpha,i} = -\frac{dM_{z,i}}{d\alpha_{ai}}$$

By substituting Equations 2.5-2.13 and 2.35 into Equation 2.46, the effective axle aligning stiffness accounting also for ply-steer and conicity effect is

$$-C_{M\alpha,i} = -2C_{M\alpha,i} + C^{*}_{eff,i}\cdot\left(C_{M\alpha,roll,i} - C_{M\alpha,susp,i} + C_{M\alpha,steer,i} + C_{M\alpha,\Delta F_z,i}\right) \tag{2.47}$$

The terms of Equation 2.47 are defined below

$$C_{M\alpha,roll,i} = \frac{2l}{l-a_i}\cdot\frac{h'}{k_{\varphi1} + k_{\varphi2} - mgh'}\left(\varepsilon_i C_{M\alpha,i0} - \tau_i C_{M\gamma,i0}\right) \tag{2.48}$$

$$C_{M\alpha,susp,i} = 2C_{M\alpha,i0}\cdot c_{sf,i} \tag{2.49}$$

$$C_{M\alpha,steer,i} = 2C_{M\alpha,i0}\cdot\frac{n_i - t_{0,i}}{c_{\Psi,i}} \tag{2.50}$$

$$C_{M\alpha,\Delta F_z,i} = \sigma_i\frac{l}{l-a_i}\left[\xi_{M\alpha,i}\left(2\Psi_{i0} - \alpha_{ply,iL} + \alpha_{ply,iR}\right) + \xi_{M\gamma,i}\left(-2\gamma_{i0} + \gamma_{con,iL} - \gamma_{con,iR}\right)\right] \tag{2.51}$$

The effective axle aligning moment can thus be expressed as

$$M_{z,i} = -C_{M\alpha,i} \cdot \alpha_{ai} + M_{z0,i} \tag{2.52}$$

where the aligning moment shift is

$$\begin{aligned} M_{z0,i} = F_{y0,i} \cdot \left(C_{M\alpha,roll,i} - C_{M\alpha,susp,i} + C_{M\alpha,steer,i} + C_{M\alpha,\Delta F_{z,i}}\right) \\ + \Psi_{i0} \cdot \left(C_{M\alpha,i0L} - C_{M\alpha,i0R}\right) - \gamma_{i0} \cdot \left(C_{M\gamma,i0L} - C_{M\gamma,i0R}\right) + M_{z0,iL} + M_{z0,iR} \end{aligned} \tag{2.53}$$

2.5.1 Effective Axle Aligning Moment with Cross-Toe and Cross-Camber

As discussed for the axle side force, it is possible to introduce cross-toe and cross-camber in the formulation of the effective axle aligning moment as well.

Equation 2.44 is slightly modified in order to accommodate different left and right wheel alignment angles

$$\begin{aligned} M_{z,iL} &= \left(-C_{M\alpha,i0L} - \xi_{M\alpha,i} \Delta F_{z,i}\right)\left(\alpha_i - \Psi_{i0,L} + \alpha_{ply,iL}\right) + \left(C_{M\gamma,i0L} + \xi_{M\gamma,i} \Delta F_{z,i}\right)\left(\gamma_i - \gamma_{i0,L} + \gamma_{con,iL}\right) \\ M_{z,iR} &= \left(-C_{M\alpha,i0R} + \xi_{M\alpha,i} \Delta F_{z,i}\right)\left(\alpha_i + \Psi_{i0,R} + \alpha_{ply,iR}\right) + \left(C_{M\gamma,i0R} - \xi_{M\gamma,i} \Delta F_{z,i}\right)\left(\gamma_i + \gamma_{i0,R} + \gamma_{con,iR}\right) \end{aligned} \tag{2.54}$$

By introducing the mean values between static toe and static camber angles (Equations 2.39-2.40) of the left and right tires, the axle aligning moment becomes

$$\begin{aligned} M_{z,i} &= M_{z,iL} + M_{z,iR} = \\ &= -2C_{M\alpha,i0} \cdot \alpha_i + 2C_{M\gamma,i0} \cdot \gamma_i + 2\xi_{M\alpha,i} \Delta F_{z,i} \cdot \Psi_{i0} - 2\xi_{M\gamma,i} \Delta F_{z,i} \cdot \gamma_{i0} - \\ &- \xi_{M\alpha,i} \Delta F_{z,i} \cdot (\alpha_{ply,iL} - \alpha_{ply,iR}) + \xi_{M\gamma,i} \Delta F_{z,i} \cdot (\gamma_{con,iL} - \gamma_{con,iR}) + \\ &+ [-C_{M\alpha,i0L} \cdot (\alpha_{ply,iL} - \Psi_{i0,L}) - C_{M\alpha,i0R} \cdot (\alpha_{ply,iR} + \Psi_{i0,R})] + \\ &+ [C_{M\gamma,i0L} \cdot (\gamma_{con,iL} - \gamma_{i0,L}) + C_{M\gamma,i0R} \cdot (\gamma_{con,iR} + \gamma_{i0,R})] \end{aligned} \tag{2.55}$$

It is highlighted that this Equation 2.55 is substantially equal to Equation 2.46; thanks to the mean values of static toe and camber angles Ψ_{i0} and γ_{i0} previously introduced, only the last two terms in brackets are different when cross-toe and cross-camber are introduced.

As a consequence, the only difference with respect to Equation 2.52 consists in the formula of the aligning moment vertical shift $M_{z0,i}$

$$\begin{aligned} M_{z0,i} &= F_{y0,i} \cdot \left(C_{M\alpha,roll,i} - C_{M\alpha,susp,i} + C_{M\alpha,steer,i} + C_{M\alpha,\Delta F_{z,i}}\right) + \\ &+ \left(\Psi_{i0,L} \cdot C_{M\alpha,i0L} - \Psi_{i0,R} \cdot C_{M\alpha,i0R}\right) - \left(\gamma_{i0,L} \cdot C_{M\gamma,i0L} - \gamma_{i0,R} \cdot C_{M\gamma,i0R}\right) + \\ &+ M_{z0,iL} + M_{z0,iR} \end{aligned} \tag{2.56}$$

2.6 Effective Axle Characteristics in a Simple Vehicle Model

A single-track (or bicycle) roll axis vehicle model [7] is presented in this section. It has four degrees of freedom (longitudinal velocity u, lateral velocity v, yaw rate r, and roll angle φ).

The number of degrees of freedom is reduced to three in case the longitudinal velocity u is considered constant (or quasi-static), as in this book. It considers roll dynamics and no pitch dynamics. It can accommodate separated tire slip forces and moments.

Steering angle δ, slip angles α, roll angle φ, and roll axis inclination θ are assumed to remain small.

The complete equations of motion as in Pacejka [7] are reported in Equation 2.57. The involved parameters can be identified by inspection of **Figure 2.6**.

FIGURE 2.6 Simple vehicle model.

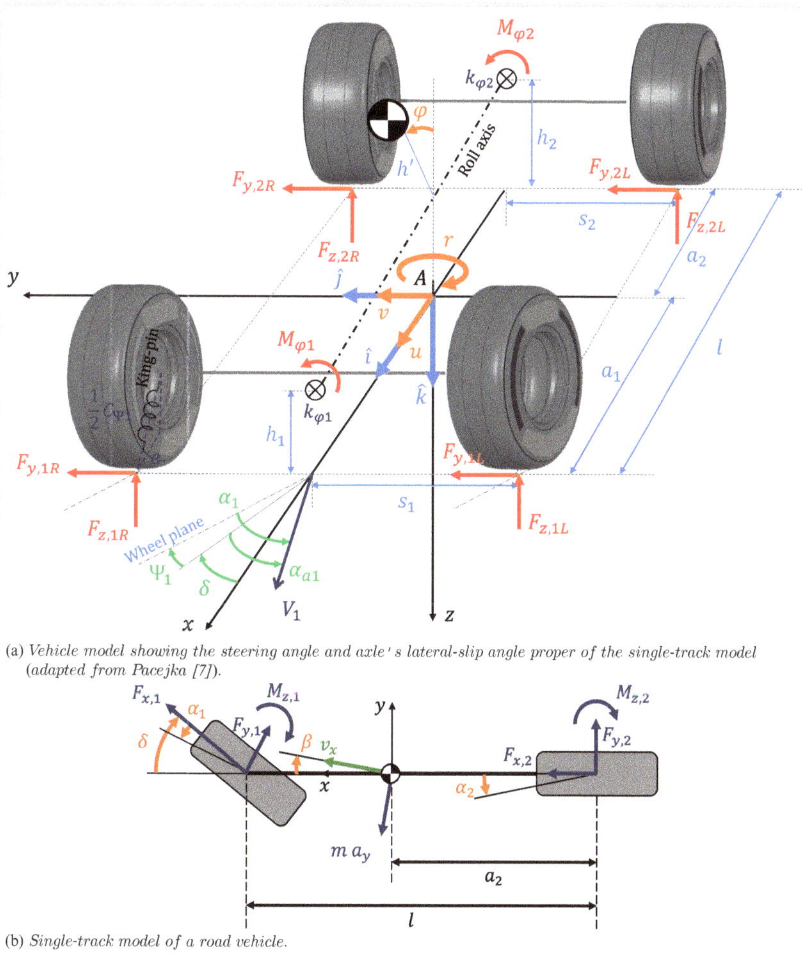

(a) *Vehicle model showing the steering angle and axle's lateral-slip angle proper of the single-track model (adapted from Pacejka [7]).*

(b) *Single-track model of a road vehicle.*

FIGURE 2.7 Effective *i-axle* cornering stiffness (adapted from [7]).

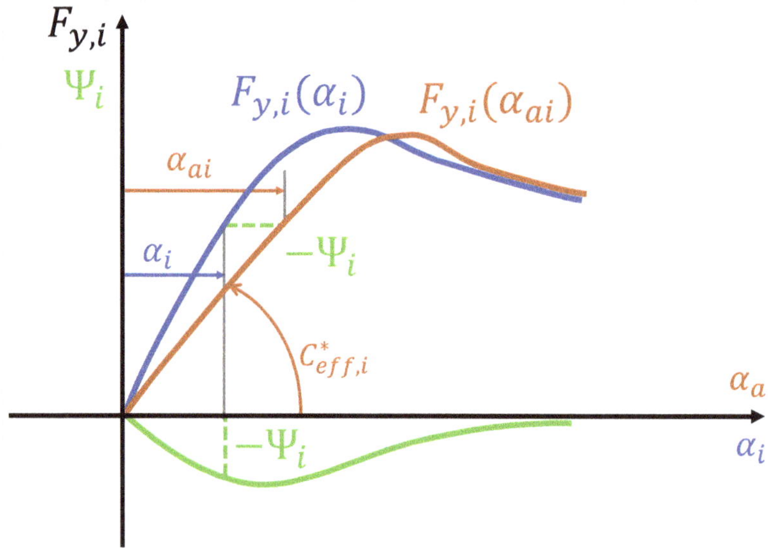

$$\begin{cases} \swarrow u: & m(\dot{u} - rv - h'\varphi\dot{r} - 2h'r\dot{\varphi}) = F_{x,1} - F_{y,1}\delta + F_{x,2} \\ \nwarrow v: & m(\dot{v} + ru + h'\ddot{\varphi} - h'r^2\varphi) = F_{x,1}\delta + F_{y,1} + F_{y,2} \\ \curvearrowright r: & I_z\dot{r} + (I_z\theta_r - I_{xz})\ddot{\varphi} - mh'(\dot{u} - rv)\varphi = \\ & a_1F_{x,1}\delta + a_1F_{y,1} + M_{z,1} - a_2F_{y,2} + M_{z,2} \\ \curvearrowleft \varphi: & (I_x + mh'^2)\ddot{\varphi} + mh'(\dot{v} + ru) + (I_z\theta_r - I_{xz})\dot{r} - \\ & -(mh'^2 + I_y - I_z)r^2\varphi + (c_{\varphi1} + c_{\varphi2})\dot{\varphi} + \\ & +(k_{\varphi1} + k_{\varphi2} - mgh')\varphi = 0 \end{cases} \tag{2.57}$$

More details pertaining to the derivation of these equations of motion are reported in Appendix A.1.

The steady-state (or the quasi-steady-state) turning behavior of the vehicle can be analyzed using the simple single-track (or bicycle) model, as discussed in [7]. The simple model does not account for elastokinematics in the suspension and steering systems, nor for the load transfer and body roll. These effects can be considered by adopting the effective axle cornering stiffness $C^*_{eff,i}$, introduced in this chapter, as represented qualitatively in Figure 2.7.

Henceforth, virtual side slip angles α_{ai} will be called α_i for sake of simplicity.

CHAPTER 3

Tire Effect on Vehicle Straight Line Motion

Developing from the study of Lee [17], the ply-steer and conicity effects previously discussed are considered together with the simple vehicle model introduced in Section 2.6. In this chapter, the effect of tire characteristics on vehicle straight motion is analyzed, with a link to the impact on the handling diagram theory.

If (from Figure 2.6) $s_i \cdot |r| \ll u$ then $v_x \approx u$. Thus, slip angles of the axles are defined as [7]

$$\begin{aligned} \alpha_1 &= \delta - \frac{v + a_1 r}{u} \\ \alpha_2 &= -\frac{v - a_2 r}{u} \end{aligned} \tag{3.1}$$

The steering angle δ is obtained from the difference between the front and the rear slip angles (defined in Equation 3.1)

$$\alpha_1 - \alpha_2 = \delta - l\frac{r}{u} = \delta - \frac{1}{R} = \delta - \delta_0 \tag{3.2}$$

where R is the curve radius when in circular motion.

When driving straight, the following conditions hold

- $r = 0,\ \dot{r} = 0$
- $\dot{v} = 0$
- $\dot{\varphi} = 0,\ \ddot{\varphi} = 0$

The lateral acceleration $a_y = \dot{v} + ru$ is therefore null. Its expression can be derived from Figure 2.6

$$\begin{aligned}\vec{a} &= \dot{u}\hat{i} + \dot{v}\hat{j} + \left(r\hat{k}\right) \wedge [u, v, 0] = \\ &= \left(\dot{u} - rv\right)\hat{i} + \left(\dot{v} + ru\right)\hat{j} = \left(a_x\right)\hat{i} + \left(a_y\right)\hat{j}\end{aligned} \tag{3.3}$$

Thus, the equations of motion of the simple vehicle model (which have been introduced in Equation 2.57) become

$$\begin{cases} \nwarrow v: \quad F_{x,1}\delta + F_{y,1} + F_{y,2} = 0 \\ \curvearrowright r: \quad a_1 F_{x,1}\delta + a_1 F_{y,1} + M_{z,1} - a_2 F_{y,2} + M_{z,2} = 0 \end{cases} \tag{3.4}$$

From Equation 2.25, the side forces expression is reported here as

$$F_{y,i} = C^*_{eff,i} \cdot \alpha_i + F_{y0,i} \tag{3.5}$$

To begin with, a simple linear formulation for the aligning moment characteristics can be considered

$$\begin{aligned} &M_{z,i} = -C_{M\alpha,i} \cdot \alpha_i + M_{z0,i} \\ &\text{where} \\ &-C_{M\alpha,i} = -\left(C_{M\alpha,iL} + C_{M\alpha,iR}\right) \\ &M_{z0,i} = M_{z0,iL} + M_{z0,iR} \end{aligned} \tag{3.6}$$

The effect of the longitudinal tire forces F_{xi} is neglected: the tire pull effect is analyzed in this case during a free rolling straight run, in neutral gear. By substituting Equations 3.5 and 3.6 into Equation 3.4, the side slip angles for front and rear axle can be obtained (the full derivation is reported in Appendix A.6)

- **Straight motion front slip angle**

$$\alpha_{1,st} = \frac{-C^*_{eff,2}\left(M_{z0,1} + M_{z0,2} - lF_{y0,1}\right) - C_{M\alpha,2}\left(F_{y0,1} + F_{y0,2}\right)}{lC^*_{eff,1}C^*_{eff,2} - C_{M\alpha,1}C^*_{eff,2} + C_{M\alpha,2}C^*_{eff,1}} \tag{3.7}$$

- **Straight motion rear slip angle**

$$\alpha_{2,st} = \frac{C^*_{eff,1}\left(M_{z0,1} + M_{z0,2} - lF_{y0,2}\right) + C_{M\alpha,1}\left(F_{y0,1} + F_{y0,2}\right)}{lC^*_{eff,1}C^*_{eff,2} - C_{M\alpha,1}C^*_{eff,2} + C_{M\alpha,2}C^*_{eff,1}} \tag{3.8}$$

Henceforth, the subscript *st* is used to indicate the *straight motion* condition.

From Equation 3.2, the steering angle δ_{st} when driving straight $\left(\delta_0 = \frac{l}{R} = 0\right)$ is computed as the difference between front and rear slip angles

- **Straight motion steering angle**

$$\delta_{st} = \alpha_{1,st} - \alpha_{2,st} = \\ = \frac{-\left(C^*_{eff,1} + C^*_{eff,2}\right)\left(M_{z0,1} + M_{z0,2}\right) - \left(C_{M\alpha,1} + C_{M\alpha,2}\right)\left(F_{y0,1} + F_{y0,2}\right)}{lC^*_{eff,1}C^*_{eff,2} - C_{M\alpha,1}C^*_{eff,2} + C_{M\alpha,2}C^*_{eff,1}} + \\ + \frac{l\left(C^*_{eff,1}F_{y0,2} - C^*_{eff,2}F_{y0,1}\right)}{lC^*_{eff,1}C^*_{eff,2} - C_{M\alpha,1}C^*_{eff,2} + C_{M\alpha,2}C^*_{eff,1}} \tag{3.9}$$

The attitude β of the vehicle can be computed by applying the sinus theorem to the scheme of Figure 3.1, assuming small angles

$$\frac{R}{a_2} = \frac{sin\left(\frac{\pi}{2} - \alpha_2\right)}{sin\left(\beta + \alpha_2\right)} \approx \frac{1}{\beta + \alpha_2} \\ \beta = -\alpha_2 + \frac{a_2}{R} \tag{3.10}$$

Thus, the expression of vehicle attitude during straight motion is

$$\beta_{st} = -\alpha_{2,st} + \frac{a_2}{R} = -\alpha_{2,st} = \\ = \frac{C^*_{eff,1}\left(M_{z0,1} + M_{z0,2} - lF_{y0,2}\right) + C_{M\alpha,1}\left(F_{y0,1} + F_{y0,2}\right)}{lC^*_{eff,1}C^*_{eff,2} - C_{M\alpha,1}C^*_{eff,2} + C_{M\alpha,2}C^*_{eff,1}} \tag{3.11}$$

FIGURE 3.1 Vehicle attitude during circular driving (scheme adapted from [24]).

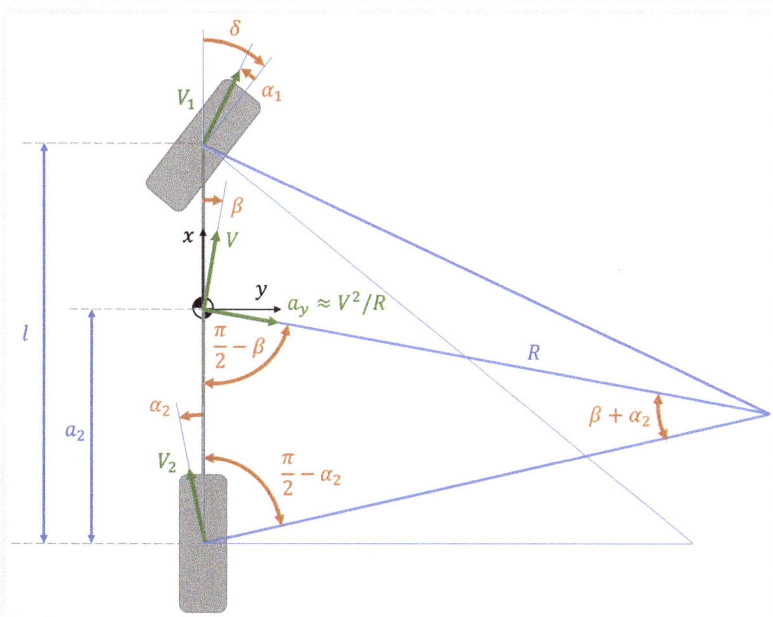

3.1 Simplifications, PRAT and CRAT Contributions

In Equations 3.7, 3.8, and 3.9, the offsets and stiffness related to aligning moment are considerably smaller (at least one order of magnitude for both) than the ones related to side force. Therefore, the side slip and steering angles can be approximated as

$$\begin{aligned}\alpha_{1,st} &\approx \frac{-C^*_{eff,2}\cdot lF_{y0,1}}{lC^*_{eff,1}C^*_{eff,2}} = -\frac{F_{y0,1}}{C^*_{eff,1}}\\ \alpha_{2,st} &\approx \frac{-C^*_{eff,1}\cdot lF_{y0,2}}{lC^*_{eff,1}C^*_{eff,2}} = -\frac{F_{y0,2}}{C^*_{eff,2}}\\ \delta_{st} &\approx \frac{l\left(C^*_{eff,1}F_{y0,2}-C^*_{eff,2}F_{y0,1}\right)}{lC^*_{eff,1}C^*_{eff,2}} = -\frac{F_{y0,1}}{C^*_{eff,1}}+\frac{F_{y0,2}}{C^*_{eff,2}}\end{aligned} \tag{3.12}$$

This solution corresponds to Equation 3.4 without the moments $M_{z,1}$ and $M_{z,2}$, which appear to provide a negligible contribution. The characteristics of each *i-axle* are qualitatively represented in **Figure 3.2**.

It is important to underline that the simplifications expressed in Equation 3.12 can be adopted also in case the effective axle aligning moment (introduced in Section 2.5) is considered: indeed, the aligning moment characteristics remain aptly smaller than the side force characteristics.

For the straight motion condition, each axle has an approximately null side force $F_{yi,st}$ and a nonnull residual aligning moment $M_{zi,st}$

$$\begin{aligned}F_{yi,st} &= C^*_{eff,i}\cdot\alpha_{i,st}+F_{y0,i}\approx 0\\ M_{zi,st} &= -C_{M\alpha,i}\cdot\alpha_{i,st}+M_{z0,i}\end{aligned} \tag{3.13}$$

In the simplest case, with equal left and right tire cornering stiffness and aligning stiffness and $C^*_{eff,i}=2C_{F\alpha,i0}$, the residual aligning moment can be computed from the

FIGURE 3.2 Axle *i* side force and aligning moment in the straight motion condition.

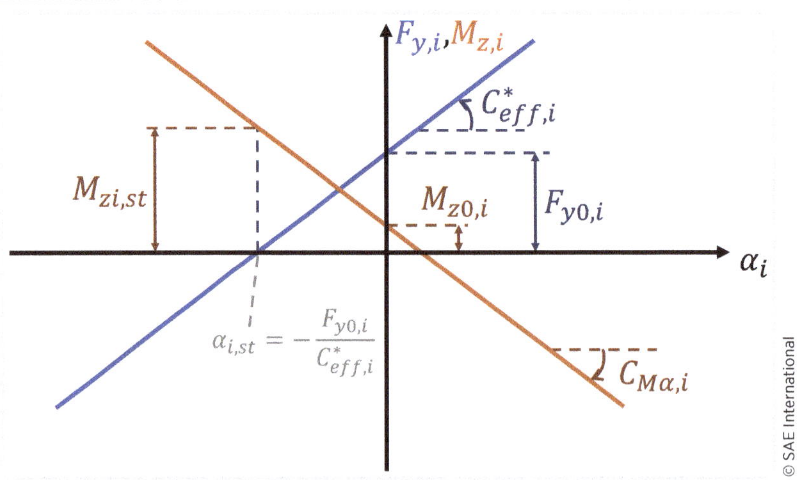

FIGURE 3.3 Axle's tires assembly configurations. Letters *r* and *l* are indicating, respectively, the *right* and the *left* tire of the default configuration (see *Configuration a*). Subscripts *out* and *in* of the symbols *r* and *l* are used to indicate whether the tire is mounted on the rim with face, respectively *outboard* or *inboard*. Letters *R* and *L* refer, respectively, to the *right* and the *left* wheel of the axle.

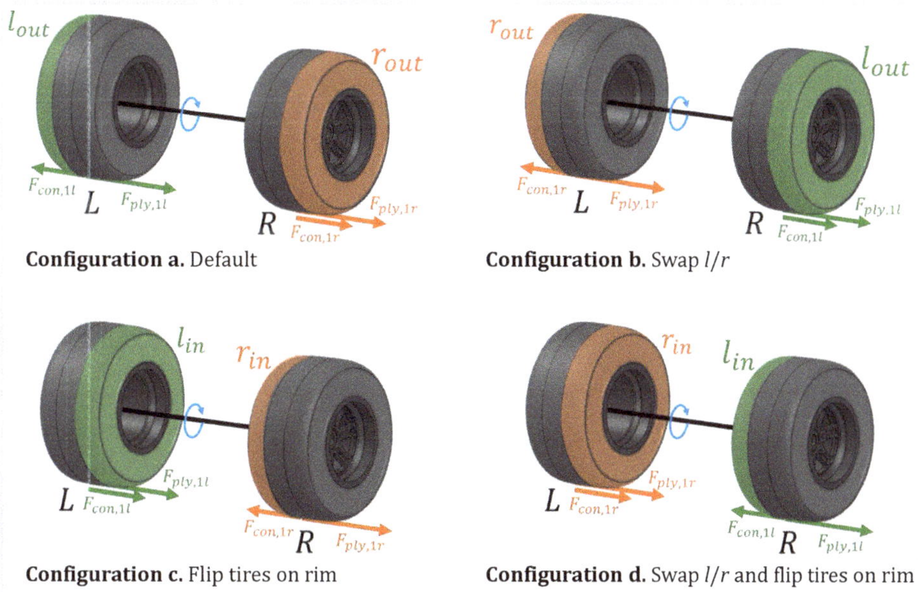

values of *PRAT* and *CRAT* of the single tires. The value depends on the tires' assembly configuration. It is assumed that left and right tire characteristics are expressed with their ply-steer and conicity components (here reported for tires mounted with face outboard, refer to Figure 1.17)

$$
\begin{aligned}
F_{y0,ir} &= F_{ply,ir} + F_{con,ir} \\
F_{y0,il} &= F_{ply,il} - F_{con,il} \\
M_{z0,ir} &= M_{ply,ir} + M_{con,ir} \\
M_{z0,il} &= M_{ply,il} - M_{con,il}
\end{aligned}
$$

The configurations represented in **Figure 3.3** are analyzed.

- **Configuration a:** In this default configuration, tires are mounted with the sides intended to be external towards the outside (*out*). The tire mounted on the right wheel of the axle is identified with letter *r*, while the other one is identified with letter *l*. The residual aligning moment on the axle is

$$
\begin{aligned}
M_{zi,st} &= \frac{C_{M\alpha,i}}{2C_{F\alpha,i0}}\left(F_{y0,iR} + F_{y0,iL}\right) + M_{z0,iR} + M_{z0,iL} = \\
&= \frac{C_{M\alpha,i}}{2C_{F\alpha,i0}}\left(F_{ply,ir} + F_{con,ir} + F_{ply,il} - F_{con,il}\right) + M_{ply,ir} + M_{con,ir} + M_{ply,il} - M_{con,il} = \\
&= PRAT_{ir} + PRAT_{il} + CRAT_{ir} - CRAT_{il}
\end{aligned}
$$

Notice that in reality seldom, if ever, do a pair of tires have the exact same *PRAT* and *CRAT*.

- **Configuration b:** In this configuration, axle's left and right tires are swapped, maintaining the sides intended to be external towards the outside (*out*). The residual aligning moment on the axle is

$$M_{zi,st} = \frac{C_{M\alpha,i}}{2C_{F\alpha,i0}}\left(F_{y0,iR} + F_{y0,iL}\right) + M_{z0,iR} + M_{z0,iL} =$$
$$= \frac{C_{M\alpha,i}}{2C_{F\alpha,i0}}\left(F_{ply,il} + F_{con,il} + F_{ply,ir} = F_{con,ir}\right) + M_{ply,il} + M_{con,il} + M_{ply,ir} - M_{con,ir} =$$
$$= PRAT_{ir} + PRAT_{il} - CRAT_{ir} + CRAT_{il}$$

In the simplest case, with left and right tire with same *PRAT* and *CRAT*, the two configurations discussed would have a same residual aligning moment $M_{zi,st} = 2 \cdot PRAT_i$, equal to two times the *PRAT* of the single tire.

As it can be observed, when $CRAT_{ir} \approx CRAT_{il}$ there is no significant difference between *default* configuration and *swap L/R* configuration: the two terms cancel out and only *PRAT* contributions remain. In reality, *CRAT* is strongly influenced by manufacturing variability, thus it is unlikely for a pair of tires to have the exact same *CRAT*. Therefore, a different residual aligning moment $M_{zi,st}$ can be observed when swapping left and right tires of an axle.

- **Configuration c:** In this configuration, axle's tires are flipped on the rim, thus with the sides intended to be external towards the inside (*in*). The residual aligning moment on the axle is

$$M_{zi,st} = \frac{C_{M\alpha,i}}{2C_{F\alpha,i0}}\left(F_{y0,iR} + F_{y0,iL}\right) + M_{z0,iR} + M_{z0,iL} =$$
$$= \frac{C_{M\alpha,i}}{2C_{F\alpha,i0}}\left(F_{ply,ir} - F_{con,ir} + F_{ply,il} + F_{con,il}\right) + M_{ply,ir} - M_{con,ir} + M_{ply,il} + M_{con,il} =$$
$$= PRAT_{ir} + PRAT_{il} - CRAT_{ir} + CRAT_{il}$$

For a vehicle with same load on the left and right tires, this configuration provides the same residual aligning moment of the *swap L/R* configuration, since the direction of conicity components is the same.

- **Configuration d:** In this configuration, axle's tires are both swapped and flipped on the rim, thus with the sides intended to be external towards the inside (*in*). The residual aligning moment on the axle is

$$M_{zi,st} = \frac{C_{M\alpha,i}}{2C_{F\alpha,i0}}\left(F_{y0,iR} + F_{y0,iL}\right) + M_{z0,iR} + M_{z0,iL} =$$
$$= \frac{C_{M\alpha,i}}{2C_{F\alpha,i0}}\left(F_{ply,ir} - F_{con,il} + F_{ply,ir} = F_{con,ir}\right) + M_{ply,il} + M_{con,il} + M_{ply,ir} + M_{con,ir} =$$
$$= PRAT_{ir} + PRAT_{il} + CRAT_{ir} - CRAT_{il}$$

For a vehicle with same load on the left and right tires, this configuration provides the same residual aligning moment of the *default* configuration, since the direction of conicity components is the same.

3.2 Handling Diagram

The results of Equations 3.7, 3.8, and 3.9 can be examined from the point of view of the handling diagram theory [7, 24], in its version including also tires' aligning moments.

With reference to Figure 3.4, it is possible to write moment equilibrium around the front and rear tire contact points as

$$\begin{cases} F_{y,1}\cdot l = ma_y \cdot a_2 - \left(M_{z,1}+M_{z,2}\right) \\ F_{y,2}\cdot l = ma_y \cdot a_1 + \left(M_{z,1}+M_{z,2}\right) \end{cases} \tag{3.14}$$

By dividing the two equations respectively for mga_2 and mga_1, and substituting the normal forces on the axles we get

$$F_{zi} = mg\left(l-a_i\right)/l \tag{3.15}$$

and the axle characteristics are obtained as

$$\begin{cases} \dfrac{F_{y,1}+\left(M_{z,1}+M_{z,2}\right)/l}{F_{z1}} = \dfrac{a_y}{g} = \mu_1 \\ \dfrac{F_{y,2}-\left(M_{z,1}+M_{z,2}\right)/l}{F_{z2}} = \dfrac{a_y}{g} = \mu_2 \end{cases} \tag{3.16}$$

By substituting Equations 3.5 and 3.6 into Equation 3.16, the slip angles for straight motion $\alpha_{1,st}$ and $\alpha_{2,st}$ (expressed in Equations 3.7 and 3.8) represent the solution of the system 3.16 for null lateral acceleration $a_y = 0$ ($\mu_i = 0$).

FIGURE 3.4 Single-track model, forces, and moments during steady-state cornering (adapted from [7] and [24]).

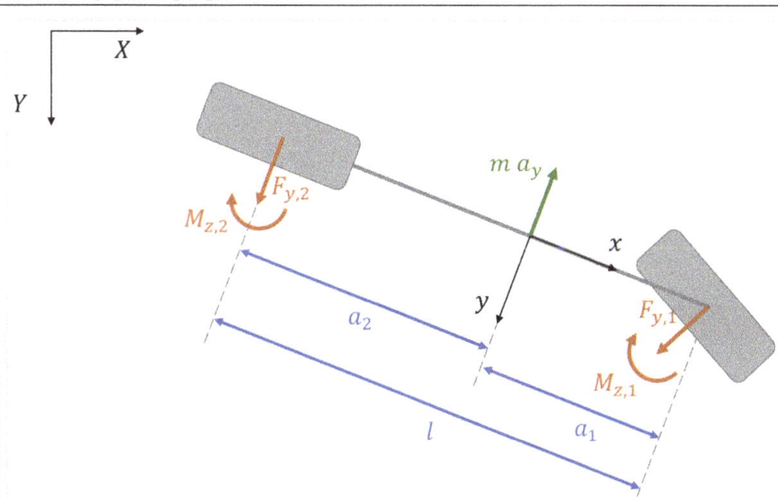

The linear handling diagram can be represented with Equation 3.16 and will have an offset with respect to the origin due to ply-steer and conicity characteristics.

In order to build the linear handling diagram (valid for small lateral accelerations), the side slip and steering angles can be computed as function of $\mu = a_y/g$. The full derivation of the handling diagram equations is reported in Appendix A.6. Again, the terms related to aligning moment play a negligible role with respect to the ones related to side force, and can thus be disregarded

- **Front side slip angle**

$$\begin{aligned}
\alpha_1 &= K_{\alpha 1} \cdot \mu + \alpha_{1,st} \\
K_{\alpha 1} &= \frac{lF_{z1}C^*_{eff,2} + C_{M\alpha,2}mg}{lC^*_{eff,1}C^*_{eff,2} - C_{M\alpha,1}C^*_{eff,2} + C_{M\alpha,2}C^*_{eff,1}} \\
&\approx \frac{F_{z1}}{C^*_{eff,1}} \\
\alpha_{1,st} &= \frac{-C^*_{eff,2}(M_{z0,1} + M_{z0,2} + lF_{y0,1}) - C_{M\alpha,2}(F_{y0,1} + F_{y0,2})}{lC^*_{eff,1}C^*_{eff,2} - C_{M\alpha,1}C^*_{eff,2} + C_{M\alpha,2}C^*_{eff,1}} \\
&\approx -\frac{F_{y0,1}}{C^*_{eff,1}}
\end{aligned} \tag{3.17}$$

- **Rear side slip angle**

$$\begin{aligned}
\alpha_2 &= K_{\alpha 2} \cdot \mu + \alpha_{2,st} \\
K_{\alpha 2} &= \frac{lF_{z2}C^*_{eff,1} + C_{M\alpha,1}mg}{lC^*_{eff,1}C^*_{eff,2} - C_{M\alpha,1}C^*_{eff,2} + C_{M\alpha,2}C^*_{eff,1}} \\
&\approx \frac{F_{z2}}{C^*_{eff,2}} \\
\alpha_{2,st} &= \frac{C^*_{eff,1}(M_{z0,1} + M_{z0,2} - lF_{y0,2}) + C_{M\alpha,1}(F_{y0,1} + F_{y0,2})}{lC^*_{eff,1}C^*_{eff,2} - C_{M\alpha,1}C^*_{eff,2} + C_{M\alpha,2}C^*_{eff,1}} \\
&\approx -\frac{F_{y0,2}}{C^*_{eff,2}}
\end{aligned} \tag{3.18}$$

- **Steering angle** The angle $\delta_0 = \frac{l}{R}$ of Equation 3.2 can be expressed as function of lateral acceleration $\mu = a_y/g$. By introducing the centripetal acceleration $a_y = \frac{u^2}{R}$

$$\delta_0 = \frac{l}{R} = \left(\frac{l \cdot g}{u^2}\right) \cdot \frac{a_y}{g} = \left(\frac{l \cdot g}{u^2}\right) \cdot \mu \tag{3.19}$$

$$\begin{aligned}
\delta &= \delta_0 + \alpha_1 - \alpha_2 = \\
&= K_\delta \cdot \mu + \delta_{st} = \\
&= \left(K_{\alpha 1} - K_{\alpha 2} + \frac{l \cdot g}{u^2}\right) \cdot \mu + \left(\alpha_{1,st} - \alpha_{2,st}\right)
\end{aligned} \tag{3.20}$$

FIGURE 3.5 Handling diagram: construction starting from the two axles' characteristics (symbols in Figure 3.1).

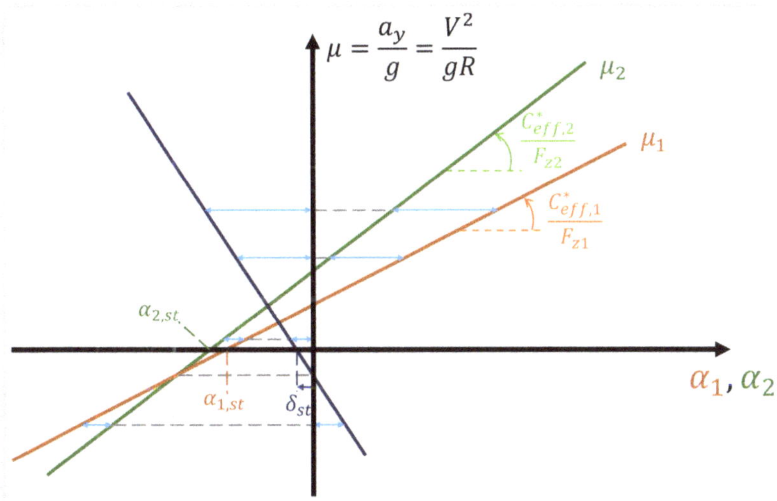

The handling diagram can be constructed starting from the two axles' characteristics (expressed in Equations 3.17 and 3.18), for all μ values. The left-hand side line in **Figure 3.5** is given by the subtraction $\alpha_1 - \alpha_2$, according to Equation 3.20.

With the handling diagram, it is possible to obtain for each steady-state cornering condition (described by the combination of lateral acceleration $a_y = \mu \cdot g$, velocity V, and curve radius R), the steering angle δ required for the vehicle with the given axles characteristics. The correspondence is qualitatively indicated in **Figure 3.6**.

Moreover, for each condition selected (parameters indicated as overlined in the following), it is possible to compute the corresponding slip angles $\overline{\alpha_i}$ as function of the lateral acceleration $\bar{\mu}$, and thus the corresponding axle side force $\overline{F_{yi}}$, and aligning moment $\overline{M_{zi}}$, as represented qualitatively in **Figure 3.7**.

FIGURE 3.6 Handling diagram: steering angle corresponding to selected steady-state cornering conditions.

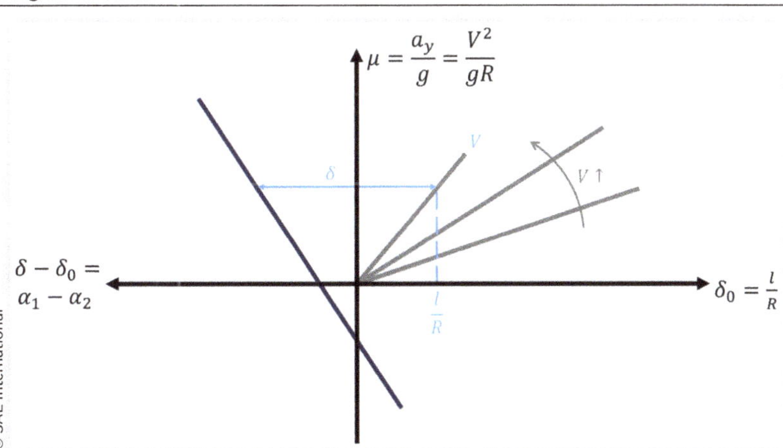

FIGURE 3.7 Handling diagram: slip angle and axle characteristics corresponding to a selected steady-state cornering condition (indicated as overlined).

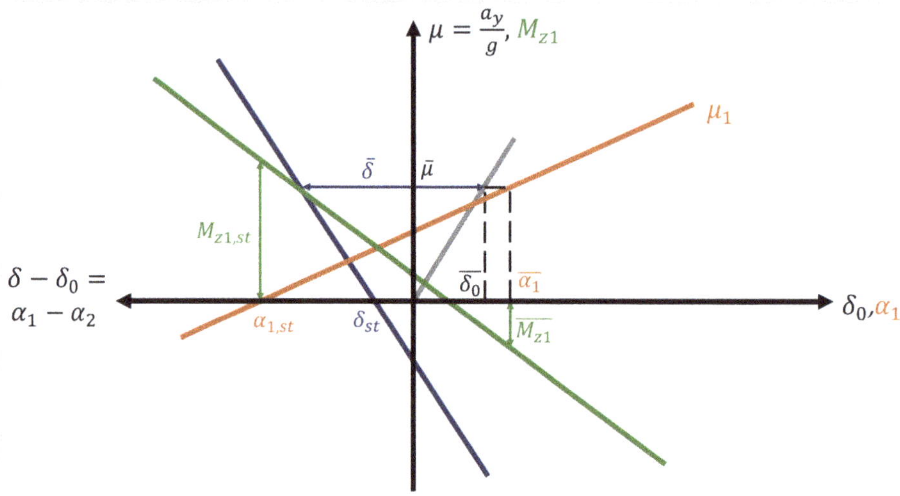

Therefore, it is possible to effectively express the axle characteristics, namely side force F_{yi} and aligning moment M_{zi}, as function of the lateral acceleration a_y. From Equations 3.6, 3.13, 3.17, and 3.18, it is possible to write

$$F_{y,i} = C^*_{eff,i} \cdot \left(\frac{a_y}{g} \cdot K_{\alpha i} + \alpha_{i,st} \right) + F_{y0,i} \approx \frac{a_y}{g} \cdot F_{zi}$$
$$M_{z,i} = -C_{M\alpha,i} \cdot \left(\frac{a_y}{g} \cdot K_{\alpha i} \right) + M_{zi,st} \tag{3.21}$$

FIGURE 3.8 Axle characteristics as function of lateral acceleration in steady-state cornering condition.

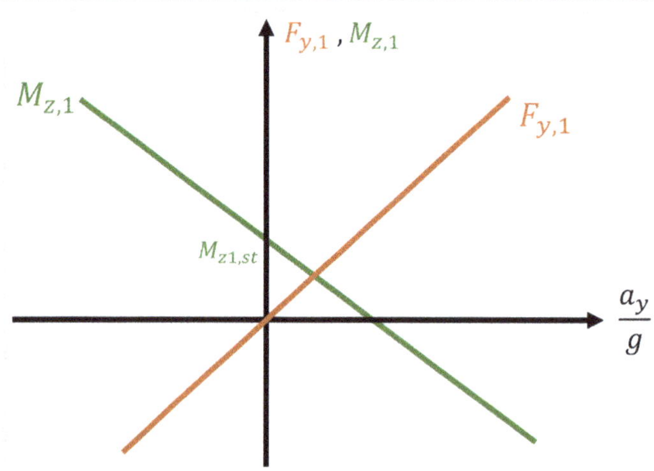

These two equations are represented qualitatively in **Figure 3.8** for the front axle case ($i = 1$). Since $\alpha_{1,st} \approx \frac{F_{y0,1}}{C^*_{eff,1}}$, the axle side force is approximately null for null lateral acceleration ($F_{y,1} \approx 0$ for $a_y = 0$). Hence, the F_{y1} line is represented passing through the origin in Figure 3.8.

CHAPTER 4

Steering Torque

The dependence of steering torque on tire characteristics is analyzed in this chapter.

Side forces and aligning moments of the front tires are sensed as a steering torque feedback by the driver.

The aligning moments $M_{z,1L}$ and $M_{z,1R}$ are transmitted to the steering system together with the side forces $F_{y,1L}$ and $F_{y,1R}$, having a moment arm given by the caster offset (or caster trail) n. This caster offset n is defined as the distance between the wheel contact patch center and the intersection between the king-pin axis line and the ground (in the side view). A scheme adapted from Mitschke [31] is represented in **Figure 4.1**.

The following work is based on the hypothesis of steady-state cornering, with no dynamic components considered. Moreover, for the situation under investigation (high speed and small steering angles), the total steering ratio n_{st} and the caster trail n can be considered constant [34].

The torque on the steering linkage side of the steering gear is

$$M_S^* = \left(F_{y,1L} + F_{y,1R}\right) \cdot n - \left(M_{z,1L} + M_{z,1R}\right) \tag{4.1}$$

Equation 4.1 can be rewritten in order to accommodate the description of *caster split*, namely different caster trail values between left and right wheel ($n_L \neq n_R$). *Caster split* is usually associated with variability in the suspension manufacturing and assembly process

FIGURE 4.1 Scheme of steering characteristics (adapted from [31, 32, 33]).

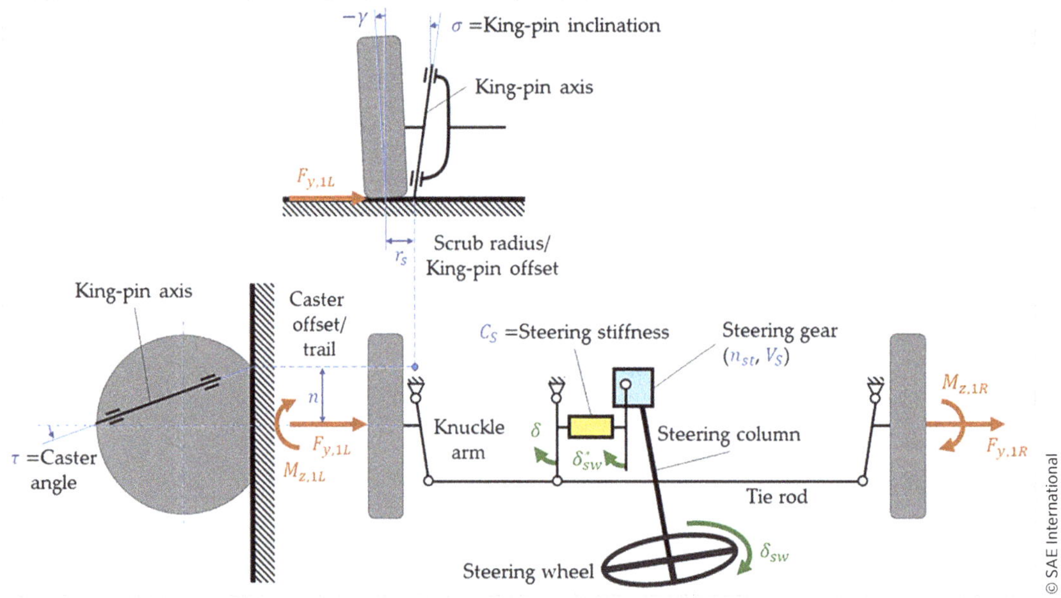

[15, 16]. Let us call n the average value of the left and right caster trail, and Δn the difference between right and left caster trail

$$n = \frac{1}{2}(n_R + n_L)$$
$$\Delta n = n_R - n_L$$

On introducing these terms into Equation 4.1, it becomes

$$\begin{aligned} M_S^* &= (F_{y,1L} \cdot n_L + F_{y,1R} \cdot n_R) - (M_{z,1L} + M_{z,1R}) = \\ &= \left[F_{y,1L} \cdot \left(n - \frac{\Delta n}{2}\right) + F_{y,1R} \cdot \left(n + \frac{\Delta n}{2}\right)\right] - (M_{z,1L} + M_{z,1R}) = \\ &= (F_{y,1R} + F_{y,1L}) \cdot n + (F_{y,1R} - F_{y,1L}) \cdot \frac{\Delta n}{2} - (M_{z,1L} + M_{z,1R}) \end{aligned} \tag{4.2}$$

It has to be underlined that longitudinal forces $F_{xi,L/R}$ are not introduced in this work, even if they might have a significant effect: the focus is on the tire pull that can be observed also during a free rolling straight run, in neutral gear.

Moreover, vertical forces $F_{zi,L/R}$ have an effect on the steering torque, which is linked to the steering system dimensions (i.e. caster trail n, caster angle τ, king-pin offset r_s, king-pin inclination σ), as it can be observed in **Figure 4.2**. This contribution can be quantified as follows

$$M_{S,Fz}^* = (-F_{z,1L} + F_{z,1R}) \cdot (tan\sigma \cdot n + tan\tau \cdot r_s) = (-F_{z,1L} + F_{z,1R}) \cdot p$$

Nevertheless, this will not be treated henceforth because the torques due to the two vertical forces nearly balance. "*Nearly*" indicates that the effect is considered negligible

FIGURE 4.2 Scheme of the effect of normal load F_z on steering torque (adapted from [24, 33]).

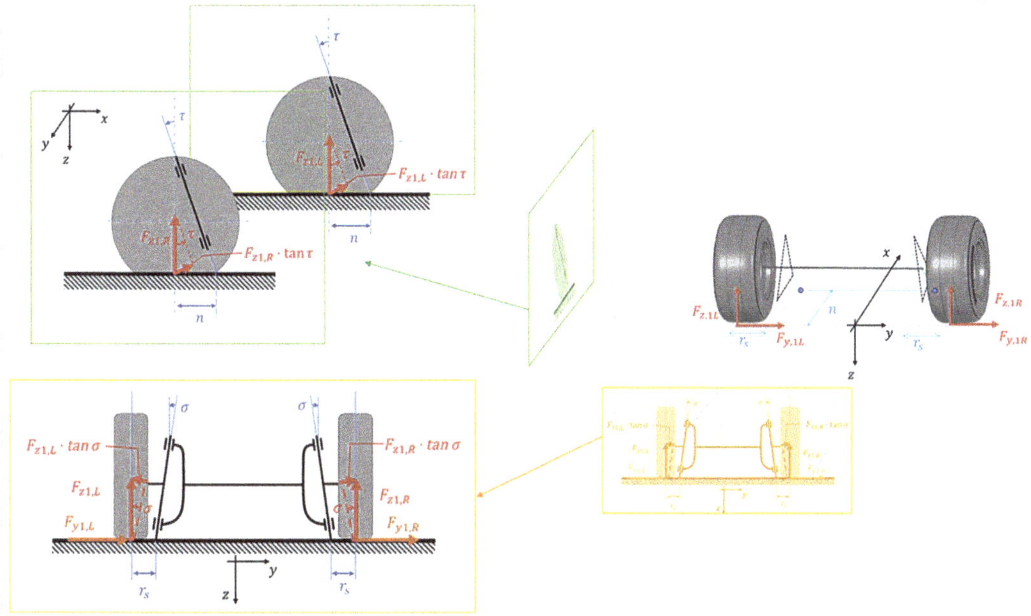

even when ordinary straight-ahead running road cross slope is introduced, where left–right weight distribution is not exactly 50/50. Including the presence of such torques in the analytical formulae of M_S seems to lead to both a complex formulation and a negligible improvement of accuracy.

The torque that the driver has to apply to steer or to hold the steering wheel gives the steering feedback, a torque transmitted from tires to steering wheel via the steering system [1]. It can be computed as

$$M_S = \frac{M_S^*}{n_{st} \cdot V_S} \tag{4.3}$$

where

- n_{st} is the total steering gear ratio, accounting both for steering gear and linkage
- V_S is the steering gain (or steering assistance ratio) in case a power steering system is employed.

This gain is inversely dependent on vehicle's velocity. For a highway driving condition, V_S can be assumed unitary [31].

4.1 Steering Assistance Ratio V_S

Pfeffer and Harrer [34] described an analytical and compact formulation for the steering assistance ratio V_S. From experimental evidence, based on subjective feedbacks, a linear relationship between V_S and lateral acceleration a_y proved to give a harmonic quasi-steady-state steering wheel torque M_S characteristic [34], which is desirable. Therefore, the steering assistance ratio can be written as

$$V_S = C_A \cdot \frac{a_y}{M_S} = C_A \cdot \left(D_A + K_A \cdot a_y\right) \tag{4.4}$$

where C_A is the vehicle-specific self-aligning coefficient, defined as

$$C_A = m \cdot \frac{l - a_1}{l} \cdot \frac{n - t_0}{n_{st}}$$

The two parameters describing the linear dependence of V_S on a_y are

- D_A: the *gradient factor*, specifying the basic boost level
- K_A: the *degressivity factor*, describing the rate at which steering wheel torque gradient $\frac{dM_S}{da_y}$ decreases, which is inversely dependent on vehicle's speed.

The steering wheel torque can hence be written as a function of lateral acceleration as

$$M_S = \frac{1}{\frac{D_A}{a_y} + K_A} \tag{4.6}$$

An example of the dependence of steering wheel torque M_S and steering assistance ratio V_S on lateral acceleration is represented in Figure 4.3.

For low lateral accelerations, expected in case of straight motion and on-center handling analyses, the assistance ratio can be considered unitary (as assumed for highway driving conditions in [31]). This, as a result, provides a "distinctive" center feel according to [34].

FIGURE 4.3 Steering wheel torque M_S and steering assistance ratio V_S as a function of lateral acceleration a_y (from [34]).

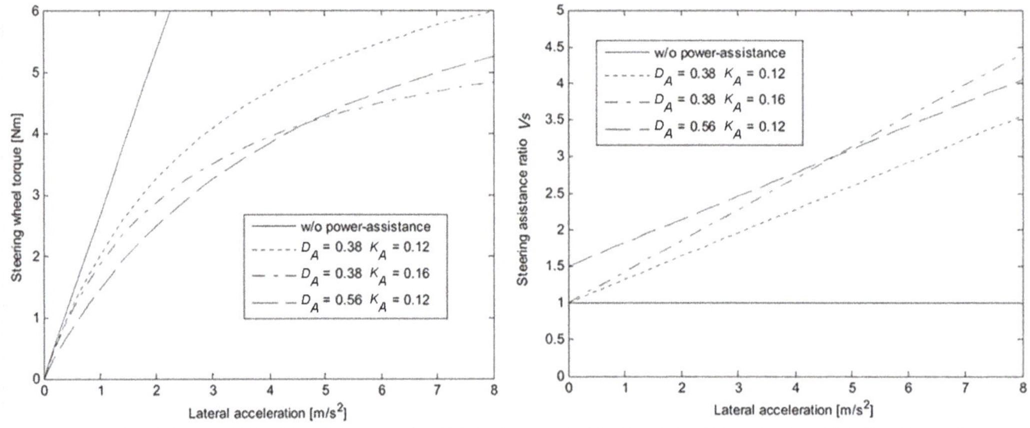

4.2 Steering Torque during On-Center Steering

By merging Equations 4.1-4.3 and 3.21, it is possible to write the steering torque as a function of the lateral acceleration $\mu = a_y/g$

$$M_S(\mu) = \frac{M_S^*(\mu)}{n_{st} \cdot V_S} = \frac{F_{y,1} \cdot n - M_{z,1}}{n_{st} \cdot V_S} =$$
$$= \frac{\mu \cdot K_{\alpha 1} \cdot \left(n \cdot C_{eff,1}^* + C_{M\alpha,1}\right) + n \cdot \left(C_{eff,1}^* + \alpha_{1,st} + F_{y0,1}\right) - M_{z1,st}}{n_{st} \cdot V_S}$$
$$\approx \frac{\mu \cdot \left(n \cdot F_{z1} + C_{M\alpha,1} \cdot K_{\alpha 1}\right) - M_{z1,st}}{n_{st} \cdot V_S} \tag{4.7}$$

Thanks to the derivation of this last formula, we see that it is possible to express *all* of the tire-induced vehicle pull quantities (namely steering torque M_S, slip angles α_i, vehicle attitude β, and steering angle δ) as a function of the same parameter, namely the normalized lateral acceleration $\mu = a_y/g$. Therefore, Equations 3.17–3.20 and 4.7 may represent an effective *set of analytical expressions* to evaluate and relate simultaneously the aforementioned quantities.

As an example, it is possible to invert Equation 4.7 to compute the value of μ for which the steering torque M_S is null. Then, it is possible to compute the values of slip angles (and thus tire side forces and aligning moments) for that μ value, with the opportunity to easily monitor the quantities affecting vehicle dynamics.

The steering torque M_S can also be expressed as a function of the steering angle δ, using the relation between δ itself and lateral acceleration shown in Equation 3.20

$$M_S(\delta) = \frac{(\delta - \delta_0 - \delta_{st}) \cdot K_{\alpha 1} \cdot \left(\frac{n \cdot C_{eff,1}^* + C_{M\alpha,1}}{K_\delta}\right) + n \cdot \left(C_{eff,1}^* \cdot \alpha_{1,st} + F_{y0,1}\right) - M_{z1,st}}{n_{st} \cdot V_S}$$
$$\approx \frac{(\delta - \delta_0 - \delta_{st}) \cdot \left(\frac{n \cdot F_{z1} + C_{M\alpha,1} \cdot K_{\alpha 1}}{K_\delta}\right) - M_{z1,st}}{n_{st} \cdot V_S} \tag{4.8}$$

As it can be observed from Equations 4.7 and 4.8, the steering torque has an offset with respect to the origin (see **Figure 4.4**). This has an effect on the so-called *on-center feel* [1]. With respect to the plot reported in **Figure 4.5**, the steering characteristic will show an offset, giving different feelings to the driver when turning to one side or to the other. This can be qualitatively observed in the different cases presented in Section 4.2.1, where both steering angle δ and driver's steering torque M_S are examined as a function of lateral acceleration $\mu = a_y/g$, according to Equations 4.7 and 3.20.

FIGURE 4.4 Steering torque M_S as a function of lateral acceleration a_y and steering angle δ.

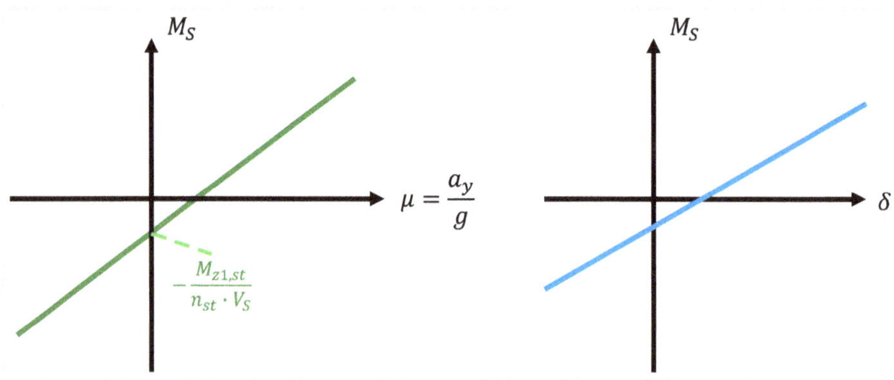

FIGURE 4.5 On-center steering torque (adapted from Gim [1]).

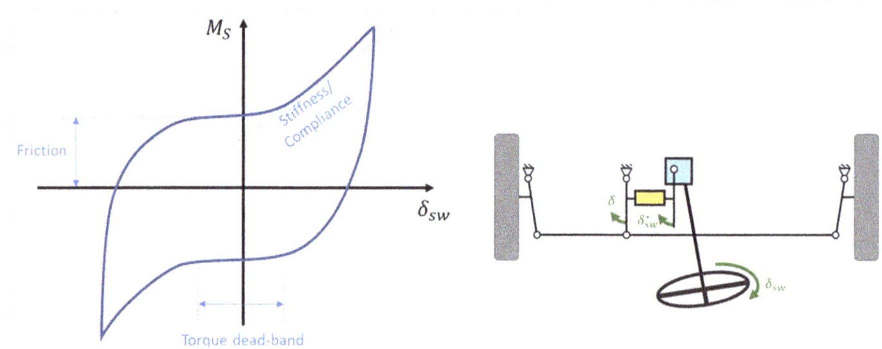

To limit the shortcoming of different steering feels for the two turning directions, the offsets to be minimized are δ_{st} and $M^*_{S,st} = M^*_S(\mu = 0)$. Let us write δ_{st} by combining Equations 3.12 and 2.36

$$
\begin{aligned}
\delta_{st} &= \alpha_{1,st} - \alpha_{2,st} \\
&\approx -\frac{F_{y0,1}}{C^*_{eff,1}} + \frac{F_{y0,2}}{C^*_{eff,2}} = \\
&= \left[-\Psi_{10} \cdot \left(\frac{C_{F\alpha,10R} - C_{F\alpha,10L}}{C_{F\alpha,10R} + C_{F\alpha,10L}} \right) - \gamma_{10} \cdot \left(\frac{C_{F\gamma,10R} - C_{F\gamma,10L}}{C_{F\alpha,10R} + C_{F\alpha,10L}} \right) - \left(\frac{F_{y0,1R} - F_{y0,1L}}{C_{F\alpha,10R} + C_{F\alpha,10L}} \right) + \right. \\
&\quad \left. + \Psi_{20} \cdot \left(\frac{C_{F\alpha,20R} - C_{F\alpha,20L}}{C_{F\alpha,20R} + C_{F\alpha,20L}} \right) + \gamma_{20} \cdot \left(\frac{C_{F\gamma,20R} - C_{F\gamma,20L}}{C_{F\alpha,20R} + C_{F\alpha,20L}} \right) + \left(\frac{F_{y0,2R} - F_{y0,2L}}{C_{F\alpha,20R} + C_{F\alpha,20L}} \right) \right]
\end{aligned}
\tag{4.9}
$$

The steering torque offset can be computed from Equations 4.7, 3.12, and 2.36

$$
\begin{aligned}
M^*_{S,st} &= M^*_S(\mu=0) = -M_{z1,st} = \\
&= C_{M\alpha,1}\cdot\alpha_{1,st} - M_{z0,1} \\
&\approx C_{M\alpha,1}\cdot\left(\frac{F_{y0,1}}{C^*_{eff,1}}\right) - M_{z0,1} = \\
&= -C_{M\alpha,1}\cdot\left[\Psi_{10}\cdot\left(\frac{C_{F\alpha,10R}-C_{F\alpha,10L}}{C_{F\alpha,10R}+C_{F\alpha,10L}}\right)+\gamma_{10}\cdot\left(\frac{C_{F\gamma,10R}-C_{F\gamma,10L}}{C_{F\alpha,10R}+C_{F\alpha,10L}}\right)+\right. \\
&\left.+\left(\frac{F_{y0,1R}+F_{y0,1L}}{C_{F\alpha,10R}+C_{F\alpha,10L}}\right)\right] - M_{z0,1R} - M_{z0,1L}
\end{aligned}
\tag{4.10}
$$

Moreover, the condition for which the vehicle has null attitude β during straight motion is pursued. From Equations 3.11 and 2.36, this corresponds to the condition

$$
\begin{aligned}
\beta_{st} &= -\alpha_{2,st} \\
&\approx \frac{F_{y0,2}}{C^*_{eff,2}} = \\
&= \Psi_{20}\cdot\left(\frac{C_{F\alpha,20R}-C_{F\alpha,20L}}{C_{F\alpha,20R}+C_{F\alpha,20L}}\right)+\gamma_{20}\cdot\left(\frac{C_{F\gamma,20R}-C_{F\gamma,20L}}{C_{F\alpha,20R}+C_{F\alpha,20L}}\right)+\left(\frac{F_{y0,2R}+F_{y0,2L}}{C_{F\alpha,20R}+C_{F\alpha,20L}}\right)
\end{aligned}
\tag{4.11}
$$

4.2.1 Examples for a Simplified Case

In this section, different cases for steering angle offset δ_{st} and steering torque offset $M_{z1,st}$ are considered to discuss their effects on the lateral deviation from the straight path and on the on-center handling properties qualitatively.

For the sake of simplicity, the case with equal left and right tire characteristics is considered:

- $C_{F\alpha,10L} = C_{F\alpha,10R} = C_{F\alpha,10}$
- $C_{M\alpha,1} = C_{M\alpha,1L} + C_{M\alpha,1R} = 2\cdot C_{M\alpha,10}$

This allows introduction of the terms for ply-steer lateral force, $F_{ply,i}$ (i = 1: front axle, i = 2: rear axle) and front tire's Ply-steer Residual Aligning Torque, $PRAT_1$ (the parameter pertains to each individual front tire). Their definition is reported below:

$$
\begin{aligned}
F_{ply,i} &= \frac{F_{y0,iL}+F_{y0,iR}}{2} \\
M_{ply,i} &= \frac{M_{z0,iL}+M_{z0,iR}}{2} \\
PRAT_1 &= M_{ply,1} + C_{M\alpha,10}\cdot\frac{F_{ply,1}}{C_{F\alpha,10}}
\end{aligned}
$$

In this simplified case, the steering angle offset δ_{st} from Equation 4.9 and the steering torque offset term $M_{z1,st}$ from Equation 4.10 can be simply rewritten as

$$\delta_{st} = \left(\frac{F_{y0,2R} + F_{y0,2L}}{C_{F\alpha,20R} + C_{F\alpha,20L}}\right) - \left(\frac{F_{y0,1R} + F_{y0,1L}}{C_{F\alpha,10R} + C_{F\alpha,10L}}\right) = \\ = \frac{F_{ply,2}}{C_{F\alpha,20}} - \frac{F_{ply,1}}{C_{F\alpha,10}} \tag{4.12}$$

$$M_{z1,st} = 2C_{M\alpha,10} \cdot \left(\frac{F_{y0,1R} + F_{y0,1L}}{2 \cdot C_{F\alpha,10}}\right) + M_{z0,1R} + M_{z0,1L} = \\ = 2C_{M\alpha,10} \cdot \left(\frac{F_{ply,1}}{C_{F\alpha,10}}\right) + 2M_{ply,1} = \\ = 2 \cdot PRAT_1 \tag{4.13}$$

The steering angle δ – a function of lateral acceleration $\mu = \frac{a_y}{g}$ – can be rewritten from Equation 3.20 expressing δ_{st} as indicated in Equation 4.12 as

$$\delta = K_\delta \cdot \mu + \delta_{st} = \\ = \left(K_{\alpha 1} - K_{\alpha 2} + \frac{l \cdot g}{u^2}\right) \cdot \mu + \left(\frac{F_{ply,2}}{C_{F\alpha,20}} - \frac{F_{ply,1}}{C_{F\alpha,10}}\right) \tag{4.14}$$

Likewise, the steering torque M_S – a function of lateral acceleration $\mu = \frac{a_y}{g}$ – can be rewritten from Equation 4.7 expressing $M_{z1,st}$ as indicated in Equation 4.13 as

$$M_S = \frac{\mu \cdot (n \cdot (F_{z1} + C_{M\alpha,1} \cdot K_{\alpha 1}) - M_{z1,st}}{n_{st} \cdot V_S} = \\ = \frac{\mu \cdot (n \cdot F_{z1} + C_{M\alpha,1} \cdot K_{\alpha 1}) - 2 \cdot PRAT_1}{n_{st} \cdot V_S} \tag{4.15}$$

Different cases are presented in **Figures 4.6–4.10**, with positive and negative values of δ_{st} and $M_{z1,st} = 2 \cdot PRAT_1$, with a qualitative representation of the steering angle and steering torque for the different lateral accelerations.

4.3 Steering Torque with Effective Axle Aligning Moment

In case the effective axle aligning moment (introduced in Section 2.5) is considered, the formulae for the steering torque M_S have to be adjusted. The effective front axle aligning moment terms (Equations 2.47–2.53) are reported here for convenience

FIGURE 4.6 Steering angle and steering torque for different lateral accelerations: $\delta_{st} = 0$, $PRAT_1 = 0$.

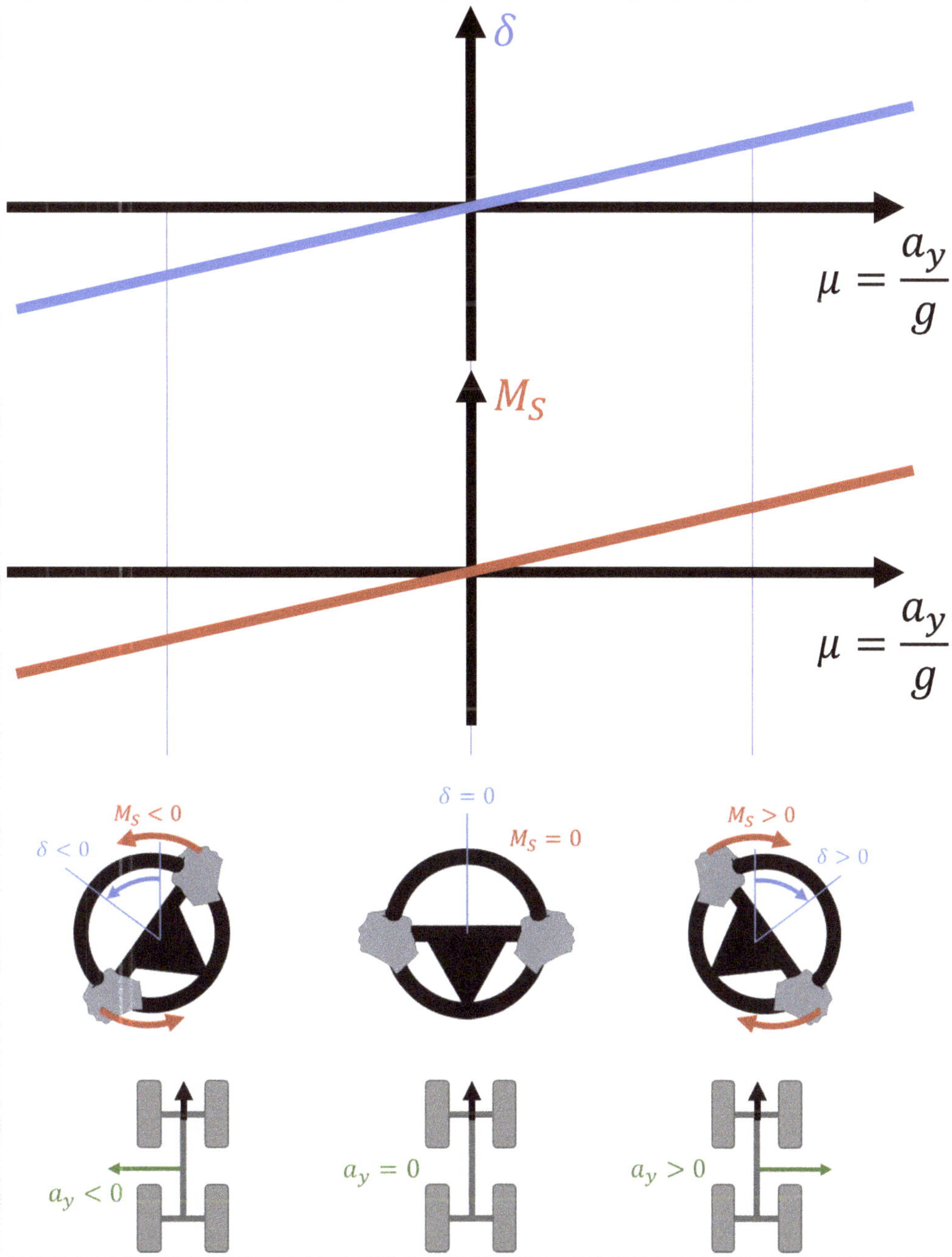

FIGURE 4.7 Steering angle and steering torque for different lateral accelerations: $\delta_{st} > 0$, $PRAT_1 > 0$.

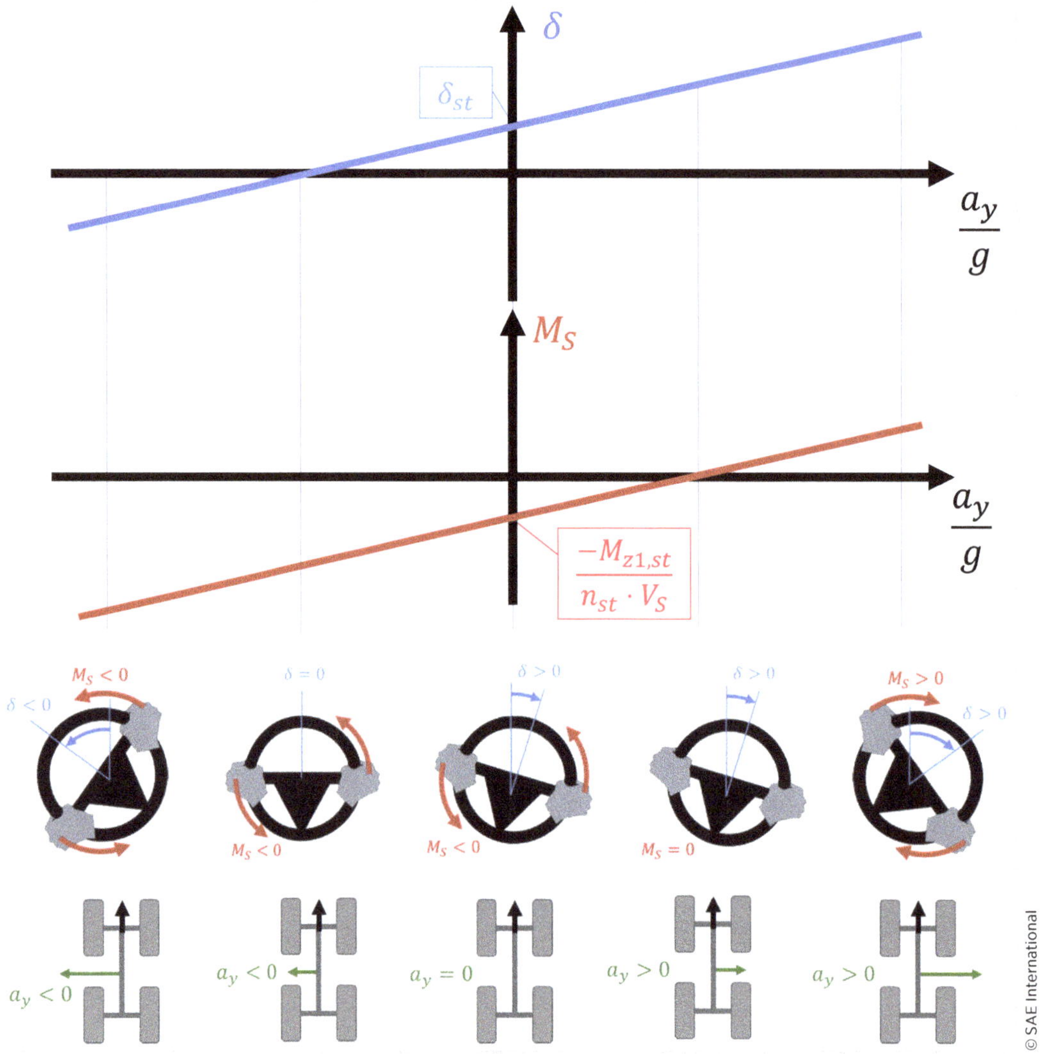

$$M_{z,1} = -C_{M\alpha,1} \cdot \alpha_1 + M_{z0,1}$$

where

$$-C_{M\alpha,1} = -2C_{M\alpha,10} + C^{*}_{eff,1} \cdot (C_{M\alpha,roll,1} - C_{M\alpha,susp,1} + C_{M\alpha,steer,1} + C_{M\alpha,\Delta F_z,1})$$

$$M_{z0,1} = F_{y0,1} \cdot (C_{M\alpha,roll,1} - C_{M\alpha,susp,1} + C_{M\alpha,Steer,1} + C_{M\alpha,\Delta F_{z,1}}) +$$
$$+\Psi_{10} \cdot (C_{M\alpha,10L} - C_{M\alpha,10R}) - \gamma_{10} \cdot (C_{M\gamma,10L} - C_{M\gamma,10R}) + M_{z0,1L} + M_{z0,1R}$$

The steering wheel torque M_S definition, namely Equations 4.3 and 4.7, is reported below as well

FIGURE 4.8 Steering angle and steering torque for different lateral accelerations: $\delta_{st} < 0$, $PRAT_1 < 0$.

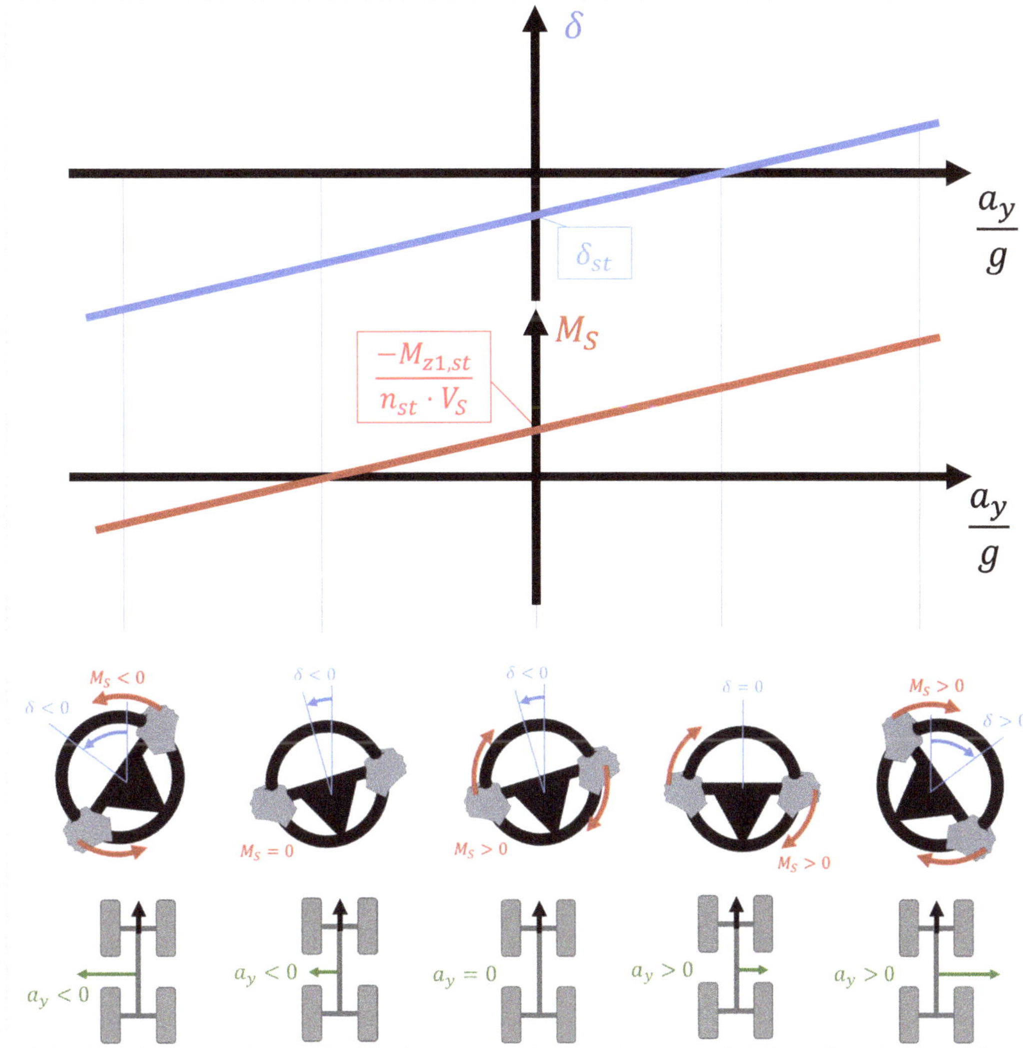

$$M_S(\mu) = \frac{M_S^*(\mu)}{n_{st} \cdot V_S} = \frac{F_{y,1} \cdot n - M_{z,1}}{n_{st} \cdot V_S} \quad (4.16)$$

On introducing into Equation 4.16, the expressions for front axle side force $F_{y,1}$, aligning moment $M_{z,1}$, and slip angle α_1, respectively, from Equations 2.35, 2.52, and 3.17, which are reported below for convenience

$$F_{y,1} = C^*_{eff,1} \cdot \alpha_1 + F_{y0,1}$$
$$M_{z,1} = -C_{M\alpha,1} \cdot \alpha_1 + M_{z0,1}$$
$$\alpha_1 = K_{\alpha 1} \cdot \mu + \alpha_{1,st}$$

FIGURE 4.9 Steering angle and steering torque for different lateral accelerations: $\delta_{st} > 0$, $PRAT_1 < 0$.

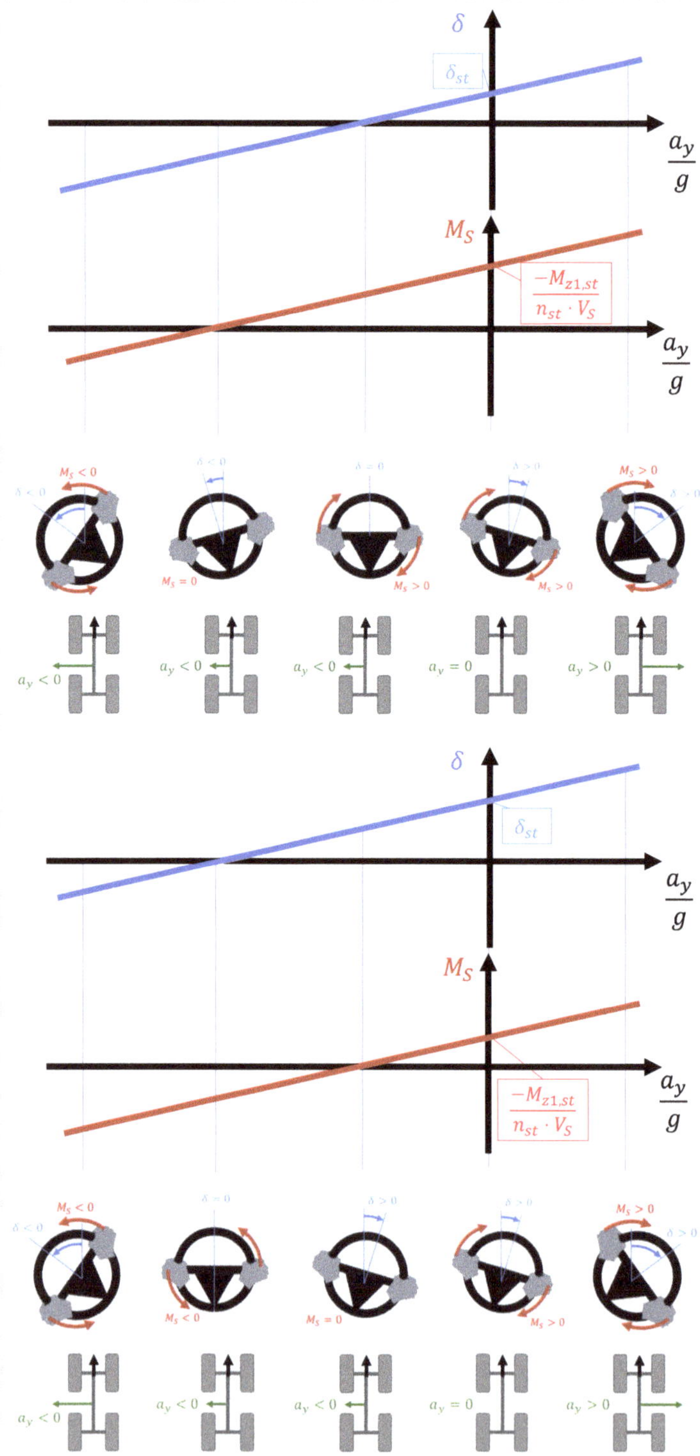

FIGURE 4.10 Steering angle and steering torque for different lateral accelerations: $\delta_{st} < 0$, $PRAT_1 > 0$.

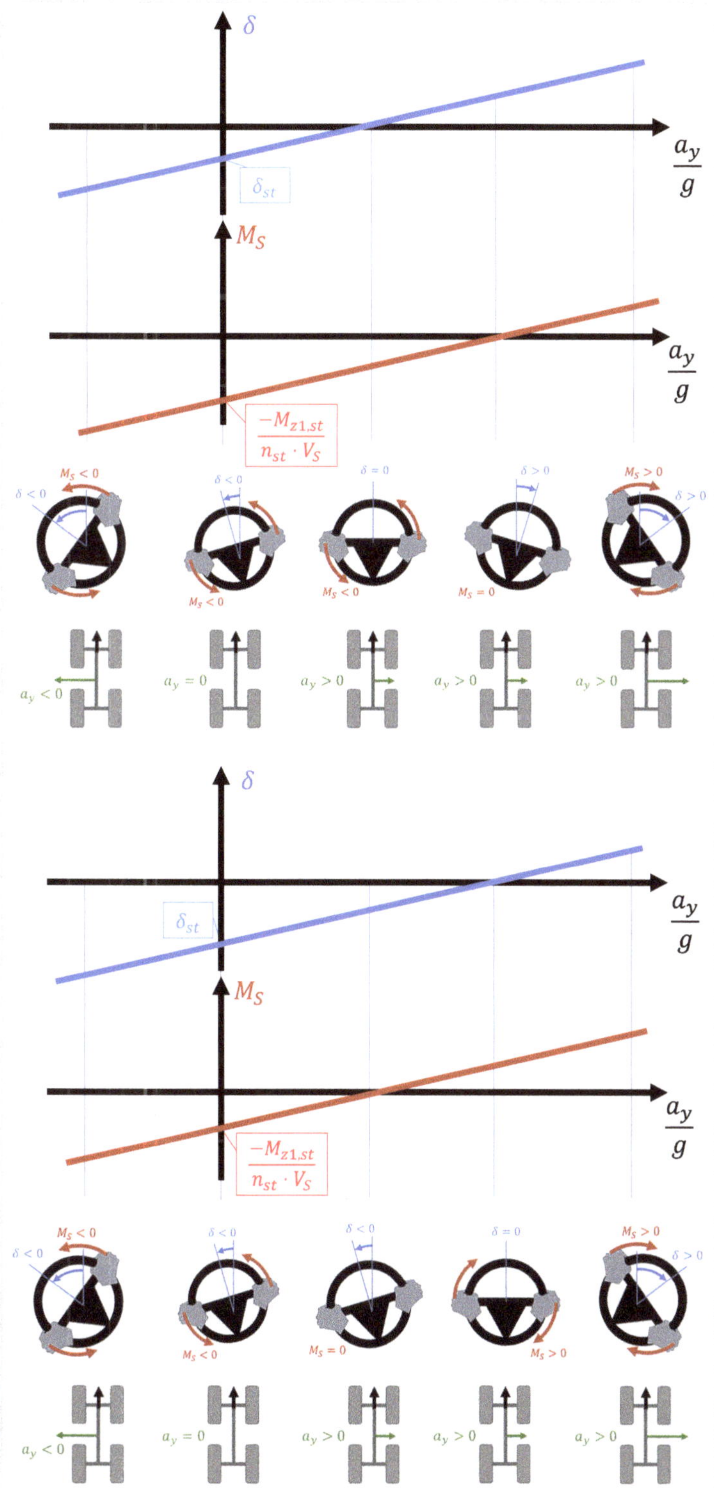

the steering wheel torque becomes

$$M_S(\mu)=\frac{M_S^*(\mu)}{n_{st}\cdot V_S}=\frac{F_{y,1}\cdot n-M_{z,1}}{n_{st}\cdot V_S}=$$
$$=\frac{\left(C_{eff,1}^*\cdot\alpha_1+F_{y0,1}\right)\cdot n-\left(-C_{M\alpha,1}\cdot\alpha_1+M_{z0,1}\right)}{n_{st}\cdot V_S}=$$
$$=\frac{\left(\mu\cdot K_{\alpha 1}+\alpha_{1,st}\right)\cdot\left(n\cdot C_{eff,1}^*+C_{M\alpha,1}\right)+F_{y0,1}\cdot n-M_{z0,1}}{n_{st}\cdot V_S} \tag{4.17}$$

It is possible to introduce the approximations expressed in Equation 3.17, namely

$$K_{\alpha 1}\approx\frac{F_{z1}}{C_{eff,1}^*}$$
$$\alpha_{1,st}\approx-\frac{F_{y0,1}}{C_{eff,1}^*}$$

With these approximations, Equation 4.17 can be rewritten as follows

$$M_S(\mu)\approx\frac{\left(\mu\cdot\frac{F_{z1}}{C_{eff,1}^*}-\frac{F_{y0,1}}{C_{eff,1}^*}\right)\cdot\left(n\cdot C_{eff,1}^*+C_{M\alpha,1}\right)+F_{y0,1}\cdot n-M_{z0,1}}{n_{st}\cdot V_S}=$$
$$=\frac{\mu\cdot F_{z1}\cdot\left(n+\frac{C_{M\alpha,1}}{C_{eff,1}^*}\right)-\frac{F_{y0,1}}{C_{eff,1}^*}\cdot C_{M\alpha,1}-M_{z0,1}}{n_{st}\cdot V_S} \tag{4.18}$$

The effective axle aligning moment terms, reported at the beginning of this section, can now be substituted into Equation 4.18. The torque on the steering linkage side of the steering gear M_S^* is thus

$$\begin{aligned}M_S^*(\mu)\approx{}&\mu\cdot F_{z1}\cdot\left[n+\frac{2C_{M\alpha,10}}{C_{eff,1}^*}-(C_{M\alpha,roll,1}-C_{M\alpha,susp,1}+C_{M\alpha,steer,1}+C_{M\alpha,\Delta F_z,1})\right]-\\&-\frac{F_{y0,1}\cdot 2C_{M\alpha,10}}{C_{eff,1}^*}+F_{y0,1}\cdot(C_{M\alpha,roll,1}-C_{M\alpha,susp,1}+C_{M\alpha,steer,1}+C_{M\alpha,\Delta F_z,1})-\\&-F_{y0,1}\cdot(C_{M\alpha,roll,1}-C_{M\alpha,susp,1}+C_{M\alpha,steer,1}+C_{M\alpha,\Delta F_z,1})-\\&-\Psi_{10}\cdot(C_{M\alpha,10L}-C_{M\alpha,10R})+\gamma_{10}\cdot(C_{M\gamma,10L}-C_{M\gamma,10R})-M_{z0,1L}-M_{z0,1R}=\\&=\mu\cdot F_{z1}\cdot\left[n+\frac{2C_{M\alpha,10}}{C_{eff,1}^*}-(C_{M\alpha,roll,1}-C_{M\alpha,susp,1}+C_{M\alpha,steer,1}+C_{M\alpha,\Delta F_z,1})\right]-\\&-\frac{F_{y0,1}\cdot 2C_{M\alpha,10}}{C_{eff,1}^*}-\Psi_{10}\cdot(C_{M\alpha,10L}-C_{M\alpha,10R})+\gamma_{10}\cdot(C_{M\gamma,10L}-C_{M\gamma,10R})-\\&-M_{z0,1L}-M_{z0,1R}\end{aligned} \tag{4.19}$$

CHAPTER 5

Tests for Studying Vehicle Pull Effects

In this chapter, different test methods that can be used to analyze various occurrences of vehicle pull and to examine on-center steering feel for road vehicles are discussed.

5.1 Trajectory Simulation

In the literature, two different test methods to evaluate vehicle pull effects are mainly reported [3, 16, 35] (please refer to **Figure 5.1** for symbols)

5.1.1 Free Control Test

- $M_S = 0$
- $\delta \neq 0$
- $\Delta y \neq 0$

The vehicle is maintained at a constant velocity (usually $u = 80$ or 100 km/h). The steering wheel is released ($M_S = 0$). The amount of vehicle lateral deviation Δy (called "*drift*" in literature [1]) after 100 m from the point of steering wheel release (origin) is an objective measure of one way of expressing the steering hands-off drift problem. Examples of the allowable ranges for the amount of lateral *drift* can be found in literature, such as ±1 m with road cross slope of $\varphi_r = 0.01$ rad [16], although no unanimity exists.

FIGURE 5.1 Qualitative representation of *null steer test* (δ = 0) trajectory simulation, global, and local (vehicle's) reference systems.

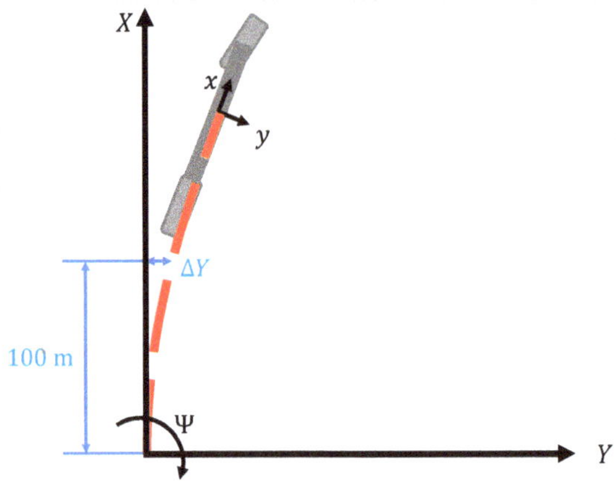

5.1.2 Fixed Control Test

- $M_S \neq 0$
- $\delta \neq 0$
- $\Delta y = 0$

The vehicle is maintained at a constant velocity (usually u = 80 or 100 km/h) on a straight path. The steering torque M_S to be applied by the driver is measured by means of a steering torque sensor. This residual steering torque (or *pull*) is another way of expressing the vehicle pull problem.

A customer clinic study of an Original Equipment Manufacturer (OEM), reported by Badiru [2], quantifies the relation between steering *pull* and customers' complaint cumulative percentage, indicating a threshold of M_S = 0.60 Nm for the straight-driving steering torque sensed at the steering wheel.

These two tests are easy to perform in theory: they just require the driver to release the steering wheel or to follow a straight path. Pottinger [3] provides practical suggestions on how to perform these basic experiments in case of cross-wind or road cross slope: for example, acquiring data back-to-back in both directions of travel allows removal of the external factors whose effect is not of interest, or to isolate them.

However, these tests are actually difficult to be physically run with sufficient repeatability and reproducibility. Therefore, most OEMs are leaning toward analytical verification process in place of physical experiments: this allows for more repeatable results and more extensive analyses (e.g., robustness to production tolerances of vehicle and tires as well as range of road cross sloped), and is more effective both in terms of cost and time.

5.1.3 Null Steer Test

A third kind of test is introduced in this book, which is easier to simulate using equations of motion or models that have the longitudinal velocity u and the steering angle δ (e.g., the single-track vehicle model represented by Equation 2.57) as inputs.

- $M_S \neq 0$
- $\delta = 0$
- $\Delta y \neq 0$

The vehicle is maintained at a constant velocity (u = 100 km/h) with steering wheel held fixed $\delta_{sw} = 0$ (input $\delta = 0$). The amount of vehicle lateral *drift* Δy after 100 m from the origin can be considered an objective measure of the null steer drift problem.

The trajectory can be simulated with the steering wheel held fixed (input δ) at the constant forward velocity u = 100 km/h.

The equations of motion adopted are a simplified version of Equation 2.57, since small angles are expected, with negligible roll angle value and dynamics and constant longitudinal velocity.

$$\begin{cases} \nwarrow v: \quad \dot{v} = \dfrac{1}{m}\left(F_{y,1} + F_{y,2}\right) - u \cdot r \\ \curvearrowleft r: \quad \dot{r} = \dfrac{1}{I_z}\left(a_1 \cdot F_{y,1} + M_{z,1} - a_2 \cdot F_{y,2} + M_{z,2}\right) \end{cases} \tag{5.1}$$

The slip angles are computed as in Equation 3.1, forces as in Equation 3.5, and moments as in Equation 3.6.

The trajectory is then obtained with Crank–Nicolson method

$$\begin{cases} X_{i+1} = X_i + \dfrac{1}{2}\left(\dot{X}_{i+1} + \dot{X}_i\right) \cdot \left(t_{i+1} - t_i\right) \\ Y_{i+1} = Y_i + \dfrac{1}{2}\left(\dot{Y}_{i+1} + \dot{Y}_i\right) \cdot \left(t_{i+1} - t_i\right) \end{cases} \tag{5.2}$$

where

$$\begin{cases} \dot{X} = u \cdot cos\Psi - v \cdot sin\Psi \\ \dot{Y} = u \cdot sin\Psi + v \cdot cos\Psi \end{cases} \tag{5.3}$$

The deviation from the straight path after 100 m (see Figure 5.1), with the fixed steering (input δ) from the origin at a constant velocity of 100 km/h, is adopted as an index of the unbalance given by ply-steer and conicity characteristics of tires.

The initial conditions can be set to simulate the situation in which the vehicle is following a straight path before the origin, and the steering is fixed to the input value δ instantaneously when passing through the origin at time $t = 0$

- $y(0) = 0$

  the lateral deviation is null at the origin.

- $v(0) = \beta_{st} \cdot u$

  the lateral velocity v is equal to the product between the straight motion vehicle attitude β_{st} (defined in Equation 3.10) and the longitudinal velocity u.

- $\Psi(0) = -\beta_{st} = \alpha_{2,st}$

  the yaw angle is opposite to the straight motion vehicle attitude β_{st}, defined in Equation 3.10.

- $r(0) = 0$

  the yaw rate is null during straight motion.

These initial conditions ensures that the slip angles α_1 and α_2 (defined in Equation 3.1) are equal to the straight motion slip angles $\alpha_{1,st}$ and $\alpha_{2,st}$, with steering angle $\delta = \delta_{st}$.

By selectively imposing a fixed steering angle $\delta = \delta|_{MS=0}$ or $\delta = 0$, it is possible to simulate the trajectory, respectively, of a *free control test* or of a *null steer test*.

5.2 On-Center Handling and Steering Feel

The steering *feel* can mainly be characterized by off-center and on-center handling properties. A good on-center handling does not require the driver to apply many corrections and gives clear feedback information on how much intervention is required, processing the driver's commands accurately [36–38]. This is particularly important for highway driving at high speed, and most of the time vehicles are driven with moderate steering corrections [39].

The ISO defines *on-center handling* as the "description of steering 'feel' and precision of a vehicle during nominally straight-line driving and in negotiating large radius bends at high speeds but low lateral accelerations," representing "that part of the straight-line directional stability characteristics of the vehicle existing at lateral acceleration levels, typically no greater than $1 m/s^2$" [40]. Steering *feel* is a subjective perception of the driver, which has noteworthy impact on automotive brand's image, and thus constitutes a fundamental aspect of vehicle dynamics development.

There are many researches in literature on how to objectify on-center handling with different testing methods and metrics [36–38, 41]. An example of a test method is the *weave test* [40], which consists in applying an oscillatory steering wheel angle input (nominally sinusoidal) with constant vehicle speed. This is a fairly standardized test method, employed among the others by *General Motors Co.* [37] and *Hyundai Motor Company* [42]. An adapted version of the *weave test* has been used in an analysis conducted by Salaani et al. [37].

Farrer [36, 37] proposed the description of three characteristics in order to inspect on-center handling

- **Steering activity**. Identified by steering wheel angle δ_{SW}, steering wheel velocity $\dot{\delta}_{SW}$, and steering wheel torque M_S. The steering activity should not be excessive for a good on-center handling.
- **Steering "feel"**. Identified by steering friction, torque dead-band, and steering stiffness. The steering feel should be informative for the driver to guarantee good on-center characteristics.
- **Vehicle response**. Identified by response dead-band, yaw rate r response, and response time lag. The response of a vehicle with good on-center handling should be precise.

Salaani et al. [37] conducted an extensive experimental investigation of steering feel performances. In particular, they proposed to characterize the on-center behavior by means of three cross plots, an example of which is represented in **Figure 5.2**. The differences between the upper and lower curves of the loop were analyzed, with their relevance to the driver feel.

The loop exhibited in the time domain cross plots is due to inherent dynamics of the vehicle, causing the output motion to have a phase lag with respect to the input signal. Further influences can be considered, coming from hysteresis, friction, and elasticities in steering and suspension systems.

Nonlinearity, on the other hand, is ascribable to unavoidable steering friction (steering gear and king-pin axis), steering system, and bushing compliances, and power steering assistance characteristics.

FIGURE 5.2 On-center handling experimental measures (from Salaani et al. [37]).

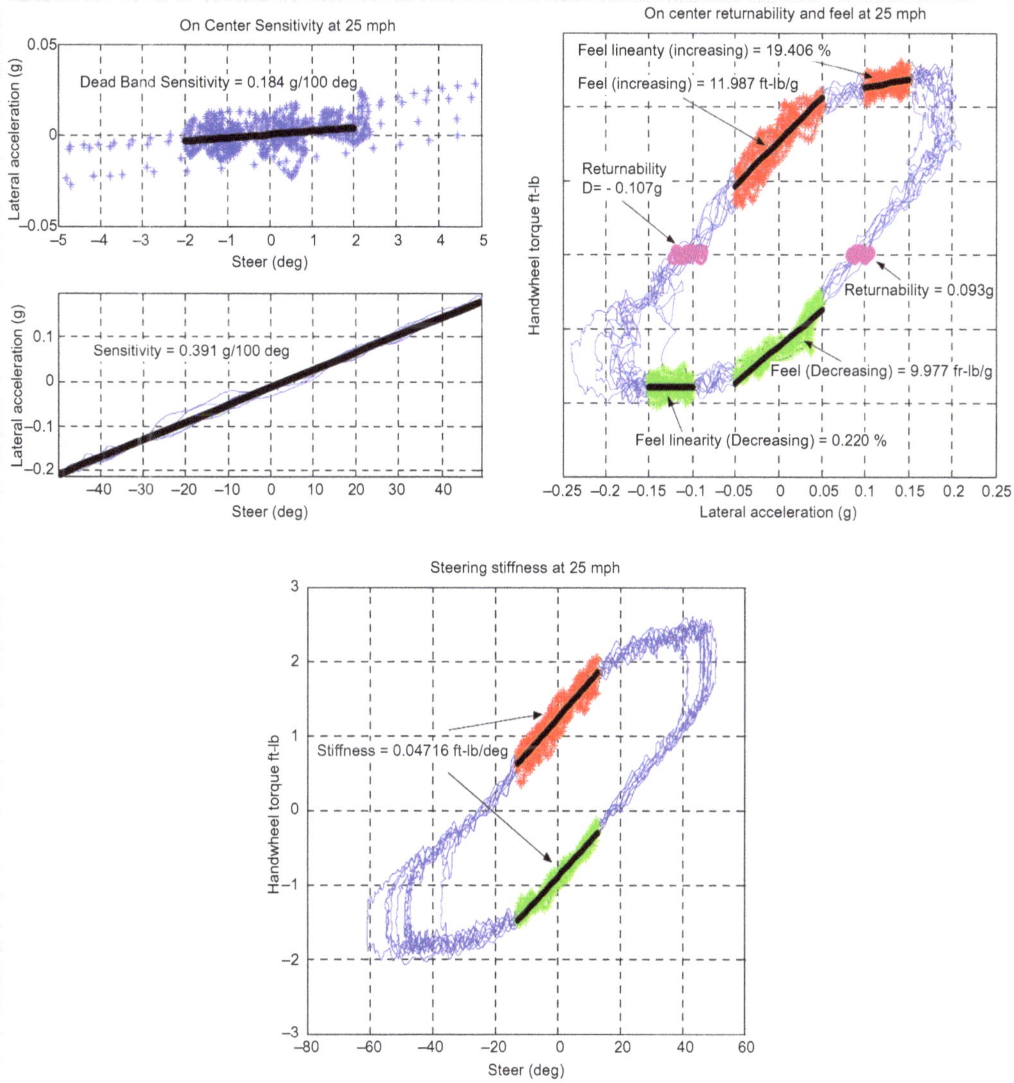

The mentioned cross plots are

1. **Steering wheel angle–vehicle response** (δ_{SW} – $\mathbf{a}_y$). This plot allows to estimate the *steering sensitivity*. Salaani et al. reported that, in the *steering dead-band* (namely $\delta_{SW} \in [-2 \text{ deg}, +2 \text{ deg}]$), the lateral acceleration a_y can be different from zero due to asymmetric properties of the vehicle, including different left and right tire characteristics, suspension kinematics (static toe and camber angles), and weight unbalance.
2. **Steering torque–vehicle response** ($\mathbf{M}_S$ – $\mathbf{a}_y$). This plot turns out to be extremely useful for the objective characterization of the on-center steering feel. Firstly, it allows to estimate the *feel* as the torque gradient at null lateral acceleration value.

Moreover, the upper and lower curves of loop – representing, respectively, right and left turn – intersect the $M_S = 0$ axis at different a_y values, and vice-versa. This difference in behavior for left and right turn can be attributed to asymmetries of vehicle and tires. Salaani et al. define *returnability* as the lateral acceleration a_y values (left turn and right turn) when the steering torque M_S is null: these represent the dynamic status of the vehicle when the driver is not feeling any torque from the steering system.

The *steering torque–vehicle response* plot is considered also by Anstalt für Verbrennungskraftmaschinen List (*AVL*) [35] for the development of a model-based optimization methodology of steering feel.

3. **Steering torque–steering wheel angle** ($\mathbf{M_S} - \delta_{SW}$). This plot provides information on the steering wheel *centering* felt by the driver as well as on *torque dead-band* and *friction* level in the steering system.

A research on the influence of tire characteristics on vehicle on-center steering was carried out by *Hyundai Motor Company* [42]. It concluded that better on-center performances are obtained with larger tire aligning moment M_z at small slip angle of $\alpha = 1$ deg. Thus, the effects of tire structural design parameters, tread pattern shape, and viscoelastic properties on aligning moment stiffness $C_{M\alpha}$ were described, with trade-off considerations with respect to cornering stiffness $C_{F\alpha}$. No tire characteristics' asymmetries were considered. Moreover, the resolution of tire measurements was probably too low, having aligning moment M_z data for slip angle intervals of $\Delta\alpha = 1°$: this prevented inspecting the effects of tire characteristic offsets considered in this book.

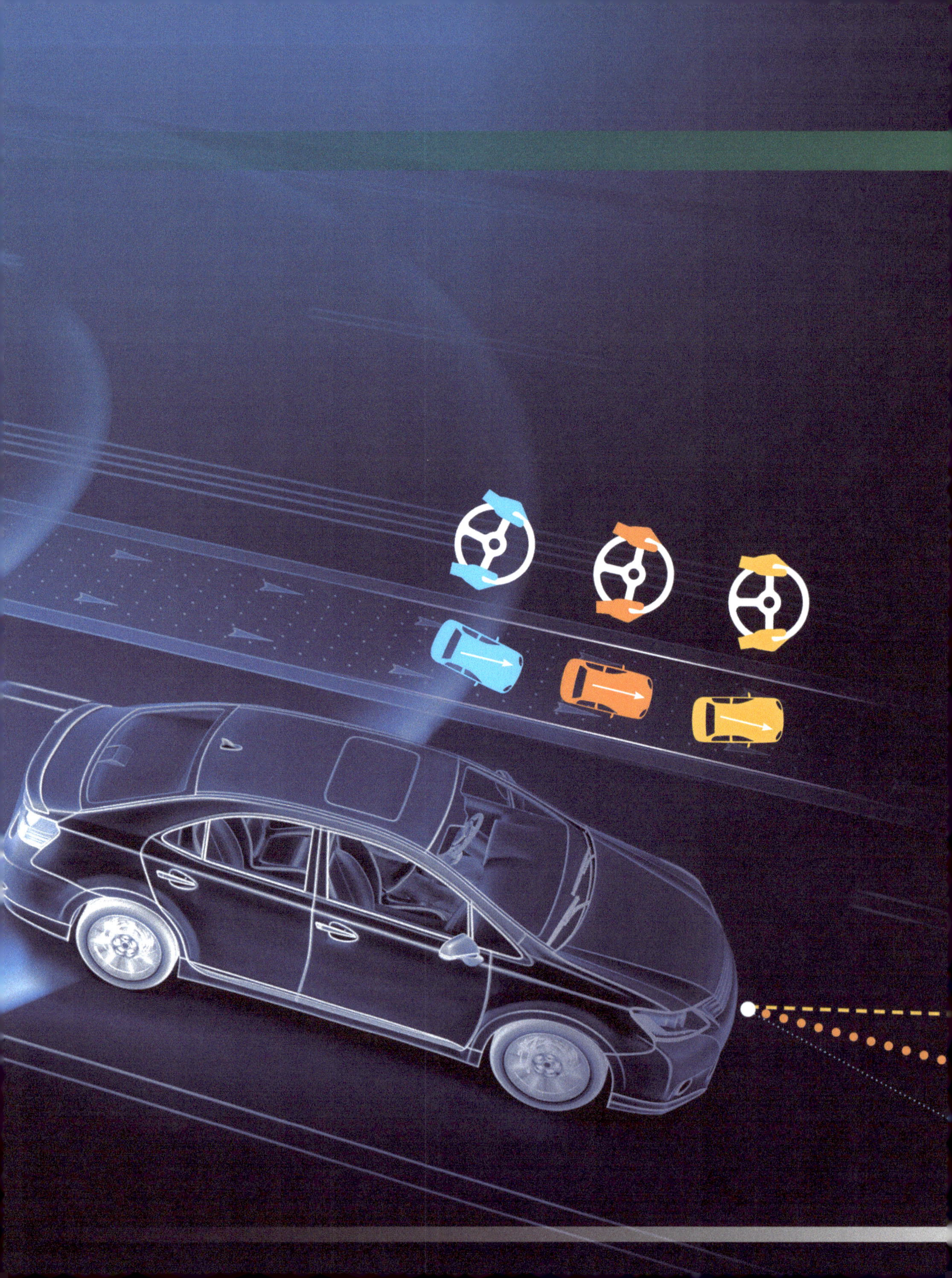

CHAPTER 6

Road Cross Slope Effect

The effect of road cross slope on the aspects discussed so far is dealt with in this chapter.

Roads are usually built with a slope in the cross section to induce rain to flow out of the traveled way. Accumulated water can cause problems such as aquaplaning. Estimations of the required slope depending on hydraulic permeability of the construction and water discharge rate can be found in literature [43].

Different configurations of road cross slope (or crown) may be found, as illustrated in the examples of **Figure 6.1**. A common one is the *normal crown* or *centerline crown*, where the road centerline is higher than the sides, so that drainage flows out of road surface into the surroundings. For road design engineering, different standards and guidelines can be found, indicating the proper geometry of different road cross sections. An example from the Italian *Ministero delle Infrastrutture e dei Trasporti* (*Ministry of Infrastructure and Transport*) is provided in **Figure 6.2**.

Cross slope is usually measured as percentage of height gain per unit of horizontal distance, as illustrated in **Figure 6.3**, and can be computed (with small-angle approximation) as

$$
\begin{aligned}
i_c &= \frac{h_r}{b_r} \cdot 100 = tan\varphi_r \cdot 100 \\
tan\varphi_r &= \frac{h_r}{b_r} \approx \varphi_r
\end{aligned}
\qquad (6.1)
$$

FIGURE 6.1 Road cross slope types (adapted from [44]).

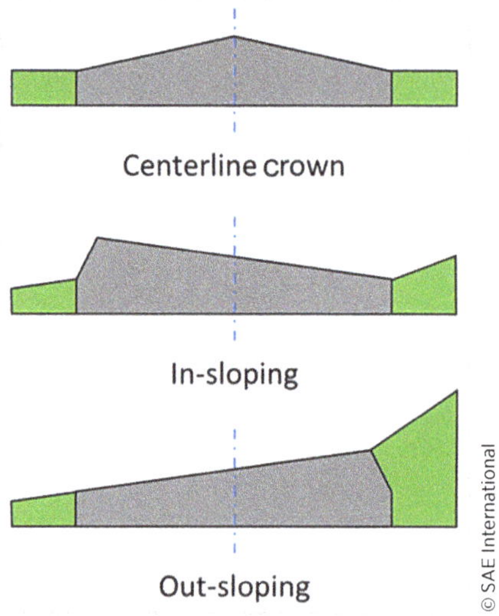

FIGURE 6.2 Road cross section geometry measures (i_C indicates cross slope), adapted from [45].

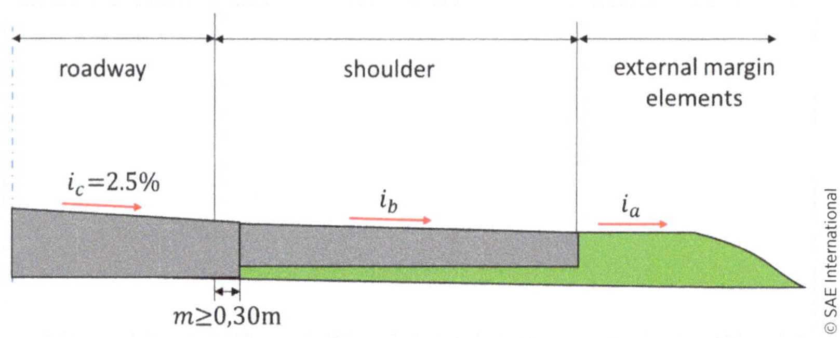

FIGURE 6.3 Normal or centerline crown, cross slope scheme.

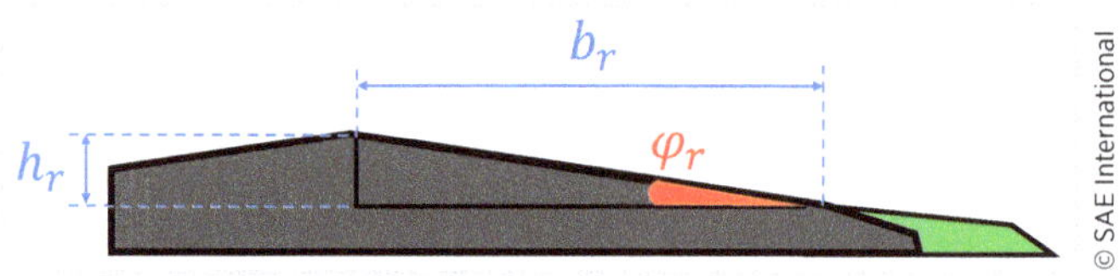

The *cross slope* (or *road camber*) angle φ_r is normally small. In the Italian regulations [45], the indicated nominal value is $i_c \approx 2.5\%$ for all roads. In the US [43, 46, 47], the most common value for highways is $i_c \leq 2.0\%$.

Road cross slope effects on the vehicle can be evaluated by considering induced roll angle φ_{rv} and static load transfer $\Delta F_{zr,i}$.

6.1 Induced Roll and Load Transfer

The cross slope angle φ_r of the road induces a roll angle φ_{rv} to the vehicle body, which can be computed via the moment equilibrium around the roll axis with a procedure similar to the one employed in Chapter 2

$$\varphi_{rv}\cdot\left(k_{\varphi 1}+k_{\varphi 2}\right)-h'sin\varphi_{rv}\cdot mgcos\varphi_r-h'cos\varphi_{rv}\cdot mgsin\varphi_r=0$$
$$\varphi_{rv}\approx\frac{h'mg\cdot\varphi_r}{k_{\varphi 1}+k_{\varphi 2}-h'mg} \tag{6.2}$$

where the small-angle approximation has been introduced

- $sin\varphi_r\to\varphi_r$
- $sin\varphi_{rv}\to\varphi_{rv}$
- $cos\varphi_r\to 1$
- $cos\varphi_{rv}\to 1$

For the i-axle, the load transfer pertaining just to road cross slope can be computed similarly to Equation 2.7. Let us consider the moment equilibrium around a wheel's contact point in the scheme represented in **Figure 6.4**

$$\Delta F_{zr,i}=\frac{1}{2s_i}\left[k_{\varphi i}\cdot\varphi_{rv}+\frac{l-a_i}{l}h_i\cdot mg\varphi_r\right]$$
$$\Rightarrow\Delta F_{zr,i}=\frac{1}{2s_i}\left[\frac{h'k_{\varphi i}}{k_{\varphi 1}+k_{\varphi 2}-h'mg}+\frac{l-a_i}{l}h_i\right]\cdot mg\varphi_r \tag{6.3}$$

FIGURE 6.4 Scheme of vehicle on a sloped road.

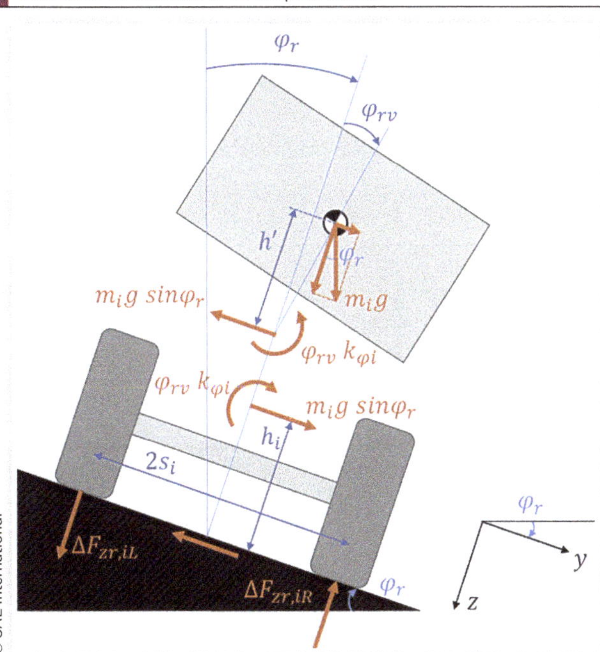

Therefore, the i-axle left (L) and right (R) wheel's static load (normal to the road surface) are

$$F_{z,iL,stat} = \frac{1}{2}\left(mg \cdot \frac{l-a_i}{l}\right) - \Delta F_{zr,i}$$
$$F_{z,iR,stat} = \frac{1}{2}\left(mg \cdot \frac{l-a_i}{l}\right) + \Delta F_{zr,i} \tag{6.4}$$

It has to be mentioned that the positive sign convention for the road cross slope φ_r is consistent with right-side driving countries. For countries following left-side driving, the road cross slope φ_r assumes a negative value.

6.2 Effect on Left and Right Tire Characteristics

A road cross slope-induced load transfer between left and right side of an axle changes the normal force applied on the tires. Therefore, it has an influence on the actual value of the tire characteristics in terms of cornering stiffness $C_{F\alpha,i0L/R}$, aligning moment stiffness $C_{M\alpha,iL/R}$, side force vertical shift $F_{y0,iL/R}$, and aligning moment shift $M_{z0,iL/R}$.

In this work, the left and right normal forces calculated according to Equation 6.4 are used to compute the actual left and right tire characteristics, with a polynomial fit of the available data for the different load conditions as explained in Chapter 9. Then, the equations introduced in Chapters 2, 3, 5, and 4 are used with the left and right characteristics computed for the actual tire loads, according to Equations 9.1–9.4.

6.3 Effect on Handling Diagram Theory and Straight Motion

Besides the modification of tire characteristics, an additional lateral force component – directed along y-*axis* (refer to Figure 6.4) – has to be introduced, due to the combined effect of road inclination and vehicle weight. Starting from Equation 3.16, an additional term $mg\varphi_r \cdot a_i$ is introduced

$$\begin{cases} F_{y,1} \cdot l + mg\varphi_r \cdot a_2 = ma_y \cdot a_2 - \left(M_{z,1} + M_{z,2}\right) \\ F_{y,2} \cdot l + mg\varphi_r \cdot a_1 = ma_y \cdot a_1 + M_{z,1} + M_{z,2}) \end{cases}$$

With the same procedure adopted in Chapter 3, observing the analogy with the lateral acceleration term ($ma_y \cdot a_i$), the system can be rewritten as

$$\begin{cases} \dfrac{F_{y,1} + \left(M_{z,1} + M_{z,2}\right)/l}{F_{z1}} + \varphi_r = \dfrac{a_y}{g} = \mu_1 \\ \dfrac{F_{y,2} + \left(M_{z,1} + M_{z,2}\right)/l}{F_{z2}} + \varphi_r = \dfrac{a_y}{g} = \mu_2 \end{cases} \tag{6.5}$$

The full derivation of the extended Handling Diagram theory is reported in Appendix A.6.

Therefore, slip angles and steering angle can be expressed adding to Equations 3.17–3.20, the lateral force term given by road slope φ_r and vehicle weight.

- **Front side slip angle**

$$\begin{aligned}\alpha_1 &= K_{\alpha 1}\cdot\mu - K_{\alpha 1}\cdot\varphi_r + \alpha_{1,st}\\ &\approx \frac{F_{z1}}{C^*_{eff,1}}\cdot\mu - \frac{F_{z1}}{C^*_{eff,1}}\cdot\varphi_r - \frac{F_{y0,1}}{C^*_{eff,1}}\end{aligned} \tag{6.6}$$

- **Rear side slip angle**

$$\begin{aligned}\alpha_2 &= K_{\alpha 2}\cdot\mu - K_{\alpha 2}\cdot\varphi_r + \alpha_{2,st}\\ &\approx \frac{F_{z2}}{C^*_{eff,2}}\cdot\mu - \frac{F_{z2}}{C^*_{eff,2}}\cdot\varphi_r - \frac{F_{y0.2}}{C^*_{eff,2}}\end{aligned} \tag{6.7}$$

- **Steering angle**

$$\begin{aligned}\delta &= \delta_0 + \alpha_1 - \alpha_2 =\\ &= K_\delta\cdot\mu - \left(K_{\alpha 1} - K_{\alpha 2}\right)\cdot\varphi_r + \delta_{st} =\\ &= \left(K_{\alpha 1} - K_{\alpha 2} + \frac{l\cdot g}{u^2}\right)\cdot\mu - \left(K_{\alpha 1} - K_{\alpha 2}\right)\cdot\varphi_r + \left(\alpha_{1,st} - a_{2,st}\right)\end{aligned} \tag{6.8}$$

The angles for straight driving are obtained for $\mu = 0$, thus the additional term with φ_r has to be considered together with the terms having the subscript *st*.

6.4 Effect on Equations of Motion

An additional force term has to be included in Equation 2.57 to the *y-axis* force equilibrium expression (roll dynamics is not considered in this analysis) and is given as

$$\nwarrow v: m\left(\dot{v} + ru + h'\ddot{\varphi} - h'r^2\varphi\right) = F_{x,1}\delta + F_{y,1} + F_{y,2} + mg\varphi_r \tag{6.9}$$

The simplified version, neglecting roll dynamics and longitudinal tire forces, is used in the differential equations for the computation of the trajectory. Equation 5.1, including the road cross slope effect, becomes

$$\begin{cases}\nwarrow v: & \dot{v} = \dfrac{1}{m}\left(F_{y,1} + F_{y,2} + mg\varphi_r\right) - ur\\ \curvearrowright r: & \dot{r} = \dfrac{1}{I_z}\left(a_1F_{y,1} + M_{z,1} - a_2F_{y,2} + M_{z,2}\right)\end{cases} \tag{6.10}$$

6.5 Effect on Straight-Driving Steering Angle and Vehicle Attitude

Equations 6.6–6.8 are used to introduce the road cross slope φ_r effect in the expressions of steering angle δ_{st} (defined in Equation 4.9) and vehicle attitude β_{st} (defined in Equation 4.11) in the straight motion condition and are given below

$$\begin{aligned}\delta_{st} &= \alpha_{1,st} - \alpha_{2,st} - (K_{\alpha 1} - K_{\alpha 2})\varphi_r \\ &\approx -\frac{F_{y0,1}}{C^*_{eff,1}} + \frac{F_{y0,2}}{C^*_{eff,2}} + \left(-\frac{F_{z1}}{C^*_{eff,1}} + \frac{F_{z2}}{C^*_{eff,2}}\right)\varphi_r = \\ &= \left[-\Psi_{10} \cdot \left(\frac{C_{F\alpha,10R} - C_{F\alpha,10L}}{C_{F\alpha,10R} + C_{F\alpha,10L}}\right) - \gamma_{10} \cdot \left(\frac{C_{F\gamma,10R} - C_{F\gamma,10L}}{C_{F\alpha,10R} + C_{F\alpha,10L}}\right) - \left(\frac{F_{y0,1R} + F_{y0,1L}}{C_{F\alpha,10R} + C_{F\alpha,10L}}\right) + \right. \\ &\quad \left. + \Psi_{20} \cdot \left(\frac{C_{F\alpha,20R} - C_{F\alpha,20L}}{C_{F\alpha,20R} + C_{F\alpha,20L}}\right) + \gamma_{20} \cdot \left(\frac{C_{F\gamma,20R} - C_{F\gamma,20L}}{C_{F\alpha,20R} + C_{F\alpha,20L}}\right) + \left(\frac{F_{y0,2R} + F_{y0,2L}}{C_{F\alpha,20R} + C_{F\alpha,20L}}\right)\right] + \\ &\quad + \left(-\frac{F_{z1}}{C^*_{eff,1}} + \frac{F_{z2}}{C^*_{eff,2}}\right)\varphi_r\end{aligned} \tag{6.11}$$

$$\begin{aligned}\beta_{st} &= -\alpha_{2,st} + K_{\alpha 2}\varphi_r \\ &\approx \frac{F_{y0,2}}{C^*_{eff,2}} + \left(\frac{F_{z2}}{C^*_{eff,2}}\right)\varphi_r = \\ &= \Psi_{20} \cdot \left(\frac{C_{F\alpha,20R} - C_{F\alpha,20L}}{C_{F\alpha,20R} + C_{F\alpha,20L}}\right) + \gamma_{20} \cdot \left(\frac{C_{F\gamma,20R} - C_{F\gamma,20L}}{C_{F\alpha,20R} + C_{F\alpha,20L}}\right) + \left(\frac{F_{y0,2R} + F_{y0,2L}}{C_{F\alpha,20R} + C_{F\alpha,20L}}\right) + \\ &\quad + \left(\frac{F_{z2}}{C^*_{eff,2}}\right)\varphi_r\end{aligned} \tag{6.12}$$

6.6 Effect on Straight-Driving Steering Torque

Equation 6.6 is used to introduce the road cross slope φ_r effect in the steering torque (expressed in Equation 4.7) and is given below

$$\begin{aligned}M_S(\mu) &= \frac{F_{y,1} \cdot n - M_{z,1}}{n_{st} \cdot V_S} = \frac{M^*_S(\mu)}{n_{st} \cdot V_S} \\ M^*_S(\mu) &= F_{y,1} \cdot n - M_{z,1} = \\ &= n \cdot [C^*_{eff,1} \cdot (K_{\alpha 1} \cdot \mu - K_{\alpha 1} \cdot \varphi_r + \alpha_{1,st}) + F_{y0,1}] - \\ &\quad -[-C_{M\alpha,1} \cdot (K_{\alpha 1} \cdot \mu - K_{\alpha 1} \cdot \varphi_r + \alpha_{1,st}) + M_{z0,1R} + M_{z0,1L}] \\ M^*_S(\mu) &= (\mu - \varphi_r) \cdot K_{\alpha 1} \cdot [C^*_{eff,1} \cdot n + C_{M\alpha,1}] + n \cdot (C^*_{eff,1} \cdot \alpha_{1,st} + F_{y0,1}) + \\ &\quad + C_{M\alpha,1} \cdot \alpha_{1,st} - M_{z0,1R} - M_{z0,1L}\end{aligned} \tag{6.13}$$

By introducing the approximations expressed in Equation 3.17, it is possible to obtain a simplified formulation which is given as

$$M_S^*(\mu) \approx (\mu - \varphi_r) \cdot \frac{F_{z1}}{C_{eff,1}^*} \cdot \left[C_{eff,1}^* \cdot n + C_{M\alpha,1} \right] + n \cdot \left(C_{eff,1}^* \cdot \frac{-F_{y0,1}}{C_{eff,1}^*} + F_{y0,1} \right) +$$

$$+ C_{M\alpha,1} \cdot \frac{-F_{y0,1}}{C_{eff,1}^*} - M_{z0,1R} - M_{z0,1L}$$

$$M_S^*(\mu) \approx (\mu - \varphi_r) \cdot F_{z1} \cdot \left[n + \frac{C_{M\alpha,1}}{C_{eff,1}^*} \right] + C_{M\alpha,1} \cdot \frac{-F_{y0,1}}{C_{eff,1}^*} - M_{z0,1R} - M_{z0,1L} \tag{6.14}$$

Thus, for null lateral acceleration, the steering torque offset $M_{S,st}^*$ reads

$$M_{S,st}^* = M_S^*(\mu = 0) =$$

$$= -\varphi_r \cdot K_{\alpha 1} \cdot \left[C_{eff,1}^* \cdot n + C_{M\alpha,1} \right] + n \cdot \left(C_{eff,1}^* \cdot \alpha_{1,st} + F_{y0,1} \right) +$$

$$+ C_{M\alpha,1} \cdot \alpha_{1,st} - M_{z0,1R} - M_{z0,1L} =$$

$$= -\varphi_r \cdot K_{\alpha 1} \cdot \left[C_{eff,1}^* \cdot n + C_{M\alpha,1} \right] + n \cdot \left(C_{eff,1}^* \cdot \alpha_{1,st} + F_{y0,1} \right) - M_{z1,st}$$

$$M_{S,st}^* \approx -\varphi_r \cdot F_{z1} \cdot \left[n + \frac{C_{M\alpha,1}}{C_{eff,1}^*} \right] - M_{z1,st} \tag{6.15}$$

where the term $M_{z1,st}$ has been defined in Equation 4.10 and is reported here as

$$M_{z1,st} = -C_{M\alpha,1} \cdot \alpha_{1,st} + M_{z0,1}$$

$$\approx C_{M\alpha,1} \cdot \left(\frac{F_{y0,1}}{C_{eff,1}^*} \right) + M_{z0,1} =$$

$$= C_{M\alpha,1} \cdot \left[\Psi_{10} \cdot \left(\frac{C_{F\alpha,10R} - C_{F\alpha,10L}}{C_{F\alpha,10R} + C_{F\alpha,10L}} \right) + \gamma_{10} \cdot \left(\frac{C_{F\gamma,10R} - C_{F\gamma,10L}}{C_{F\alpha,10R} + C_{F\alpha,10L}} \right) + \right.$$

$$\left. + \left(\frac{F_{y0,1R} + F_{y0,1L}}{C_{F\alpha,10R} + C_{F\alpha,10L}} \right) \right] + M_{z0,1R} + M_{z0,1L} \tag{6.16}$$

6.6.1 Effect on Straight-Driving Steering Torque with Effective Axle Aligning Moment

The aforementioned equations, pertaining to the steering wheel torque, can be rewritten considering the effective axle aligning moment (introduced in Section 2.5).

Equation 6.6 is used to introduce the road cross slope φ_r effect in the steering torque (expressed in Equation 4.17) and is given as

$$\begin{aligned} M_S(\mu) &= \frac{F_{y,1}\cdot n - M_{z,1}}{n_{st}\cdot V_S} = \frac{M_S^*(\mu)}{n_{st}\cdot V_S} \\ M_S^*(\mu) &= F_{y,1}\cdot n - M_{z,1} = \\ &= \left(C_{eff,1}^*\cdot\alpha_1 + F_{y0,1}\right)\cdot n - \left(-C_{M\alpha,1}\cdot\alpha_1 + M_{z0,1}\right) = \\ &= \left[(\mu-\varphi_r)\cdot K_{\alpha 1} + \alpha_{1,st}\right]\cdot\left(n\cdot C_{eff,1}^* + C_{M\alpha,1}\right) + F_{y0,1}\cdot n - M_{z0,1} \end{aligned} \tag{6.17}$$

By introducing the approximations expressed in Equation 3.17, namely

$$K_{\alpha 1} \approx \frac{F_{z1}}{C_{eff,1}^*}$$

$$\alpha_{1,st} \approx -\frac{F_{y0,1}}{C_{eff,1}^*}$$

it is possible to obtain a simplified formulation for the steering torque (refer to Equation 4.19 to compare the following expression with the one with null road cross slope)

$$M_S^*(\mu) \approx (\mu-\varphi_r)\cdot F_{z1}\cdot\left(n + \frac{C_{M\alpha,1}}{C_{eff,1}^*}\right) - \frac{F_{y0,1}}{C_{eff,1}^*}\cdot C_{M\alpha,1} - M_{z0,1} \tag{6.18}$$

Thus, for null lateral acceleration, the steering torque offset $M_{S,st}^*$ reads

$$\begin{aligned} M_{S,st}^* &= M_S^*(\mu=0) = \\ &= (-\varphi_r\cdot K_{\alpha 1} + \alpha_{1,st})\cdot(n\cdot C_{eff,1}^* + C_{M\alpha,1}) + F_{y0,1}\cdot n - M_{z0,1} \\ M_{S,st}^* &\approx -\varphi_r\cdot F_{z1}\cdot\left(n + \frac{C_{M\alpha,1}}{C_{eff,1}^*}\right) - \frac{F_{y0,1}}{C_{eff,1}^*}\cdot C_{M\alpha,1} - M_{z0,1} = \\ &= -\varphi_r\cdot F_{z1}\cdot\left[n + \frac{2C_{M\alpha,10}}{C_{eff,1}^*} - (C_{M\alpha,roll,1} - C_{M\alpha,susp,1} + C_{M\alpha,steer,1} + C_{M\alpha,\Delta F_z,1})\right] - \\ &\quad -\frac{F_{y0,1}\cdot 2C_{M\alpha,10}}{C_{eff,1}^*} - \Psi_{10}\cdot(C_{M\alpha,10L} - C_{M\alpha,10R}) + \gamma_{10}\cdot(C_{M\gamma,10L} - C_{M\gamma,10R}) - \\ &\quad - M_{z0,1L} - M_{z0,1R} \end{aligned} \tag{6.19}$$

Substituting also the expression of axle side force shift $F_{y0,1}$, according to Equation 2.36, it is possible to rewrite Equation 6.19 in an expanded fashion as

$$\begin{aligned} M_{S,st}^* &\approx -\varphi_r\cdot F_{z1}\cdot\left[n + \frac{2C_{M\alpha,10}}{C_{eff,1}^*} - (C_{M\alpha,roll,1} - C_{M\alpha,susp,1} + C_{M\alpha,steer,1} + C_{M\alpha,\Delta F_z,1})\right] - \\ &\quad -2C_{M\alpha,10}\cdot\left[\Psi_{10}\cdot\left(\frac{C_{F\alpha,10R} - C_{F\alpha,10L}}{C_{F\alpha,10R} + C_{F\alpha,10L}}\right) + \gamma_{10}\cdot\left(\frac{C_{F\gamma,10R} - C_{F\gamma,10L}}{C_{F\alpha,10R} + C_{F\alpha,10L}}\right) + \right. \\ &\quad \left. + \left(\frac{F_{y0,1R} + F_{y0,1L}}{C_{F\alpha,10R} + C_{F\alpha,10L}}\right)\right] - \Psi_{10}\cdot(C_{M\alpha,10L} - C_{M\alpha,10R}) + \\ &\quad +\gamma_{10}\cdot(C_{M\gamma,10L} - C_{M\gamma,10R}) - M_{z0,1L} - M_{z0,1R} \end{aligned} \tag{6.20}$$

CHAPTER 7

Straight Motion Parameter Setting: Wheel Alignment and Tire Characteristics

In this chapter, we will define how the wheel alignment parameters influence the straight-driving steering angle, steering torque and vehicle attitude offsets. In particular, the relationships between these parameters aimed at obtaining null straight-driving offsets are presented. Analytical solutions are given which does not guarantee an engineering actual feasibility.

Equations 6.11–6.12 and 6.20 exhibit the contributions belonging to the offsets of steering angle, vehicle attitude and steering torque when driving straight ($a_y = 0$). The equations are here reported in their full forms, i.e., taking into account road cross slope and the related load transfer. The expressions are written after the introduction of the approximations already discussed in Chapter 6

- **Steering angle**

$$\begin{aligned}\delta_{st} \approx & \left[-\Psi_{10}\cdot\left(\frac{C_{F\alpha,10R}-C_{F\alpha,10L}}{C_{F\alpha,10R}+C_{F\alpha,10L}}\right)-\gamma_{10}\cdot\left(\frac{C_{F\gamma,10R}-C_{F\gamma,10L}}{C_{F\alpha,10R}+C_{F\alpha,10L}}\right)-\right.\\ & -\left(\frac{F_{y0,1R}+F_{y0,1L}}{C_{F\alpha,10R}+C_{F\alpha,10L}}\right)+\left(\frac{F_{y0,2R}+F_{y0,2L}}{C_{F\alpha,20R}+C_{F\alpha,20L}}\right)+\\ & \left.+\Psi_{20}\cdot\left(\frac{C_{F\alpha,20R}-C_{F\alpha,20L}}{C_{F\alpha,20R}+C_{F\alpha,20L}}\right)+\gamma_{20}\cdot\left(\frac{C_{F\gamma,20R}-C_{F\gamma,20L}}{C_{F\alpha,20R}+C_{F\alpha,20L}}\right)\right]+\\ & +\left(-\frac{F_{z1}}{C^{*}_{eff,1}}+\frac{F_{z2}}{C^{*}_{eff,2}}\right)\varphi_r \qquad (7.1)\end{aligned}$$

- **Steering torque**

$$
\begin{aligned}
M^{*}_{S,st} \approx & -\varphi_r \cdot F_{z1} \cdot \left[n + \frac{2C_{M\alpha,10}}{C^{*}_{eff,1}} - (C_{M\alpha,roll,1} - C_{M\alpha,susp,1} + C_{M\alpha,steer,1} + C_{M\alpha,\Delta F_z,1}) \right] - \\
& -2C_{M\alpha,10} \cdot \left[\Psi_{10} \cdot \left(\frac{C_{F\alpha,10R} - C_{F\alpha,10L}}{C_{F\alpha,10R} + C_{F\alpha,10L}} \right) + \gamma_{10} \cdot \left(\frac{C_{F\gamma,10R} - C_{F\gamma,10L}}{C_{F\alpha,10R} + C_{F\alpha,10L}} \right) + \right. \\
& \left. + \left(\frac{F_{y0,1R} + F_{y0,1L}}{C_{F\alpha,10R} + C_{F\alpha,10L}} \right) \right] - \Psi_{10} \cdot (C_{M\alpha,10L} - C_{M\alpha,10R}) + \\
& + \gamma_{10} \cdot (C_{M\gamma,10L} - C_{M\gamma,10R}) - M_{z0,1L} - M_{z0,1R}
\end{aligned}
\tag{7.2}
$$

- **Vehicle attitude**

$$
\begin{aligned}
\beta_{st} \approx & \Psi_{20} \cdot \left(\frac{C_{F\alpha,20R} - C_{F\alpha,20L}}{C_{F\alpha,20R} + C_{F\alpha,20L}} \right) + \gamma_{20} \cdot \left(\frac{C_{F\gamma,20R} - C_{F\gamma,20L}}{C_{F\alpha,20R} + C_{F\alpha,20L}} \right) + \\
& + \left(\frac{F_{y0,2R} + F_{y0,2L}}{C_{F\alpha,20R} + C_{F\alpha,20L}} \right) + \left(\frac{F_{z2}}{C^{*}_{eff,2}} \right) \varphi_r
\end{aligned}
\tag{7.3}
$$

7.1 Sorting Out the Offsets Expressions

In order to efficaciously analyze the contribution of wheel alignment parameters to the straight-driving offsets, it is necessary to cluster the coefficients multiplying wheel alignment angles (namely static toe angle Ψ_{i0} and static camber angle γ_{i0}). Therefore, some terms pertaining to effective axle cornering stiffness and aligning stiffness have to be rewritten in order to isolate the terms related to wheel alignment angles.

First of all, the **effective axle cornering stiffness** $C^{*}_{eff,i}$ depends on static toe Ψ_{i0} and static camber γ_{i0} angles. This dependence can be emphasized by rewriting Equation 2.30

$$
\begin{aligned}
\frac{1}{C^{*}_{eff,i}} = & \frac{1 + C_{roll,i} - C_{susp,i} + C_{steer,i} + C^{*}_{\Delta F_z,i}}{2C_{F\alpha,i0}} = \\
& = \frac{1}{2C_{F\alpha,i0}} \cdot \left\{ 1 + C_{roll,i} - C_{susp,i} + C_{steer,i} + \sigma_i \frac{l}{l - a_i} \cdot \left[\xi_{\alpha,i}(-\alpha_{ply,iL} + \alpha_{ply,iR}) + \right. \right. \\
& \left. \left. + \xi_{\gamma,i}(-\gamma_{con,iL} + \gamma_{con,iR}) \right] + \sigma_i \frac{1}{l - a_i} \cdot \left[\xi_{\alpha,i} \cdot 2\Psi_{i0} + \xi_{\gamma,i} \cdot 2\gamma_{i0} \right] \right\} = \\
& = \frac{C_{el,AL,i}}{2C_{F\alpha,i0}} + \frac{1}{2C_{F\alpha,i0}} \cdot \sigma_i \frac{l}{l - a_i} \cdot \left[\xi_{\alpha,i} \cdot 2\Psi_{i0} + \xi_{\gamma,i} \cdot 2\gamma_{i0} \right] = \\
& = \frac{C_{el,AL,i}}{2C_{F\alpha,i0}} + \frac{C_{el,\Psi,i}}{2C_{F\alpha,i0}} \cdot \Psi_{i0} + \frac{C_{el,\gamma,i}}{2C_{F\alpha,i0}} \cdot \gamma_{i0}
\end{aligned}
\tag{7.4}
$$

where the elastokinematics coefficients have been separated into the different contributions:

- $C_{el,AL,i}$ when wheels are aligned
- $C_{el,\Psi,i}$ multiplying static toe angle Ψ_{i0}
- $C_{el,\gamma,i}$ multiplying static camber angle γ_{i0}

and are defined below as

$$\begin{aligned} C_{el,AL,i} &= 1 + C_{roll,i} - C_{susp,i} + C_{steer,i} + \\ &+ \sigma_i \frac{l}{l-a_i} \cdot \left[\xi_{\alpha,i}\left(-\alpha_{ply,iL} + \alpha_{ply,iR}\right) + \xi_{\gamma,i}\left(-\gamma_{con,iL} + \gamma_{con,iR}\right)\right] \\ C_{el,\Psi,i} &= \sigma_i \frac{l}{l-a_i} \cdot 2 \cdot \xi_{\alpha,i} \\ C_{el,\gamma,i} &= \sigma_i \frac{l}{l-a_i} \cdot 2 \cdot \xi_{\gamma,i} \end{aligned} \tag{7.5}$$

It is highlighted that $\frac{1}{C^*_{eff,i,AL}} = \frac{C_{el,AL,i}}{2C_{F\alpha,i0}}$ corresponds to the reciprocal of the effective axle cornering stiffness when the wheels are aligned ($\Psi_{i0} = 0$, $\gamma_{i0} = 0$)

$$\begin{aligned} \frac{1}{C^*_{eff,i,AL}} &= \frac{1}{2C_{F\alpha,i0}} \cdot \Big\{1 + C_{roll,i} - C_{susp,i} + C_{steer,i} + \\ &+ \sigma_i \frac{l}{l-a_i} \cdot \left[\xi_{\alpha,i}\left(-\alpha_{ply,iL} + \alpha_{ply,iR}\right) + \xi_{\gamma,i}\left(-\gamma_{con,iL} + \gamma_{con,iR}\right)\right]\Big\} \\ &= \frac{C_{el,AL,i}}{2C_{F\alpha,i0}} \end{aligned} \tag{7.6}$$

Then, the same dependency is highlighted also for the **effective axle aligning stiffness** $C_{M\alpha,i}$. Rewriting the terms of Equation 2.47, we get

$$-C_{M\alpha,i} = -2C_{M\alpha,i0} + C^*_{eff,i} \cdot \left(C_{M\alpha,roll,i} - C_{M\alpha,susp,i} + C_{M\alpha,steer,i} + C_{M\alpha,\Delta F_z,i}\right)$$

the elastokinematics coefficient $C_{M\alpha,\Delta Fz,i}$ can be split into its different contributions:

- $C_{M\alpha,el,AL,i}$ when wheels are aligned
- $C_{M\alpha,el,\Psi,i}$ multiplying static toe angle Ψ_{i0}
- $C_{M\alpha,el,\gamma,i}$ multiplying static camber angle γ_{i0}

and are defined below as

$$
\begin{aligned}
C_{M\alpha,\Delta F_z,i} &= \sigma_i \frac{l}{l-a_i}[\xi_{M\alpha,i}(2\Psi_{i0} - \alpha_{ply,iL} + \alpha_{ply,iR}) + \\
&\quad + \xi_{M\gamma,i}(-2\gamma_{i0} + \gamma_{con,iL} - \gamma_{con,iR})] = \\
&= C_{M\alpha,el,AL,i} + C_{M\alpha,el,\Psi,i} \cdot \Psi_{i0} + C_{M\alpha,el,\gamma,i} \cdot \gamma_{i0} \\
&\text{where} \\
C_{M\alpha,el,AL,i} &= \sigma_i \frac{l}{l-a_i}[\xi_{M\alpha,i}(-\alpha_{ply,iL} + \alpha_{ply,iR}) + \xi_{M\gamma,i}(\gamma_{con,iL} - \gamma_{con,iR})] \\
C_{M\alpha,el,\Psi,i} &= \sigma_i \frac{l}{l-a_i} \cdot 2 \cdot \xi_{M\alpha,i} \\
C_{M\alpha,el,\gamma,i} &= -\sigma_i \frac{l}{l-a_i} \cdot 2 \cdot \xi_{M\gamma,i}
\end{aligned}
\tag{7.7}
$$

It is now possible to rewrite in a more compact fashion the three offsets expressed in Equations 7.1–7.3 as follows:

- **Steering angle**

$$
\begin{aligned}
\delta_{st} &= -\Psi_{10} \cdot A_1 - \gamma_{10} \cdot B_1 - C_1 + \\
&\quad + \Psi_{20} \cdot A_2 + \gamma_{20} \cdot B_2 + C_2 + \\
&\quad + (-D_1 + D_2)\varphi_r
\end{aligned}
\tag{7.8}
$$

- **Steering torque**

$$
M^*_{S,st} = -2C_{M\alpha,10} \cdot \left[\Psi_{10} \cdot (A_1 + A_{1M}) + \gamma_{10} \cdot (B_1 + B_{1M}) + C_1 + C_{1M} + E_{1M} \cdot \varphi_r\right] \tag{7.9}
$$

- **Vehicle attitude**

$$
\beta_{st} = \Psi_{20} \cdot A_2 + \gamma_{20} \cdot B_2 + C_2 + D_2\varphi_r \tag{7.10}
$$

where the following **coefficients** have been used

A: coefficients multiplying toe (front Ψ_{10}, rear Ψ_{20})

$$
\begin{aligned}
A_1 &= \left(\frac{C_{F\alpha,10R} - C_{F\alpha,10L}}{C_{F\alpha,10R} + C_{F\alpha,10L}}\right) + \frac{F_{z1}}{2C_{F\alpha,10}} \cdot C_{el,\Psi,1} \cdot \varphi_r \\
A_2 &= \left(\frac{C_{F\alpha,20R} - C_{F\alpha,20L}}{C_{F\alpha,20R} + C_{F\alpha,20L}}\right) + \frac{F_{z2}}{2C_{F\alpha,20}} \cdot C_{el,\Psi,2} \cdot \varphi_r \\
A_{1M} &= -\left(\frac{C_{M\alpha,10R} - C_{M\alpha,10L}}{C_{M\alpha,10R} + C_{M\alpha,10L}}\right) - \frac{F_{z1}}{C_{M\alpha,10R} + C_{M\alpha,10L}} \cdot C_{M\alpha,el,\Psi,1} \cdot \varphi_r
\end{aligned}
\tag{7.11}
$$

B: coefficients multiplying camber (front γ_{10}, rear γ_{20})

$$
\begin{aligned}
B_1 &= \left(\frac{C_{F\gamma,10R} - C_{F\gamma,10L}}{C_{F\alpha,10R} + C_{F\alpha,10L}}\right) + \frac{F_{z1}}{2C_{F\alpha,10}} \cdot C_{el,\gamma,1} \cdot \varphi_r \\
B_2 &= \left(\frac{C_{F\gamma,20R} - C_{F\gamma,20L}}{C_{F\alpha,20R} + C_{F\alpha,20L}}\right) + \frac{F_{z2}}{2C_{F\alpha,20}} \cdot C_{el,\gamma,2} \cdot \varphi_r \\
B_{1M} &= \left(\frac{C_{M\gamma,10R} - C_{M\gamma,10L}}{C_{M\alpha,10R} + C_{M\alpha,10L}}\right) - \frac{F_{z1}}{C_{M\alpha,10R} + C_{M\alpha,10L}} \cdot C_{M\alpha,el,\gamma,1} \cdot \varphi_r
\end{aligned}
\tag{7.12}
$$

C: coefficients related to tire characteristics' offsets ($F_{y0,i}$ and $M_{z0,i}$), yielded by ply-steer and conicity

$$
\begin{aligned}
C_1 &= \left(\frac{F_{y0,1R} + F_{y0,1L}}{C_{F\alpha,10R} + C_{F\alpha,10L}}\right) \\
C_2 &= \left(\frac{F_{y0,2R} + F_{y0,2L}}{C_{F\alpha,20R} + C_{F\alpha,20L}}\right) \\
C_{1M} &= \left(\frac{M_{z0,1R} + M_{z0,1L}}{C_{M\alpha,10R} + C_{M\alpha,10L}}\right)
\end{aligned}
\tag{7.13}
$$

D, E: coefficients multiplying road cross slope φ_r

$$
\begin{aligned}
D_1 &= \frac{F_{z1}}{C^*_{eff,1,AL}} = F_{z1} \cdot \frac{C_{el,AL,1}}{2C_{F\alpha,10}} \\
D_2 &= \frac{F_{z2}}{C^*_{eff,2,AL}} = F_{z2} \cdot \frac{C_{el,AL,2}}{2C_{F\alpha,20}} \\
E_{1M} &= \frac{F_{z1}}{2C_{M\alpha,10}} \cdot \left(n - C_{M\alpha,roll,1} + C_{M\alpha,susp,1} - C_{M\alpha,steer,1} - C_{M\alpha,el,AL,1} + \frac{2C_{M\alpha,10}}{C^*_{eff,1,AL}}\right)
\end{aligned}
\tag{7.14}
$$

More details pertaining to the derivation of Equation 7.9 are reported in Appendix A.7.

7.2 Wheel Alignment Setting for Null Straight Motion Offsets

The condition for which all the three offsets are simultaneously null is pursued. Actually, when this condition is verified, for a determined road cross slope φ_r the vehicle moves in a straight path ($a_y = 0$) with

- $\beta = 0$: null attitude (no "*dog-tracking*" straight motion).
- $M_S = 0$: the driver does not have to apply any torque (or "*pull*") on the steering wheel to maintain the straight motion.
- $\delta = 0$: the steering wheel does not have an angular displacement (or "*misalignment*").

Once the road cross slope φ_r is determined, the parameters remaining in the equations are related to

- **Wheel alignment**. Static toe angles (front Ψ_{10}, rear Ψ_{20}) and static camber (front γ_{10}, rear γ_{20}).
- **Tires**. Tire characteristics are differentiated between left and right side of the same axle, to take into account manufacturing heterogeneity and load transfer effect (provoked by road cross slope φ_r).
- **Vehicle**. The weight distribution of the vehicle is affecting directly the axles' loads (front F_{z1}, rear F_{z2}).
  Suspension elastokinematics has impact on the effective axle cornering stiffness $\left(C^*_{eff,1}, C^*_{eff,2}\right)$ and aligning stiffness ($C_{M\alpha,1}$). It is important to underline that elastokinematics is only affecting the terms related to road cross slope φ_r contribution.

Therefore, for a specific vehicle and road cross slope, it is possible to discuss analytically which wheel alignment and tire characteristics setting is theoretically required to cancel out the straight-driving offsets of steering angle, steering torque, and vehicle attitude.

7.2.1 Wheel Alignment Setting, Fixed Tire Characteristics

In case the vehicle parameters and the tires' characteristics are given, it is theoretically possible to have simultaneously $\delta_{st} = 0$, $M^*_{S,st} = 0$, and $\beta_{st} = 0$. Actually, we have three linear equations (7.8–7.10) and four unknowns, namely Ψ_{10}, Ψ_{20}, γ_{10}, and γ_{20}. Nevertheless, the resulting condition pertaining to wheel alignment setting turns out to be numerically (hence practically) unattainable. This will be discussed in Chapter 13.

We deal with specific cases, the full derivation of the solutions presented below is reported in Appendix A.7.

$M^*_{S,st} = 0$, $\beta_{st} = 0$, $\delta_{st} \neq 0$ One possibility is to set wheel alignment in order to have null steering torque and vanishing vehicle attitude when driving straight. To nullify the steering wheel angular offset, on the other hand, a third condition is required. Otherwise, the steering wheel will generally have a nonnull offset, as shown in **Figure 7.1**.

FIGURE 7.1 Straight motion with $M^*_{S,st} = 0$, $\beta_{st} = 0$, $\delta_{st} \neq 0$.

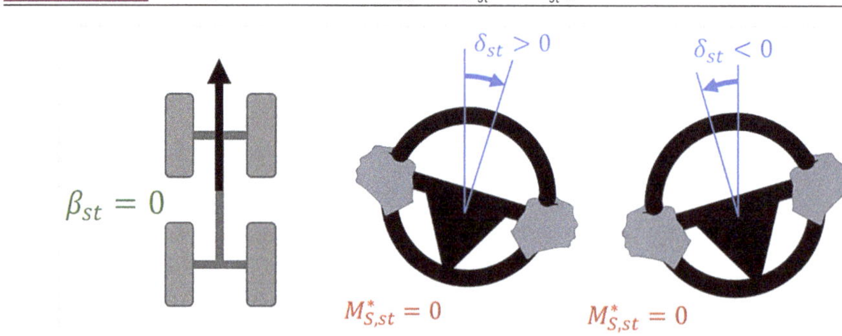

$$\begin{cases} \beta_{st} = 0: & \Psi_{20} \cdot A_2 + \gamma_{20} \cdot B_2 = -C_2 - D_2\varphi_r \\ M^*_{S,st} = 0: & \Psi_{10} \cdot (A_1 + A_{1M}) + \gamma_{10} \cdot (B_1 + B_{1M}) + C_1 + C_{1M} + E_{1M} \cdot \varphi_r = 0 \end{cases} \tag{7.15}$$

$$\begin{aligned} \delta_{st} &= -\Psi_{10} \cdot A_1 - \gamma_{10} \cdot B_1 - C_1 - D_1\varphi_r = \\ &= \Psi_{10} \cdot A_{1M} + \gamma_{10} \cdot B_{1M} + C_{1M} + (E_{1M} - D_1)\varphi_r \end{aligned} \tag{7.16}$$

$M^*_{S,st} = 0, \delta_{st} = 0, \beta_{st} \neq 0$ Another possibility is to set wheel alignment in order to have null steering torque and steering angle when driving straight. The vehicle attitude, on the other hand, would generally be nonnull, as shown in Figure 7.2, unless a third condition is fulfilled.

$$\begin{cases} \delta_{st} = 0: & \Psi_{20} \cdot A_2 + \gamma_{20} \cdot B_2 + \Psi_{10} \cdot A_{1M} + \gamma_{10} \cdot B_{1M} = \\ & = -C_2 - C_{1M} - (E_{1M} + D_2 - D_1)\varphi_r \\ M^*_{S,st} = 0: & \Psi_{10} \cdot (A_1 + A_{1M}) + \gamma_{10} \cdot (B_1 + B_{1M}) + C_1 + C_{1M} + E_{1M} \cdot \varphi_r = 0 \end{cases} \tag{7.17}$$

$$\begin{aligned} \beta_{st} &= \Psi_{20} \cdot A_2 + \gamma_{20} \cdot B_2 + C_2 + D_2\varphi_r = \\ &= -\Psi_{10} \cdot A_{1M} - \gamma_{10} \cdot B_{1M} - C_{1M} - (E_{1M} - D_1)\varphi_r \end{aligned} \tag{7.18}$$

$\delta_{st} = 0, \beta_{st} = 0, M^*_{S,st} \neq 0$ Another possibility is to adjust wheel alignment in order to have null steering angle and vehicle attitude when driving straight. The steering torque, on the other hand, would generally be nonnull, as shown in Figure 7.3, unless a third condition is fulfilled.

$$\begin{cases} \beta_{st} = 0: & \Psi_{20} \cdot A_2 + \gamma_{20} \cdot B_2 = -C_2 - D_2\varphi_r \\ \delta_{st} = 0: & \Psi_{10} \cdot A_1 + \gamma_{10} \cdot B_1 = -C_1 - D_1\varphi_r \end{cases} \tag{7.19}$$

$$M^*_{S,st} = -2C_{M\alpha,10} \cdot \left[\Psi_{10} \cdot A_{1M} + \gamma_{10} \cdot B_{1M} + C_{1M} + (E_{1M} - D_1)\varphi_r\right] \tag{7.20}$$

FIGURE 7.2 Straight motion with $M^*_{S,st} = 0, \delta_{st} = 0, \beta_{st} \neq 0$.

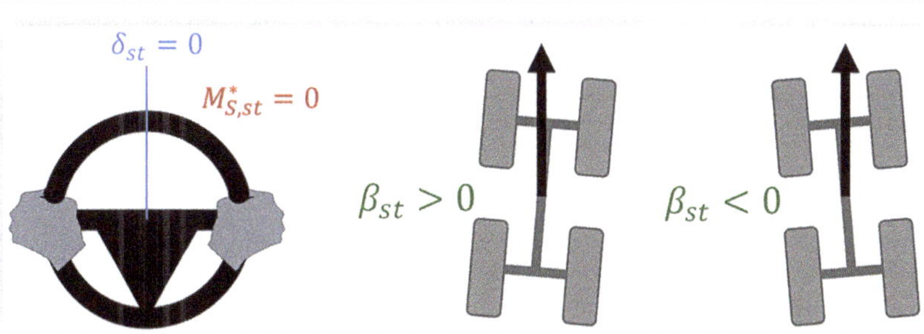

FIGURE 7.3 Straight motion with $\delta_{st} = 0$, $\beta_{st} = 0$, $M^*_{S,st} \neq 0$.

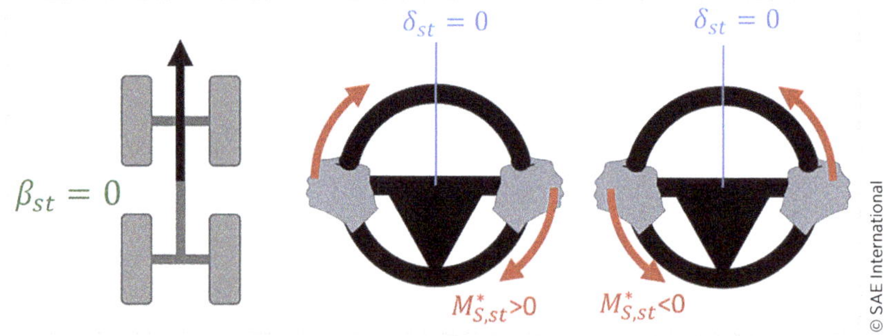

7.2.2 Wheel Alignment Setting, Adjustable Tire Characteristics

In case the tire characteristics are not fixed, but sufficient degree of freedom is available (e.g., during the design phase of a new tire or with slight changes on an existing one), it is theoretically possible to have simultaneously $\delta_{st} = 0$, $M^*_{S,st} = 0$, and $\beta_{st} = 0$, as represented in **Figure 7.4**.

The **necessary condition** involves both wheel alignment angles and tire characteristics

$$\Psi_{10} \cdot A_{1M} + \gamma_{10} \cdot B_{1M} = -C_{1M} - \left(E_{1M} - D_1\right)\varphi_r$$

and on expanding the coefficients, we get

$$\Psi_{10} \cdot \left[-\left(\frac{C_{M\alpha,10R} - C_{M\alpha,10L}}{C_{M\alpha,10R} + C_{M\alpha,10L}} \right) - \frac{F_{z1}}{C_{M\alpha,10R} + C_{M\alpha,10L}} \cdot C_{M\alpha,el,\Psi,1} \cdot \varphi_r \right] +$$
$$+\gamma_{10} \cdot \left[\left(\frac{C_{M\gamma,10R} - C_{M\gamma,10L}}{C_{M\alpha,10R} + C_{M\alpha,10L}} \right) - \frac{F_{z1}}{C_{M\alpha,10R} + C_{M\alpha,10L}} \cdot C_{M\alpha,el,\gamma,1} \cdot \varphi_r \right] =$$
$$= \frac{-M_{z0,1R} - M_{z0,1L} - F_{z1} \cdot \varphi_r \cdot \left(n - C_{M\alpha,roll,1} + C_{M\alpha,susp,1} - C_{M\alpha,steer,1} - C_{M\alpha,et,AL,1}\right)}{C_{M\alpha,10R} + C_{M\alpha,10L}} \tag{7.21}$$

With this condition, Equation 7.9 of the steering torque offset becomes

$$M^*_{S,st} = -2C_{M\alpha,10} \cdot \left(\Psi_{10} \cdot A_1 + \gamma_{10} \cdot B_1 + C_1 + D_1 \cdot \varphi_r\right) \tag{7.22}$$

FIGURE 7.4 Straight motion with $\delta_{st} = 0$, $\beta_{st} = 0$, $M^*_{S,st} = 0$.

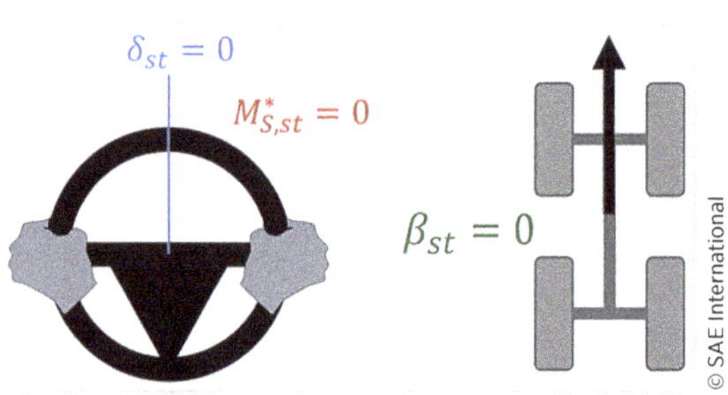

Therefore, imposing $M^*_{S,st} = 0$ is equivalent to imposing that the term inside the brackets (which is the front slip angle for straight motion) as null. This condition, together with the one for null attitude $\beta_{st} = 0$, implies $\delta_{st} = 0$ (see Equation 7.8)

$$\begin{cases} M^*_{S,st} = 0: & \Psi_{10} \cdot A_1 + \gamma_{10} \cdot B_1 = -C_1 - D_1\varphi_r \\ \beta_{st} = 0: & \Psi_{20} \cdot A_2 + \gamma_{20} \cdot B_2 = -C_2 - D_2\varphi_r \end{cases} \tag{7.23}$$

$$\begin{aligned} \Rightarrow \delta_{st} &= -\Psi_{10} \cdot A_1 - \gamma_{10} \cdot B_1 - C_1 + \\ &+ \Psi_{20} \cdot A_2 + \gamma_{20} \cdot B_2 + C_2 + \\ &+ \left(-D_1 + D_2\right)\varphi_r \\ &= 0 \end{aligned}$$

7.2.3 A Reference to the Compensation of Road Cross Slope with PRAT

In literature [3, 16], it is stated that usually OEMs deal with the steering pull issue by specifying an acceptable range of PRAT (Ply-steer Residual Aligning Torque) for the tires. This range is determined so that the effect of a specific road cross slope φ_r is balanced. This, together with a determined wheel alignment setting, causes the steering torque to be reduced when driving straight [3].

Nevertheless, balancing the road cross slope effect with the PRAT of the front tires does not allow having simultaneously $\delta_{st} = 0$, $M^*_{S,st} = 0$, and $\beta_{st} = 0$.

As discussed in Section 4.2.1, for the simplified case with equal left and right tire characteristics, the $PRAT_1$ of the front tires can be introduced in the equation of the steering torque offset contribution (see Equation 4.13)

$$\begin{aligned} &2C_{M\alpha,10} \cdot \left(\frac{F_{y0,1R} + F_{y0,1L}}{2 \cdot C_{F\alpha,10}} \right) + M_{z0,1R} + M_{z0,1L} = \\ &= 2C_{M\alpha,10} \cdot \left(\frac{F_{ply,1}}{C_{F\alpha,10}} \right) + 2M_{ply,1} = \\ &= 2 \cdot PRAT_1 \end{aligned} \tag{7.24}$$

It is to be underlined that the front tire's $PRAT_1$, usually considered in literature with reference to the vehicle pull [3], is related to the factors C_1 and C_{1M} appearing in Equation 7.9. Indeed, in case the left and right tire characteristics are not equal (e.g., due to load transfer caused by road cross slope or manufacturing heterogeneity), the following terms of Equations 7.2 and 7.9 are not exactly equal to $(2 \cdot PRAT_1)$, but are essentially related to this term

$$2C_{M\alpha,10} \cdot \left(\frac{F_{y0,1R} + F_{y0,1L}}{C_{F\alpha,10R} + C_{F\alpha,10L}} \right) + M_{z0,1R} + M_{z0,1L} = 2C_{M\alpha,10} \cdot (C_1 + C_{1M}) \leftrightsquigarrow 2 \cdot PRAT_1$$

Thus, balancing with the $PRAT_1$ value the road cross slope effect on the steering torque may mean

- balancing the whole contribution as

$$\begin{aligned} &-2 \cdot PRAT_1 \approx \\ &\approx -2C_{M\alpha,10} \cdot (C_1 + C_{1M}) = 2C_{M\alpha,10} \cdot \left[\Psi_{10} \cdot (A_1 + A_{1M}) + \gamma_{10} \cdot (B_1 + B_{1M}) + E_{1M} \cdot \varphi_r \right] \\ &\Rightarrow -\Psi_{10} \cdot A_1 - \gamma_{10} \cdot B_1 - C_1 = \Psi_{10} \cdot A_{1M} + \gamma_{10} \cdot B_{1M} + C_{1M} + E_{1M} \cdot \varphi_r \end{aligned} \tag{7.26}$$

As a consequence, the steering torque offset is null

$$M^*_{S,st} = 0$$

- balancing the contribution of the term multiplying road cross slope to get

$$\begin{aligned} -2 \cdot PRAT_1 \approx -2C_{M\alpha,10} \cdot (C_1 + C_{1M}) = 2C_{M\alpha,10} \cdot E_{1M} \cdot \varphi_r \\ \Rightarrow -C_1 = C_{1M} + E_{1M} \cdot \varphi_r \end{aligned} \tag{7.26}$$

As a consequence, to have null steering torque offset, it is necessary to impose

$$M^*_{S,st} = 0 : \Psi_{10} \cdot A_1 + \gamma_{10} \cdot B_1 = -\Psi_{10} \cdot A_{1M} - \gamma_{10} \cdot B_{1M}$$

For all the listed cases, it is generally not possible to have both $\delta_{st} = 0$ and $\beta_{st} = 0$ simultaneously, since they read

$$\begin{cases} \beta_{st} = \Psi_{20} \cdot A_2 + \gamma_{20} \cdot B_2 + C_2 + D_2\varphi_r \\ \delta_{st} = \Psi_{20} \cdot A_2 + \gamma_{20} \cdot B_2 + C_2 + D_2\varphi_r + \Psi_{10} \cdot A_{1M} + \gamma_{10} \cdot B_{1M} + C_{1M} + \left(E_{1M} - D_1\right) \cdot \varphi_r \end{cases} \tag{7.27}$$

$$\begin{cases} \beta_{st} = 0 \\ \delta_{st} = \Psi_{10} \cdot A_{1M} + \gamma_{10} \cdot B_{1M} + C_{1M} + \left(E_{1M} - D_1\right) \cdot \varphi_r \end{cases}$$

or

$$\begin{cases} \beta_{st} = -\Psi_{10} \cdot A_{1M} - \gamma_{10} \cdot B_{1M} - C_{1M} - \left(E_{1M} - D_1\right) \cdot \varphi_r \\ \delta_{st} = 0 \end{cases}$$

The only condition for which it is possible to have all the three offsets simultaneously null is the necessary condition already expressed in Equation 7.21.

7.3 A Brief Remark on the Effect of Wheel Alignment on Motion Resistance

Wheel alignment, in particular for what concerns static toe angle Ψ_{i0}, has an influence on vehicle's motion resistance. It is possible to have a hint on wheel alignment's effect by following the work of Schuring [48] that deals with energy loss of pneumatic tires. The *SAE* sign convention adopted in his work is replaced here with the *Pacejka's Modified SAE* sign convention (see Appendix A.2). Please refer to **Figure 7.5** for symbols.

FIGURE 7.5 Rolling tire scheme.

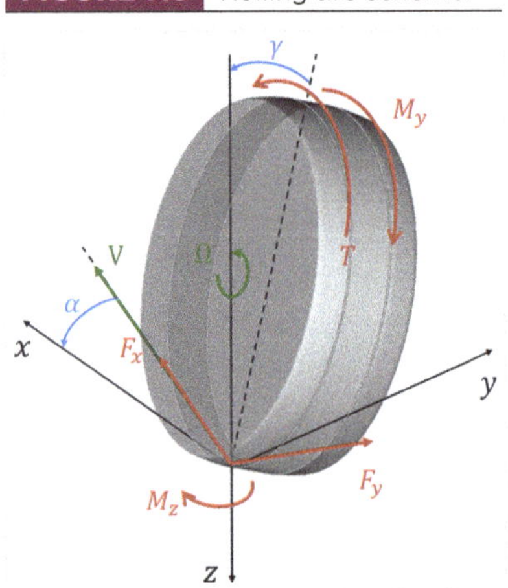

Schuring defines the *rolling resistance* e_R of a single tire as the "*energy dissipated (ΔE) per unit distance traveled (ΔL)*", which is also the power loss ΔP divided by the wheel's velocity V, namely

$$e_R \triangleq \frac{\Delta E}{\Delta L} = \frac{\Delta P}{V} \tag{7.28}$$

The power loss ΔP is obtained from the power balance, i.e., the input power minus the output power, given as

$$\Delta P = T \cdot \Omega - \left(F_x \cdot cos\alpha - F_y \cdot sin\alpha\right) \cdot V - \Delta P' \tag{7.29}$$

where the product between wheel torque T and spin velocity Ω represents the input power conveyed through the wheel axle to the tire. The term $(F_x cos\alpha - F_y sin\alpha)\ V$, on the other hand, is the output power, i.e., the traction force projected in the direction of motion multiplied by velocity. Thus, it represents the power effectively expended for vehicle propulsion. $\Delta P'$ is power loss due to aerodynamic effects and other hysteresis factors. The slip angle α is directly affected by the static toe angle Ψ_{i0} setting.

Neglecting $\Delta P'$, with moment equilibrium around wheel axis (see Figure 7.5), it is possible to express the torque T as [48]

$$T = F_x \cdot r_L + M_y \cdot cos\gamma + M_z \cdot sin\gamma \tag{7.30}$$

where r_L is the wheel's loaded radius. Moreover, the moment of rolling resistance M_y can be expressed as a function of longitudinal force F_x [48]

$$M_y = M_{yf} \cdot \left(1 + \frac{k}{r_{Lf}}\right) + k \cdot F_x \tag{7.31}$$

where the coefficient k is small ($k < 0.05 \cdot r_{Lf}$), and M_{yf} and r_{Lf} indicate, respectively, the resistance moment and loaded radius under free rolling.

Substituting Equation 7.31 into Equation 7.30 yields an expression of F_x as a function of T

$$\begin{gathered} T = F_x \cdot r_L + \left[M_{yf} \cdot \left(1 + \frac{k}{r_{Lf}}\right) + k \cdot F_x\right] \cdot cos\gamma + M_z \cdot sin\gamma \\ \Rightarrow F_x = \frac{T}{r_L + k \cdot cos\gamma} - \frac{M_{yf} \cdot \left(1 + \frac{k}{r_{Lf}}\right)}{r_L + k \cdot cos\gamma} \cdot cos\gamma - M_z \cdot \frac{sin\gamma}{r_L + k \cdot cos\gamma} \end{gathered} \tag{7.32}$$

It is thus possible to substitute Equation 7.32 into Equations 7.29 and 7.28 to obtain Schuring's expression of rolling resistance e_R

$$\begin{aligned} e_R = \frac{\Delta P}{V} = T \cdot \left(\frac{\Omega}{V} - \frac{cos\alpha}{r_L + k \cdot cos\gamma}\right) + \frac{M_{yf} \cdot \left(1 + \frac{k}{r_{Lf}}\right)}{r_L + k \cdot cos\gamma} \cdot cos\gamma \cdot cos\alpha + \\ + \frac{M_z}{r_L + k \cdot cos\gamma} \cdot sin\gamma \cdot cos\alpha + F_y \cdot sin\alpha \end{aligned} \tag{7.33}$$

If the wheel is freely rolling, the input torque is null ($T = 0$). Moreover, under the assumption that the rolling resistance moment M_y is unchanged for small α and γ, it is possible to indicate as F_{xf} the approximately constant free rolling resistance term

$$F_{xf} = \frac{M_{yf} \cdot \left(1 + \frac{k}{r_{Lf}}\right)}{r_L + k \cdot cos\gamma} \cdot cos\gamma \cdot cos\alpha$$

Finally, introducing the small-angle approximation for α and γ – namely for small wheel alignment adjustments, respectively, of static toe angle Ψ_{i0} and camber γ_{i0} – and with the linearized expressions of tire side force and aligning moment (see Equations 2.1 and 3.6), for a free rolling tire, Equation 7.33 can be rewritten as

$$\begin{aligned} e_R &\approx F_{xf} + \frac{-C_{M\alpha} \cdot \alpha + M_{z0}}{r_L + k} \cdot \gamma + \left(C_{F\alpha} \cdot \alpha + C_{F\gamma} \cdot \gamma + F_{y0}\right) \cdot \alpha = \\ &= C_{F\alpha} \cdot \alpha^2 + \left(F_{y0} + C_{F\gamma} \cdot \gamma - \frac{C_{M\alpha}}{r_L + k} \cdot \gamma\right) \cdot \alpha + \frac{M_{z0}}{r_L + k} \cdot \gamma + F_{xf} \end{aligned} \tag{7.34}$$

The trend of the rolling resistance e_R is qualitatively represented in **Figure 7.6**, using realistic values for tire parameters.

The impact of energy losses can be analyzed by means of standardized driving cycles. The *New European Driving Cycle* (*NEDC* or *MVEG-95*, *Motor Vehicle Emission Group-95*, represented in **Figure 7.7**) can be used to have a glance at potential energy savings. According to [49], the energy E required to travel 100 km can be computed (in case of no energy recovery) accounting for the three contributions to resitance of motion: aerodynamic drag, rolling, and acceleration resistances. An expression for the average energy $\overline{E}[\text{kJ}/100\ \text{km}]$ can be obtained with uniform time discretization of the *NEDC* driving cycle, calling $h = t_i - t_{i-1}$ the time intervals [49]

FIGURE 7.6 Normalized tire rolling resistance e_R as a function of side slip angle α and camber angle γ. As it can be observed, fractions of degree of toe angle Ψ_{i0} (thus α in this plot) may actually increase the rolling resistance of the single tire by up to a few percent points (F_{xf} represents the rolling resistance when the tire is aligned).

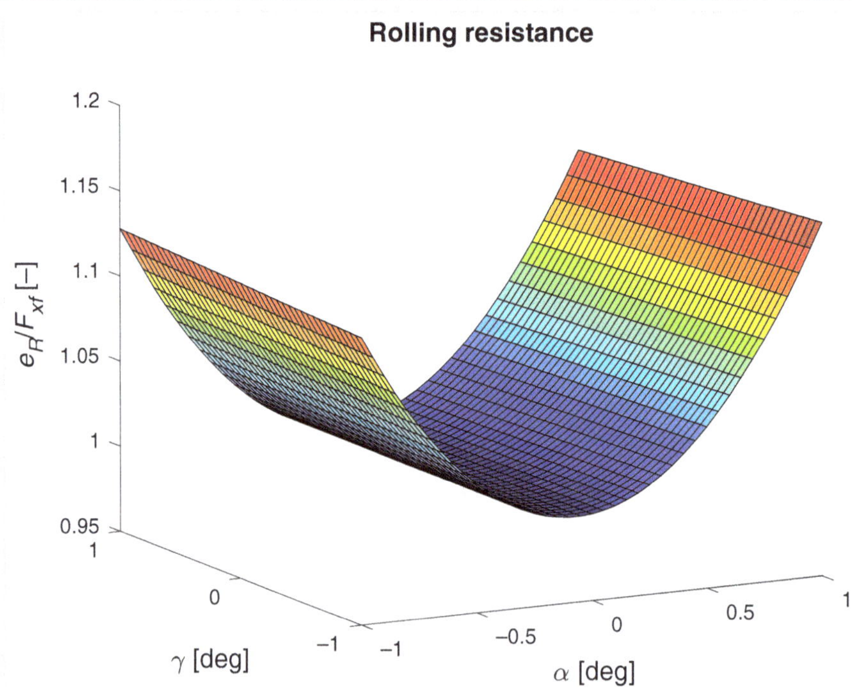

FIGURE 7.7 European test cycle *NEDC* (adapted from [49]).

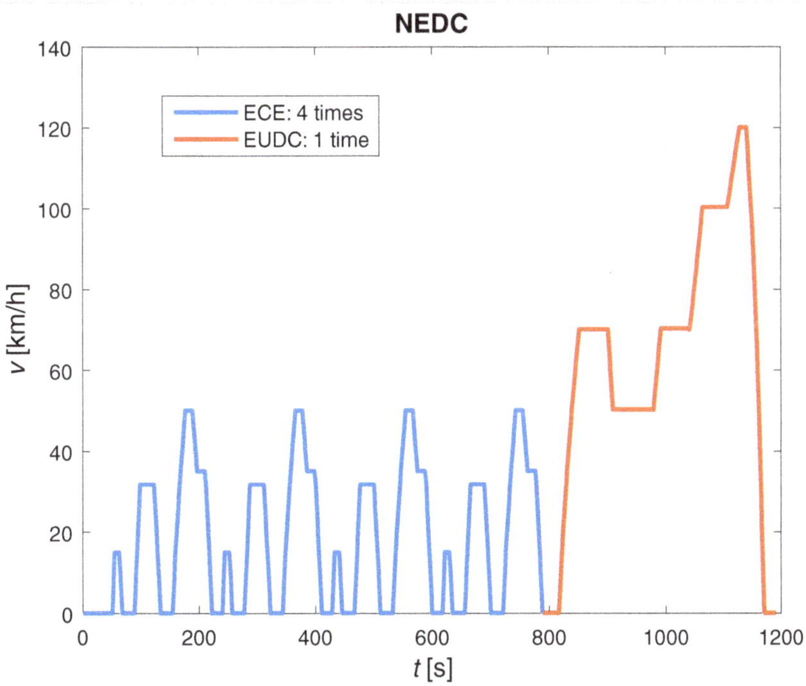

$$\overline{E}\left[\frac{\text{kJ}}{100\ \text{km}}\right] = 100 \cdot \overline{F}_{trac} = 100 \cdot \left(\overline{F}_{aero} + \overline{F}_{roll} + \overline{F}_{acc}\right) =$$

$$= 100 \cdot \left[\frac{1}{2}\rho_a \cdot A \cdot c_x \cdot \frac{1}{x_{tot}} \cdot \Sigma_i \overline{V}_i^3 \cdot h + m_{tot} \cdot f_R \cdot \frac{1}{x_{tot}} \cdot \Sigma_i \overline{V}_i \cdot h + \right.$$

$$\left. + m_{tot} \cdot \Sigma_i \overline{a}_i \cdot \overline{V}_i \cdot h \right] \approx$$

$$\approx 19000 \cdot A \cdot c_x + 840 \cdot m_{tot} \cdot f_R + 10 \cdot m_{tot} \left[\frac{\text{kJ}}{100\ \text{km}}\right] \tag{7.35}$$

It is important to specify that the constant coefficients reported in the last line of Equation 7.35 are valid for the *NEDC* only, and that air density ρ_a (at sea level) and acceleration g have been included in the constants [49].

The tires' rolling resistance coefficient f_R is directly affected by the e_R of the single tires, expressed in Equation 7.34. The effect of a 2% reduction of tire rolling resistance f_R is shown in **Table 7.1**, with realistic mid-range vehicle parameters. It should be stressed that the indicated slight energy saving can be obtained by designing tires with suitable characteristics, so that wheel alignment setting is not required as a correction for vehicle pull. That would reduce the consequent motion resistance, and the energy saving – despite being small for a single vehicle – may have a substantial role in the reduction of emissions when considering the overall number of vehicles circulating.

A further aspect that might be investigated is the effect of straight-driving vehicle attitude β_{st} on the effective aerodynamic drag coefficient c_x.

TABLE 7.1 An example of the variation of energy $\bar{E}$ for *NEDC* reference cycle due to a reduction of tires' rolling resistance coefficient f_R.

f_R [-]	m_{tot}	A [m²]	c_x [-]	$\bar{E}$[kJ / 100 km]
0.012	1488	2.2	0.27	41165
-2%	1488	2.2	0.27	40865 (-0.73%)

Another approach that can be used to estimate fuel savings can be found in [50]. A return factor R is defined in that work to relate fuel consumption improvements to relative reduction in tires' rolling loss

$$R = \frac{\Delta \text{ fuel consumption } [\%]}{\Delta \text{ Rolling loss } [\%]} \tag{7.36}$$

Based on experimental analyses, the estimated value of $R = 0.137$ is indicated for cars and light trucks, valid only for *EPA* (*U.S. Environmental Protection Agency*) tests with 55% urban and 45% highway driving. The indicated value of return factor R means that a rolling resistance reduction of 2% causes on average a fuel consumption drop of −0.27%. The actual significance of this measure might be realized in light of what is stated by Schuring [50]: given that just a fraction of the energy contained in the fuel reaches the crankshaft, "*each unit of energy saved beyond the crankshaft corresponds to several energy units back in the fuel tank*".

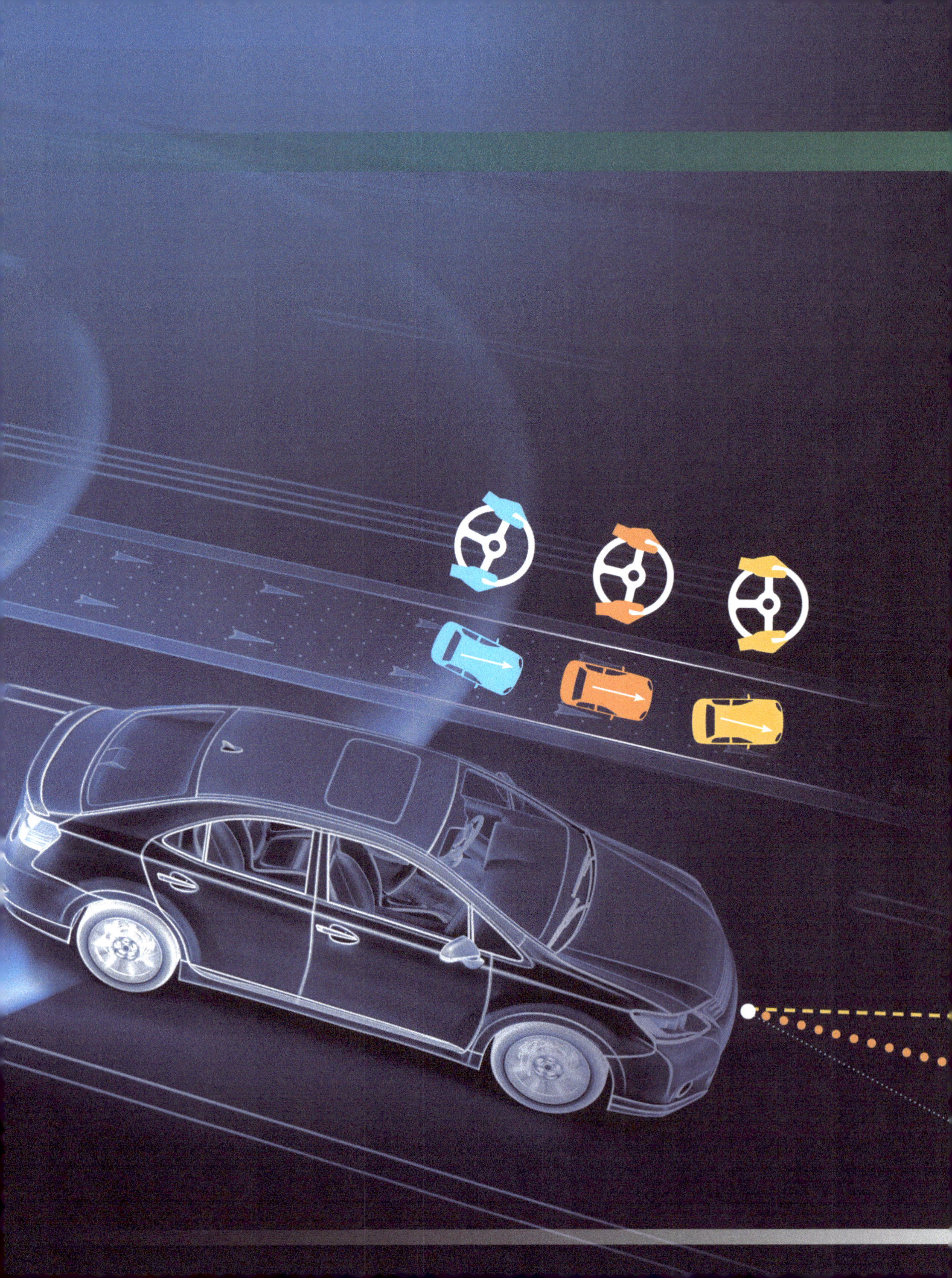

CHAPTER 8

Road Irregularity Effect

Road irregularity (or unevenness) effect on vehicle dynamics and tire characteristics is analyzed in this chapter. By introducing single-track and then two-track ground excitation, the variance of normal load is computed for the four wheels, including contributions from heave and roll motions. The actual normal load can then be computed and used to estimate the tire characteristics variation. Next, the variance of straight-driving slip angles is estimated. Finally, the influence of wheel suspension is considered to describe the effect of road irregularity on lateral displacement of the vehicle's body.

8.1 Stochastic Road Irregularity

As reported in [51], road unevenness can be described as a height profile $h(x)$ along the x-coordinate of the longitudinal path. It can be modeled as a periodic excitation of the type

$$h(x) = \sum_{k=1}^{n} h_k e^{jk\Omega x} \qquad (8.1)$$
$$\text{where } h_k \text{ is complex}: h_k = |h_k| e^{j\varepsilon_k}$$

This path-dependent height profile, being a periodic function, has a discrete spectrum. Moving to an infinitely long period, the finite sum of periodic functions becomes an integral, and the path-dependent height profile has a continuous spectrum

$$h(x)=\int_{-\infty}^{+\infty} h(\Omega)e^{j\Omega x}d\Omega \quad [\mathrm{m}]$$
$$h(\Omega)=\frac{1}{2\pi}\int_{-\infty}^{+\infty} h(x)e^{-j\Omega x}dx \quad \left[\frac{\mathrm{m}}{\mathrm{rad/m}}\right] \tag{8.2}$$

It is possible to obtain an expression for the time-dependent height profile, since spatial angular frequency Ω and temporal angular frequency ω are related via vehicle speed V

$$x=V\cdot t,\quad \omega=V\cdot\Omega$$
$$h(\Omega)\cdot d\Omega=h(\omega)\cdot d\omega \tag{8.3}$$
$$\Rightarrow h(\omega)=\frac{h(\Omega)}{V}\left[\frac{\mathrm{m}}{\mathrm{rad/s}}\right]$$

thus

$$h(t)=\int_{-\infty}^{+\infty} h(\omega)e^{j\omega t}d\omega \ [\mathrm{m}] \tag{8.4}$$

Calling $q(t)$ a generic variable describing an output of the system (e.g., wheel's load variation), its time history can be expressed by means of the frequency response function $K_{q,h}(\omega)=\frac{q(\omega)}{h(\omega)}$

$$q(t)=\int_{-\infty}^{+\infty} q(\omega)e^{j\omega t}d\omega=\int_{-\infty}^{+\infty} K_{q,h}(\omega)\ h(\omega)e^{j\omega t}d\omega \tag{8.5}$$

8.1.1 Statistical Indicators

Road irregularity can be treated as an ergodic stochastic process. Therefore, statistical indicators are more of interest than specific time instant realizations. Hence, the following statistical indicators are introduced, for the generic variable $q(t)$ [51]

- **Temporal linear mean value.**

$$\bar{q}=\frac{1}{T}\int_0^T q(t)dt \tag{8.6}$$

- **Temporal quadratic mean value.**

$$\tilde{q}=\sqrt{\frac{1}{T}\int_0^T [q(t)]^2 dt} \tag{8.7}$$

- **Standard deviation.**

$$\sigma_q=\sqrt{\frac{1}{T}\int_0^T [q(t)-\bar{q}]^2 dt}=\sqrt{\tilde{q}^2-\bar{q}^2} \tag{8.8}$$

- **Power spectral density.** By combining Equations 8.5 and 8.7, the single-sided power spectral density $\Phi_q(\omega)$ can be defined as

$$\begin{aligned}\tilde{q}^2 &= \int_0^{+\infty} \lim_{T\to+\infty} \frac{4\pi}{T}\left|q(\omega)\right|^2 d\omega \\ \Phi_q(\omega) &= \lim_{T\to+\infty} \frac{4\pi}{T}\left|q(\omega)\right|^2\end{aligned} \tag{8.9}$$

In terms of spatial angular frequency Ω, the power spectral density is

$$\Phi_q(\Omega) = \lim_{X\to+\infty} \frac{4\pi}{X}\left|q(\Omega)\right|^2 \tag{8.10}$$

where $X = V \cdot T$. Using the relationships indicated in Equation 8.3, the two power spectral densities can be linked by vehicle speed V

$$\Phi_q(\omega) = \frac{\Phi_q(\Omega)}{V} \tag{8.11}$$

The generic system variable $q(t)$ (e.g., wheel's load variation) can be rewritten by splitting a static part (constant and equal to the temporal mean $\overline{q}$) and a dynamic part, varying in time due to the road irregularity excitation

$$q(t) = q_{stat} + q_{dyn}(t) = \overline{q} + q_{dyn}(t) \tag{8.12}$$

The variance can be computed as

$$\begin{aligned}\sigma_q^2 &= \frac{1}{T}\int_0^T \left[q(t) - \overline{q}\right]^2 dt = \frac{1}{T}\int_0^T \left[q_{stat} + q_{dyn}(t) - \overline{q}\right]^2 dt = \\ &= \frac{1}{T}\int_0^T \left[q_{dyn}(t)\right]^2 dt = \tilde{q}_{dyn}^2\end{aligned} \tag{8.13}$$

Therefore, considering Equation 8.9, the variance of any generic variable q can be computed, as a function of the road unevenness power spectral density, via the integral

$$\begin{aligned}\sigma_q^2 &= \int_0^{+\infty} \Phi_q(\omega) d\omega = \int_0^{+\infty} \lim_{T\to+\infty} \frac{4\pi}{T}\left|q(\omega)\right|^2 d\omega = \\ &= \int_0^{+\infty} \lim_{T\to+\infty} \frac{4\pi}{T}\left|\frac{q(\omega)}{h(\omega)}\right|^2 \left|h(\omega)\right|^2 d\omega = \\ &= \int_0^{+\infty} \left|K_{q,h}(\omega)\right|^2 \Phi_h(\omega) d\omega\end{aligned} \tag{8.14}$$

The computational process of the variance is qualitatively illustrated in **Figure 8.1**.

FIGURE 8.1 Computation of variance σ_q^2 of a generic variable q, from road unevenness power spectral density $\Phi_h(\omega)$ and frequency response function $K_{q,h}(\omega)$ (adapted from [51]).

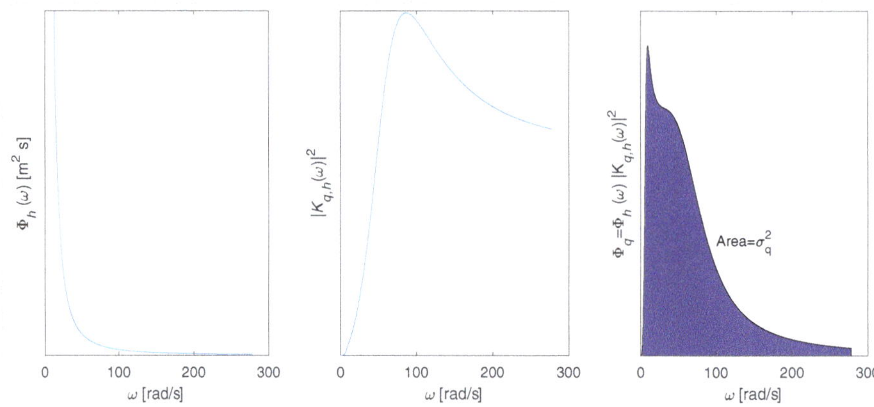

8.2 Power Spectral Density of Road Unevenness

Road unevenness in longitudinal direction is classified and described by the International Standard ISO8608 [52–54]: Roads are divided into eight classes (from A - the smoothest, to H - the roughest). Each of them has a characteristic single-sided power spectral density $\Phi_h(n)$ (a function of the spatial frequency n), or equivalently $\Phi_h(\Omega)$ (a function of the spatial angular frequency $\Omega = 2\pi n$). The unevenness can be approximated as a straight line in the double-logarithmic scale, as represented in **Figure 8.2**. The equations for the power spectral densities are given by

$$\Phi_h(n) = \Phi_h(n_0) \cdot \left(\frac{n}{n_0}\right)^{-w} \left[\frac{\text{m}^2}{1/\text{m}}\right]$$
$$\Phi_h(\Omega) = \Phi_h(\Omega_0) \cdot \left(\frac{\Omega}{\Omega_0}\right)^{-w} \left[\frac{\text{m}^2}{\text{rad}/\text{m}}\right] \tag{8.15}$$

where $-w$ is the slope of the straight line in the log-log diagram (usually $w = 2$). The waviness w of the road determines whether the long wavelengths have pronounced spectral densities (large w) or short wavelengths have pronounced spectral densities (small w).

The characteristic values of $\Phi_h(n_0)$ (or $\Phi_h(\Omega_0)$, equivalently) represent a measure of the unevenness, at reference spatial frequency $n_0 = 0.1\frac{1}{\text{m}}$ or at spatial angular frequency $\Omega_0 = 1\frac{\text{rad}}{\text{m}}$. Their values, according to ISO8608 classification, are reported in **Table 8.1**.

The power spectral density in terms of spatial frequency (n or Ω) conveys only information regarding the road. Frequencies in space and time are linked by vehicle

FIGURE 8.2 Classification of road roughness according to ISO8608 [52] (adapted from [53]).

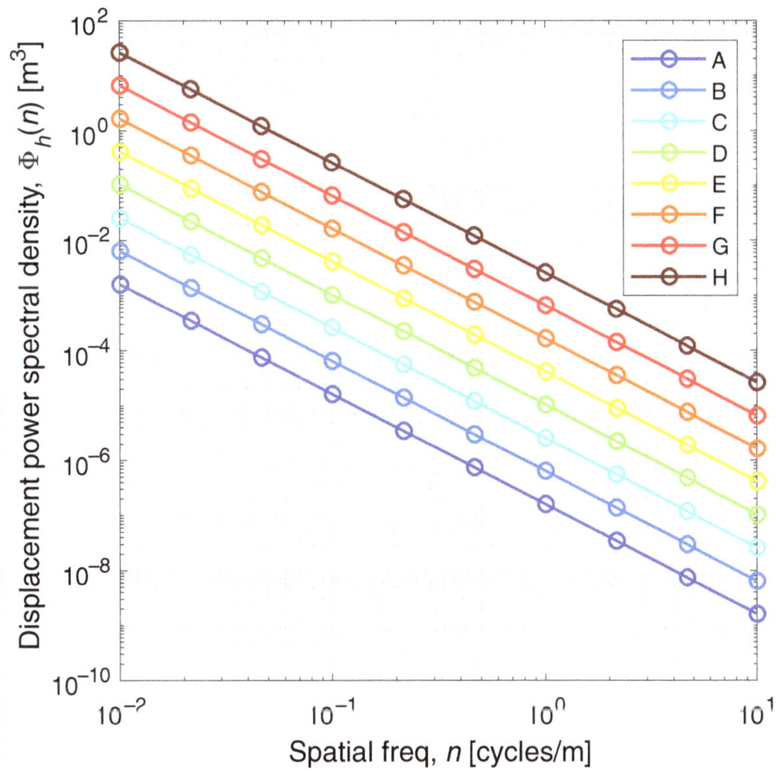

TABLE 8.1 ISO8608 road roughness classification values [52].

	$\Phi_h(n_0)$ [10^{-6} m^3]			$\Phi_h(\Omega_0)$ [10^{-6} m^3]		
Road class	**Lower limit**	**Geometric mean**	**Upper limit**	**Lower limit**	**Geometric mean**	**Upper limit**
A	-	16	32	-	1	2
B	32	64	128	2	4	8
C	128	256	512	8	16	32
D	512	1024	2048	32	64	128
E	2048	4096	8192	128	256	512
F	8192	16384	32768	512	1024	2048
G	32768	65536	131072	2048	4096	8192
H	131072	262144	-	8192	16384	-

velocity V; it is therefore possible to introduce the information on speed in the power spectral density, expressing the latter in terms of time domain frequency (f or ω)

$$\begin{gathered} f=\frac{V}{\lambda}=V\cdot n \\ \omega=2\pi f=2\pi V\cdot n=V\cdot\Omega \end{gathered} \tag{8.16}$$

For energy conservation (Parseval's theorem)

$$\Phi_h(\omega)d\omega = \Phi_h(\Omega)d\Omega \tag{8.17}$$

Therefore, the power spectral density of ground vertical displacement in terms of time frequency is

$$\Phi_h(\omega) = \frac{1}{V}\cdot\Phi_h(\Omega)\left[\frac{\mathrm{m}^2}{\mathrm{rad/s}}\right] \tag{8.18}$$

The relationship between the two power spectral densities is represented in **Figure 8.3**.

Henceforward, a road with reference power spectral density $\Phi_h(\Omega_0) = 4\cdot 10^{-6}\,m^3$ and waviness $w = 2$ is considered, corresponding to a fairly good highway (class B).

8.2.1 Colored Noise Unevenness

The power spectral density expressed by Equation 8.15 is often assumed to have a bounded frequency range [55]: according to [56–58], wavelengths above 100 m are associated with landscape variability and do not affect vehicle dynamics, while wavelengths below 0.01 m are filtered out by tires and thus excluded from the spectrum. More refined road models have different waviness w for different frequency ranges: this gives a better correspondence between the analytical expression of the road power spectral density and experimental measurements [55].

Nevertheless, for the limit case of null frequency $\Omega \to 0$, both power spectral density $\Phi_h(0) \to \infty$ and road roughness variance $\sigma_h^2 = \int_0^\infty \Phi_h(\omega)d\omega \to \infty$, which is unrealistic. Therefore, extended road unevenness models defined on the overall frequency range are characterized by the colored noise power spectral density [51, 55]

$$\Phi_h(\Omega) = \frac{\alpha_h}{\beta_h^2 + \Omega^w} \tag{8.19}$$

FIGURE 8.3 Relationship between $\Phi_h(\Omega)$ and $\Phi_h(\omega)$ power spectral densities [51] (V = 100 km/h, $\Phi_h(\Omega_0) = 4\cdot 10^{-6}$ m³, w = 2).

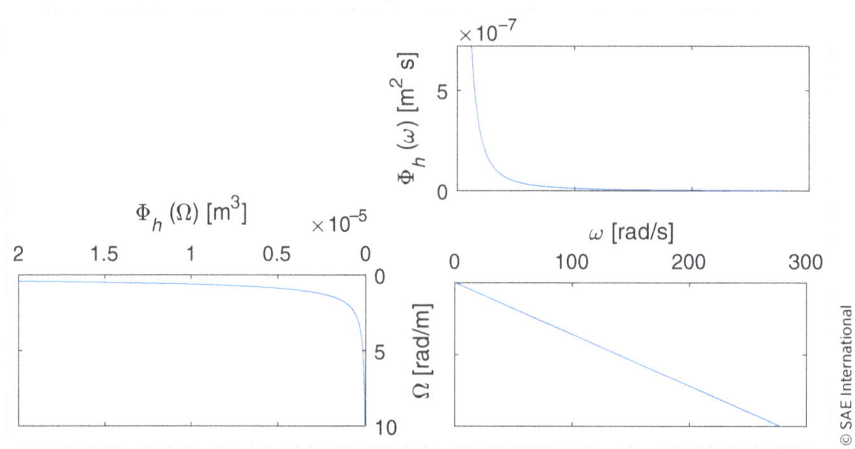

The constants α_h and β_h^2 are defined in this case so that the colored noise power spectral density has the following two characteristic values

- $\dfrac{\alpha_h}{\beta_h^2 + \Omega_0^w} = \Phi_h(\Omega_0)$: reference classification value according to ISO8608.
- $\dfrac{\alpha_h}{\beta_h^2 + 0} = \Phi_h(\Omega_{min}) = \Phi_h(\Omega_0) \cdot \left(\dfrac{\Omega_{min}}{\Omega_0}\right)^{-w}$: maximum value of the colored noise power spectral density equal to the value of the ISO power spectral density at the minimum spatial frequency (wavelengths above 100 m: $\Omega_{min} = 2\pi \dfrac{1}{100 \text{ m}}$).

Thus, the two constants are computed here as

$$\beta_h^2 = \frac{\Omega_0^w}{\Phi_h(\Omega_{min})/\Phi_h(\Omega_0) - 1} = \frac{\Omega_0^w}{\left(\dfrac{\Omega_{min}}{\Omega_0}\right)^{-w} - 1} \tag{8.20}$$

$$\alpha_h = \left(\beta_h^2 + \Omega_0^w\right) \cdot \Phi_h(\Omega_0)$$

A comparison between the power spectral densities of Equations 8.15 and 8.19 is reported in **Figure 8.4**.

FIGURE 8.4 ISO and colored noise power spectral densities $\Phi_h(\Omega)$ ($\Phi_h(\Omega_0) = 4 \cdot 10^{-6}$ m^3, $w = 2$, $\Omega_{min} = 2\pi/100$ m).

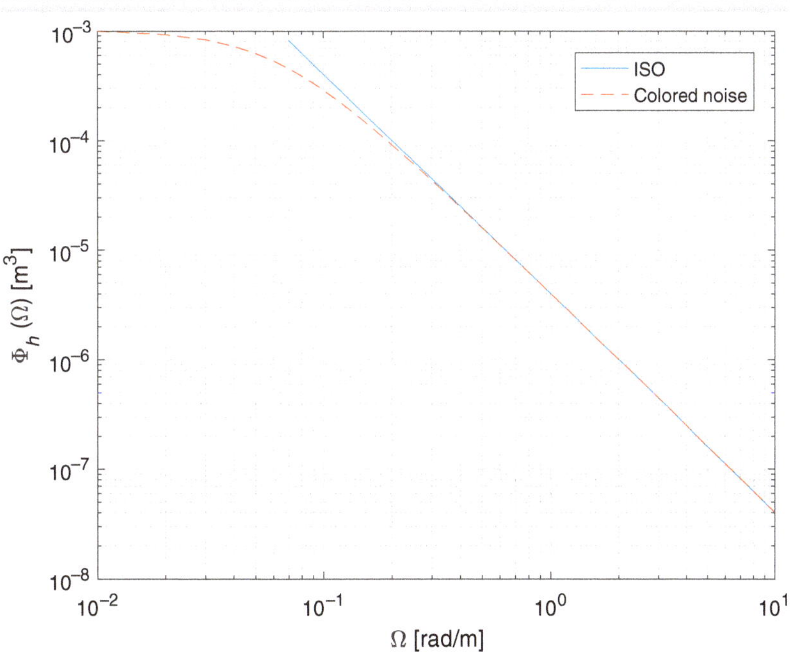

The colored noise power spectral density can be expressed in terms of time angular frequency ω as

$$\Phi_h(\omega) = \frac{\alpha_h}{\beta_h^2 + \left(\frac{\omega}{V}\right)^w} \cdot \frac{1}{V} = \frac{\alpha_h \cdot V^{w-1}}{V^w \beta_h^2 + \omega^w} \tag{8.21}$$

8.3 Vehicle Model for Road Irregularity Excitation

As considered in the analysis of Mitschke [51], a vehicle model as the one represented in Figure 8.5 can be used to evaluate the ground excitation effect on wheel's load variation.

The values of vehicle model parameters can be found in Appendix A.8, where they are listed in Tables A.1–A.3.

Tire stiffness is approximated by means of the equation reported in [24, 59, 60]

$$k_t = 3.45 + 0.00028 \cdot p_i[kPa] \cdot \sqrt{D_{out}[\mathrm{mm}] \cdot w_{tire}[\mathrm{mm}]} \left[\frac{\mathrm{kg}}{\mathrm{mm}}\right] \tag{8.22}$$

with p_i as the tire inflation pressure, D_{out} as the tire outer diameter, and w_{tire} as the tire footprint width. Further details on the computation are reported in Appendix A.8.

$$k_{t1} = 259884 \ \mathrm{N/m}$$
$$k_{t2} = 232759 \ \mathrm{N/m}$$

FIGURE 8.5 Vehicle model scheme for ground excitation analysis (adapted from [51]).

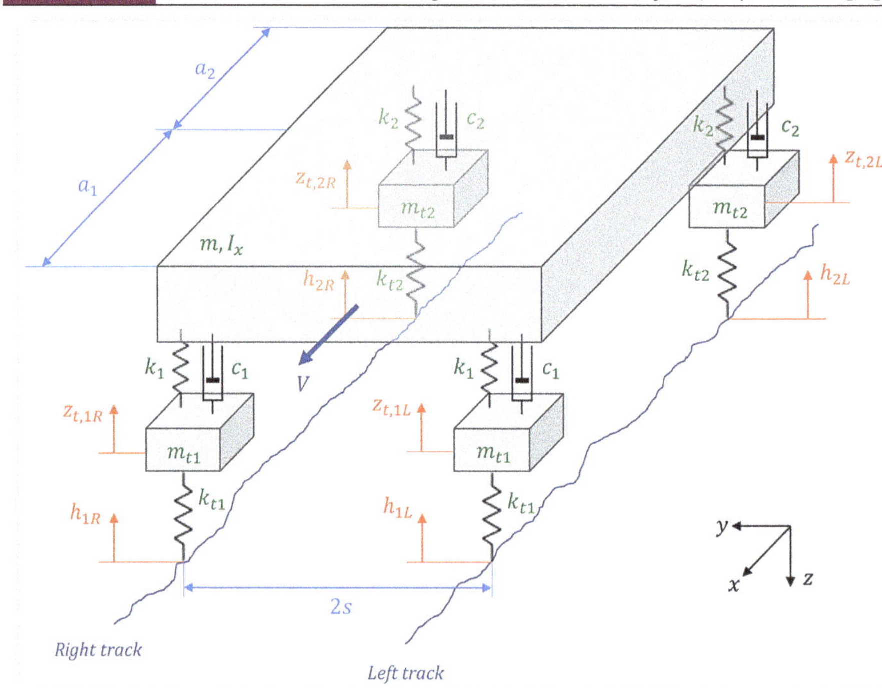

Rolling moment of inertia I_x of the vehicle's body is estimated according to [61] with the formula

$$I_x = C_x \cdot m_{tot}^{q_x} \cdot Length^{r_x} \cdot Width^{s_x} = 534.8 \text{ kg m}^2 \tag{8.23}$$

where m_{tot} is the total vehicle mass, *Length* and *Width* are the vehicle's dimensions. Details on the computation are reported in Appendix A.9.

The following hypotheses apply

- tire internal damping is neglected ($c_{ti} = 0$)
- all $z_{t,iL/R} = 0$ at static rest position, thus spring deflection under static load has already been considered
- passengers are sitting symmetrically with respect to the longitudinal axis of the vehicle, and without modifying the position of the center of gravity. Their mass is included in the vehicle's sprung mass
- the center of gravity is positioned along the longitudinal axis of the vehicle, in fact the latter is constructed symmetrically with respect to longitudinal axis
- wheels move just vertically (camber variation effect is negligible [51])
- heave and roll motions can be decoupled without losing validity [51]
- rear wheels are in line with front wheels, thus the front and rear track widths are equal ($2s_1 = 2s_2 = 2s$). As a consequence, when driving straight ahead, front and rear wheels are subjected to the same road unevenness excitation, just shifted by the wheelbase length l.

8.3.1 Heave Motion

The heave motion is analyzed considering equal left and right track excitations (the so-called *stroke excitation* in literature [51]). Therefore, a vehicle model as the one schematized in **Figure 8.6** can be considered.

It is assumed that the total sprung mass is divided into m_F (front) and m_R (rear) where

$$\begin{aligned} m_F &= \frac{F_{z,1L,stat} + F_{z,1R,stat}}{g} - 2m_{t1} \\ m_R &= \frac{F_{z,2L,stat} + F_{z,2R,stat}}{g} - 2m_{t2} \end{aligned} \tag{8.24}$$

FIGURE 8.6 Vehicle model for heave motion analysis (adapted from [51]).

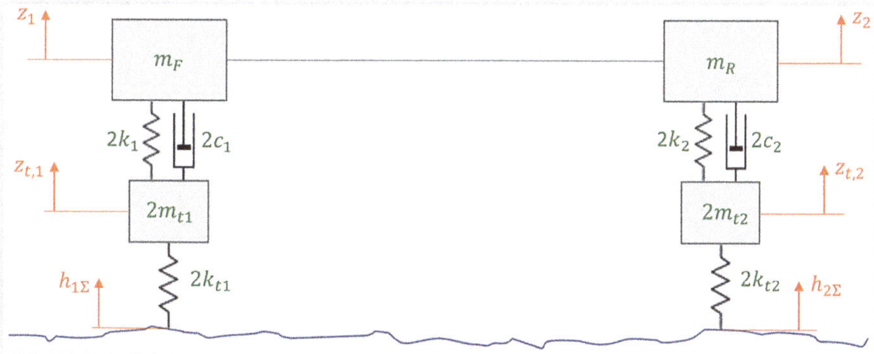

The coupling mass between front and rear is neglected. As a consequence, the vehicle model has decoupled motions of the front and rear parts.

The equations of motion are reported in matrix form [51]

$$M_z\ddot{\underline{z}}+C_z\dot{\underline{z}}+K_z\underline{z}=R_z\underline{h_\Sigma} \tag{8.25}$$

$$M_z=\begin{bmatrix} m_F & 0 & 0 & 0 \\ 0 & m_R & 0 & 0 \\ 0 & 0 & 2m_{t1} & 0 \\ 0 & 0 & 0 & 2m_{t2} \end{bmatrix}$$

$$C_z=\begin{bmatrix} 2c_1 & 0 & -2c_1 & 0 \\ 0 & 2c_2 & 0 & -2c_2 \\ -2c_1 & 0 & 2c_1 & 0 \\ 0 & -2c_2 & 0 & 2c_2 \end{bmatrix}$$

$$K_z=\begin{bmatrix} 2k_1 & 0 & -2k_1 & 0 \\ 0 & 2k_2 & 0 & -2k_2 \\ -2k_1 & 0 & 2(k_1+k_{t1}) & 0 \\ 0 & -2k_2 & 0 & 2(k_2+k_{t2}) \end{bmatrix}$$

$$R_z\underline{h_\Sigma}=\begin{bmatrix} 0 & 0 \\ 0 & 0 \\ 2k_{t1} & 0 \\ 0 & 2k_{t2} \end{bmatrix}\begin{Bmatrix} h_{1\Sigma} \\ h_{2\Sigma} \end{Bmatrix},\quad \underline{z}=\begin{Bmatrix} z_1 \\ z_2 \\ z_{t,1} \\ z_{t,2} \end{Bmatrix}$$

Vehicle's symmetry is exploited by expressing left and right side displacements as

$$\begin{aligned} z_{t,1}&=\frac{1}{2}\left(z_{t,1L}+z_{t,1R}\right) \\ z_{t,2}&=\frac{1}{2}\left(z_{t,2L}+z_{t,2R}\right) \end{aligned} \tag{8.26}$$

$$\begin{aligned} h_{1\Sigma}&=\frac{1}{2}\left(h_{1L}+h_{1R}\right) \\ h_{2\Sigma}&=\frac{1}{2}\left(h_{2L}+h_{2R}\right) \end{aligned} \tag{8.27}$$

The frequency response functions for the wheels' movement are thus derived

$$\begin{aligned} K_{zt1,h1\Sigma}&=\frac{z_{t,1}(\omega)}{h_{1\Sigma}(\omega)}= \\ &=\frac{2k_{t1}\left(-\omega^2 m_F+j2\omega c_1+2k_1\right)}{\begin{array}{l}\omega^4\left(2m_{t1}m_F\right)-j\omega^3 2c_1\left(2m_{t1}+m_F\right)- \\ -\omega^2\left[2m_{t1}2k_1+m_F\left(2k_1+2k_{t1}\right)\right]+j\omega 2c_1 2k_{t1}+2k_1\cdot 2k_{t1}\end{array}} \end{aligned} \tag{8.28}$$

$$k_{zt2,h2\Sigma} = \frac{z_{t,2}(\omega)}{h_{2\Sigma}(\omega)} =$$

$$= \frac{2k_{t2}\left(-\omega^2 m_R + j2\omega c_2 + 2k_2\right)}{\begin{array}{l}\omega^4\left(2m_{t2}m_R\right) - j\omega^3 2c_2\left(2m_{t2} + m_R\right) - \\ -\omega^2\left[2m_{t2}2k_2 + m_R\left(2k_2 + 2k_{t2}\right)\right] + j\omega 2c_2 2k_{t2} + 2k_2 \cdot 2k_{t2}\end{array}} \tag{8.29}$$

The frequency response functions for the wheels' load variation (considering $\Delta F_z = k_t(h - z_t)$) read

$$K_{Fz1,h1\Sigma} = \frac{\Delta F_{z,1}(\omega)}{h_{1\Sigma}(\omega)} = k_{t1}\left(1 - K_{zt1,h1\Sigma}\right) \tag{8.30}$$

$$K_{Fz2,h2\Sigma} = \frac{\Delta F_{z,2}(\omega)}{h_{2\Sigma}(\omega)} = k_{t2}\left(1 - K_{zt2,h2\Sigma}\right) \tag{8.31}$$

Being front and rear heave motions decoupled, $K_{Fz1\Sigma,h2\Sigma} = K_{Fz2\Sigma,h1\Sigma} = 0$.

8.3.2 Roll Motion

Roll motion is now analyzed by considering different road unevenness excitations for left and right track (*roll excitation*). The model is extended as represented in **Figure 8.7**:

FIGURE 8.7 Vehicle model for roll motion analysis (adapted from [51]).

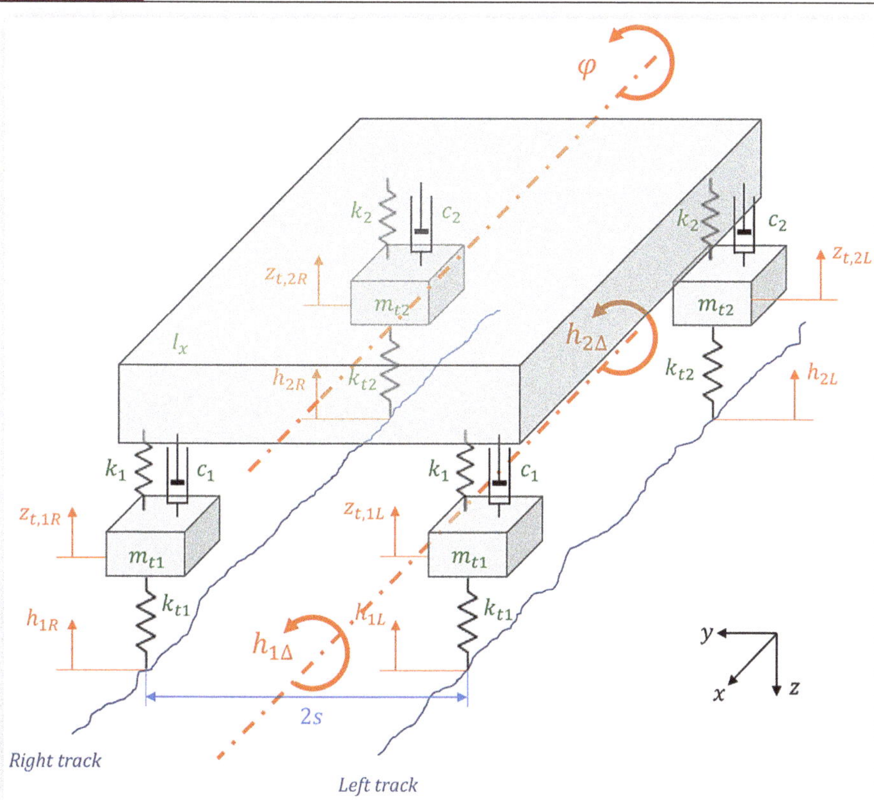

it considers fictitious roll movement of axles (φ_1 and φ_2) as difference between left and right wheel motions divided by track width

$$\varphi_1 = \frac{1}{2s}\left(z_{t,1L} - z_{t,1R}\right)$$
$$\varphi_2 = \frac{1}{2s}\left(z_{t,2L} - z_{t,2R}\right) \tag{8.32}$$

Vehicle's body roll is indicated with the coordinate φ.

Road unevenness *roll excitation* is expressed by

$$h_{1\Delta} = \frac{1}{2s}\left(h_{1L} - h_{1R}\right)$$
$$h_{2\Delta} = \frac{1}{2s}\left(h_{2L} - h_{2R}\right) \tag{8.33}$$

The equations of motion are reported in matrix form [51] as

$$M_\varphi \ddot{\underline{\varphi}} + C_\varphi \dot{\underline{\varphi}} + K_\varphi \underline{\varphi} = R_\varphi \underline{h}_\Delta \tag{8.34}$$

$$M_\varphi = \begin{bmatrix} I_x & 0 & 0 \\ 0 & 2m_{t1}\cdot s^2 & 0 \\ 0 & 0 & 2m_{t2}\cdot s^2 \end{bmatrix}$$

$$C_\varphi = \begin{bmatrix} 2(c_1+c_2)\cdot s^2 & -2c_1\cdot s^2 & -2c_2\cdot s^2 \\ -2c_1\cdot s^2 & 2c_1\cdot s^2 & 0 \\ -2c_2\cdot s^2 & 0 & 2c_2\cdot s^2 \end{bmatrix}$$

$$K_\varphi = \begin{bmatrix} 2(k_1+k_2)\cdot s^2 & -2k_1\cdot s^2 & -2k_2\cdot s^2 \\ -2k_1\cdot s^2 & 2(k_1+k_{t1})\cdot s^2 & 0 \\ -2k_2\cdot s^2 & 0 & 2(k_2+k_{t2})\cdot s^2 \end{bmatrix}$$

$$R_\varphi \underline{h}_\Delta = \begin{bmatrix} 0 & 0 \\ 2k_{t1}\cdot s^2 & 0 \\ 0 & 2k_{t2}\cdot s^2 \end{bmatrix} \begin{Bmatrix} h_{1\Delta} \\ h_{2\Delta} \end{Bmatrix}, \quad \underline{\varphi} = \begin{Bmatrix} \varphi \\ \varphi_1 \\ \varphi_2 \end{Bmatrix}$$

The frequency response functions can be derived by considering the corresponding expressions to Equation 8.34 into the Fourier domain

$$\begin{cases} A_{roll}\cdot\varphi(\omega) + B_{roll}\cdot\varphi_1(\omega) + C_{roll}\cdot\varphi_2(\omega) = 0 \\ D_{roll}\cdot\varphi_1(\omega) + E_{roll}\cdot\varphi(\omega) = F_{roll}\cdot h_{1\Delta}(\omega) \\ G_{roll}\cdot\varphi_2(\omega) + H_{roll}\cdot\varphi(\omega) = I_{roll}\cdot h_{2\Delta}(\omega) \end{cases} \tag{8.35}$$

where

$$
\begin{aligned}
A_{roll} &= -\omega^2 I_x + j\omega 2s^2(c_1 + c_2) + 2s^2(k_1 + k_2) \\
B_{roll} &= -2s^2(j\omega c_1 + k_1) \\
C_{roll} &= -2s^2(j\omega c_2 + k_2) \\
D_{roll} &= 2s^2(-\omega^2 m_{t1} + j\omega c_1 + k_1 + k_{t1}) \\
E_{roll} &= -2s^2(j\omega c_1 + k_1) \\
F_{roll} &= 2k_{t1}s^2 \\
G_{roll} &= 2s^2(-\omega^2 m_{t2} + j\omega c_2 + k_2 + k_{t2}) \\
H_{roll} &= -2s^2(j\omega c_2 + k_2) \\
I_{roll} &= 2k_{t2}s^2
\end{aligned}
$$

The frequency response functions between the wheels' motion and roll excitation are

$$K_{\varphi 1,h1\Delta} = \frac{\varphi_1(\omega)}{h_{1\Delta}(\omega)} = \frac{F_{roll}}{D_{roll} - \dfrac{E_{roll} \cdot B_{roll} \cdot G_{roll}}{A_{roll} \cdot G_{roll} - C_{roll} \cdot H_{roll}}} \tag{8.36}$$

$$K_{\varphi 1,h2\Delta} = \frac{\varphi_1(\omega)}{h_{2\Delta}(\omega)} = \frac{E_{roll} \cdot C_{roll} \cdot I_{roll}}{G_{roll} \cdot (A_{roll} \cdot D_{roll} - B_{roll} \cdot E_{roll}) - H_{roll} \cdot C_{roll} \cdot D_{roll}} \tag{8.37}$$

$$K_{\varphi 2,h2\Delta} = \frac{\varphi_2(\omega)}{h_{2\Delta}(\omega)} = \frac{I_{roll}}{G_{roll} - \dfrac{H_{roll} \cdot C_{roll} \cdot D_{roll}}{A_{roll} \cdot D_{roll} - B_{roll} \cdot E_{roll}}} \tag{8.38}$$

$$K_{\varphi 2,h1\Delta} = \frac{\varphi_2(\omega)}{h_{1\Delta}(\omega)} = \frac{H_{roll} \cdot B_{roll} \cdot F_{roll}}{D_{roll} \cdot (A_{roll} \cdot G_{roll} - C_{roll} \cdot H_{roll}) - E_{roll} \cdot B_{roll} \cdot G_{roll}} \tag{8.39}$$

The frequency response functions for the wheels' load variation (considering $\Delta F_z = k_t(h-z_t)$) are

$$K_{Fz1L,h1\Delta} = \frac{\Delta F_{z,1L}(\omega)}{h_{1\Delta}(\omega)} = k_{t1} \cdot s(1 - K_{\varphi 1,h1\Delta}) \tag{8.40}$$

$$K_{Fz1L,h2\Delta} = \frac{\Delta F_{z,1L}(\omega)}{h_{2\Delta}(\omega)} = k_{t1} \cdot s(-K_{\varphi 1,h2\Delta}) \tag{8.41}$$

$$K_{Fz1R,h1\Delta} = \frac{\Delta F_{z,1R}(\omega)}{h_{1\Delta}(\omega)} = k_{t1} \cdot s(-1 + K_{\varphi 1,h1\Delta}) \tag{8.42}$$

$$K_{Fz1R,h2\Delta} = \frac{\Delta F_{z,1R}(\omega)}{h_{2\Delta}(\omega)} = k_{t1} \cdot s(+K_{\varphi 1,h2\Delta}) \tag{8.43}$$

$$K_{Fz2L,h1\Delta} = \frac{\Delta F_{z,2L}(\omega)}{h_{1\Delta}(\omega)} = k_{t2} \cdot s(-K_{\varphi 2,h1\Delta}) \tag{8.44}$$

$$K_{Fz2L,h2\Delta} = \frac{\Delta F_{z,2L}(\omega)}{h_{2\Delta}(\omega)} = k_{t2} \cdot s\left(1 - K_{\varphi 2,h2\Delta}\right) \tag{8.45}$$

$$K_{Fz2R,h1\Delta} = \frac{\Delta F_{z,2R}(\omega)}{h_{1\Delta}(\omega)} = k_{t2} \cdot s\left(+K_{\varphi 2,h1\Delta}\right) \tag{8.46}$$

$$K_{Fz2L,h2\Delta} = \frac{\Delta F_{z,2R}(\omega)}{h_{2\Delta}(\omega)} = k_{t2} \cdot s\left(-1 + K_{\varphi 2,h2\Delta}\right) \tag{8.47}$$

8.4 Variance of Wheels' Normal Load

The work of Mitschke [51] illustrates how to compute the variance σ_q^2 of a generic variable q from the double-track excitation, caused by road unevenness with given power spectral density $\Phi_h(\Omega)$.

As expressed by Equation 8.5, a generic complex variable q of the system can be computed from the frequency response function and road unevenness

$$q(\omega) = K_{q,h}(\omega)h(\omega) \tag{8.48}$$

Its variance can be computed according to Equation 8.14. The magnitude of $q(\omega)$ can be expressed as the product with its complex conjugate (indicated with $q^*(\omega)$), namely $|q(\omega)| = q(\omega)\, q^*(\omega)$

$$\sigma_q^2 = \int_0^{+\infty} \lim_{T\to+\infty} \frac{4\pi}{T}\left[q(\omega)q*(\omega)\right]d\omega \tag{8.49}$$

Introducing the two-track excitation in heave ($h_{1\Sigma}$, $h_{2\Sigma}$) and in roll ($h_{1\Delta}$, $h_{2\Delta}$), the generic output reads

$$q(\omega) = K_{q,h1\Sigma}h_{1\Sigma} + K_{q,h2\Sigma}h_{2\Sigma} + K_{q,h1\Delta}h_{1\Delta} + K_{q,h2\Delta}h_{2\Delta} \tag{8.50}$$

Assuming the same track width $2s_1 = 2s_2$, when driving straight front and rear, axle excitations are linked by a time delay Δt, which depends on wheelbase l and velocity V, i.e., $\Delta t = \frac{l}{V}$. Therefore, front and rear road excitations can be related as

$$\begin{aligned} h_{2\Sigma} &= h_{1\Sigma}e^{-j\omega\Delta t} \\ h_{2\Delta} &= h_{1\Delta}e^{-j\omega\Delta t} \end{aligned} \tag{8.51}$$

Introducing them into Equation 8.50, the generic variable can be expressed in a more compact fashion

$$\begin{aligned} q(\omega) &= \underbrace{\left(K_{q,h1\Sigma} + K_{q,h2\Sigma}\ e^{-j\omega\Delta t}\right)}_{K_{q,h\Sigma}}\underbrace{h_{1\Sigma}}_{h_\Sigma} + \underbrace{\left(K_{q,h1\Delta} + K_{q,h2\Delta}\ e^{-j\omega\Delta t}\right)}_{K_{q,h\Delta}}\underbrace{h_{1\Delta}}_{h_\Delta} \\ \Rightarrow q(\omega) &= K_{q,h\Sigma}\ h_\Sigma + K_{q,h\Delta}\ h_\Delta \end{aligned} \tag{8.52}$$

Hence, the magnitude of $q(\omega)$ reads

$$
\begin{aligned}
q(\omega)\ q*(\omega) = &\left|K_{q,h\Sigma}\right|^2 \left|h_\Sigma\right|^2 + \left|K_{q,h\Delta}\right|^2 \left|h_\Delta\right|^2 + \\
&+ K_{q,h\Sigma} K^*_{q,h\Delta} h_\Sigma h^*_\Delta + K^*_{q,h\Sigma} K_{q,h\Delta} h^*_\Sigma h_\Delta
\end{aligned}
\tag{8.53}
$$

It is possible to introduce the power spectral densities ($\Phi_{h\Sigma\Sigma}$, $\Phi_{h\Delta\Delta}$) and the cross spectral densities ($\Phi_{h\Sigma\Delta}$, $\Phi_{h\Delta\Sigma}$)

$$
\begin{aligned}
\Phi_{h\Sigma\Sigma}(\omega) &= \lim_{T\to+\infty} \frac{4\pi}{T} \left(\left|h_\Sigma(\omega)\right|\right)^2 \\
\Phi_{h\Delta\Delta}(\omega) &= \lim_{T\to+\infty} \frac{4\pi}{T} \left(\left|h_\Delta(\omega)\right|\right)^2 \\
\Phi_{h\Sigma\Delta}(\omega) &= \lim_{T\to+\infty} \frac{4\pi}{T} \left[h_\Sigma(\omega) h^*_\Delta(\omega)\right]^2 \\
\Phi_{h\Delta\Sigma}(\omega) &= \lim_{T\to+\infty} \frac{4\pi}{T} \left[h^*_\Sigma(\omega) h_\Delta(\omega)\right]^2
\end{aligned}
\tag{8.54}
$$

According to [51], measurements confirm that roll and heave road unevenness excitations are not correlated, thus the cross spectral densities are null

$$
\Phi_{h\Sigma\Delta} = \Phi_{h\Delta\Sigma} = 0
$$

As a consequence, the expression of variance σ_q^2 simplifies as

$$
\begin{aligned}
\sigma_q^2 = &\int_0^\infty \left|K_{q,h\Sigma}\right|^2 \Phi_{h\Sigma\Sigma} d\omega + \int_0^\infty \left|K_{q,h\Delta}\right|^2 \Phi_{h\Delta\Delta} d\omega + \\
&+ \int_0^\infty K_{q,h\Sigma} K^*_{q,h\Delta} \Phi_{h\Sigma\Delta} d\omega + \int_0^\infty K^*_{q,h\Sigma} K_{q,h\Delta} \Phi_{h\Delta\Sigma} d\omega \\
\sigma_q^2 = &\int_0^\infty \left|K_{q,h\Sigma}\right|^2 \Phi_{h\Sigma\Sigma} d\omega + \int_0^\infty \left|K_{q,h\Delta}\right|^2 \Phi_{h\Delta\Delta} d\omega
\end{aligned}
\tag{8.55}
$$

8.4.1 Irregularity Power Spectral Densities and Coherence Computation

The magnitude of heave and roll road unevenness excitation can be expressed from Equations 8.27 and 8.33 in terms of the product of complex conjugates

$$
\begin{aligned}
\left|h_\Sigma\right|^2 = h_\Sigma h^*_\Sigma &= \frac{1}{4}\left(h_L + h_R\right)\left(h^*_L + h^*_R\right) \\
&= \frac{1}{4}\left(\left|h_L\right|^2 + \left|h_R\right|^2 + h_L h^*_R + h^*_L h_R\right) \\
\left|h_\Delta\right|^2 = h_\Delta h^*_\Delta &= \frac{1}{(2s)^2}\left(h_L - h_R\right)\left(h^*_L - h^*_R\right) \\
&= \frac{1}{4s^2}\left(\left|h_L\right|^2 + \left|h_R\right|^2 - h_L h^*_R - h^*_L h_R\right)
\end{aligned}
\tag{8.56}
$$

It is possible to express them in terms of power spectral density and cross spectral density of left and right track excitations

$$\begin{aligned}
\Phi_{hLL}(\omega) &= \lim_{T\to+\infty}\frac{4\pi}{T}\left(|h_L(\omega)|\right)^2 \\
\Phi_{hRR}(\omega) &= \lim_{T\to+\infty}\frac{4\pi}{T}\left(|h_R(\omega)|\right)^2 \\
\Phi_{hLR}(\omega) &= \lim_{T\to+\infty}\frac{4\pi}{T}\left(\left[h_L(\omega)h_R^*(\omega)\right]\right)^2 \\
\Phi_{hRL}(\omega) &= \lim_{T\to+\infty}\frac{4\pi}{T}\left(\left[h_R(\omega)h_L^*(\omega)\right]\right)^2
\end{aligned} \tag{8.57}$$

where the cross spectral densities are complex conjugates ($\beta(\Omega) = \beta(\omega)$)

$$\begin{aligned}
h_L(\omega)h_R^*(\omega) &= |h_L||h_R|e^{j\beta(\omega)} \\
h_R(\omega)h_L^*(\omega) &= |h_R||h_L|e^{-j\beta(\omega)} \\
\Rightarrow \Phi_{hLR}(\omega) = |\Phi_{hLR}|e^{j\beta(\omega)}, &\quad \Phi_{hRL}(\omega) = |\Phi_{hLR}|e^{-j\beta(\omega)}
\end{aligned}$$

A coherence function $\gamma_h(\Omega) = \gamma_h(\omega)$ can be introduced [51, 55, 58] in order to relate the cross spectral densities to the power spectral densities of left and right tracks

$$\gamma_h(\omega) = \frac{|\Phi_{hLR}(\omega)|}{\sqrt{\Phi_{hLL}(\omega)\Phi_{hRR}(\omega)}} \qquad \gamma_h : \mathbb{R} \in [0,1] \tag{8.58}$$

For $\gamma_h = 1$, left and right track irregularities are fully correlated. For $\gamma_h = 0$, instead, no correlation exists between left and right track irregularities. Measured coherence functions γ_h (see **Figure 8.8**) show the following relations

- **with spatial angular frequency Ω**: for low frequency Ω waves (very large wavelength L, e.g., landscape variability), left and right tracks are correlated ($\gamma_h \to 1$). For high frequency Ω waves (very small wavelength L, e.g., road roughness), left and right tracks are not correlated ($\gamma_h \to 0$).
- **with track width 2s**: as the track width increases ($\uparrow 2s$), the distance between left and right track increases and consequently the correlation between them decreases ($\downarrow \gamma_h$).

Measurement results from [51] allow the following simplifications

- power spectral densities are approximately equal for left and right track

$$\Phi_{hLL}(\Omega) \approx \Phi_{hRR}(\Omega) = \Phi_h(\Omega)$$

- the phase angles of cross spectral densities Φ_{hLR} and Φ_{hRL} are almost null

$$\beta(\omega) = \beta(\Omega) \approx 0$$

Therefore, cross spectral densities are real as the power spectral densities. With these simplifications, the coherence function can be rewritten as

$$\gamma_h \approx \frac{|\Phi_{hLR}(\Omega)|}{\Phi_h(\Omega)} \tag{8.59}$$

FIGURE 8.8 Measured coherence functions between left and right track road excitations [58]. as function of track width $2s$.

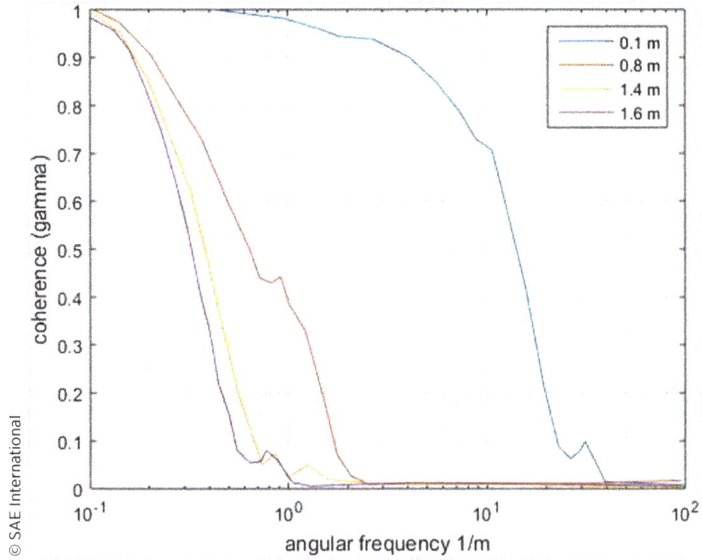

An approximated formula for the computation of coherence γ_h as function of angular frequency and track width is provided by Ammon [51, 58, 62] as

$$\gamma_h(\omega,2s) = \gamma_h(\Omega,2s) = \left\{1 + \left[\frac{\Omega}{\Omega_{p\gamma}}\left(\frac{2s}{1\ \mathrm{m}}\right)^{\alpha_\gamma}\right]^w\right\}^{-p_\gamma} \tag{8.60}$$

The values of the constants in Equation 8.60 are indicated in **Table 8.2**. The function's plot is reported in **Figure 8.9**.

The coherence can be used to compute the power spectral densities of heave and roll road excitations. From Equations 8.54, 8.57, and 8.59, it is possible to write

$$\begin{aligned}\Phi_{h\Sigma\Sigma}(\omega) &= \frac{1}{4}(\Phi_{hLL} + \Phi_{hRR} + \Phi_{hLR} + \Phi_{hRL}) \\ &\approx \frac{1}{2}(1+\gamma_h)\Phi_h(\omega)\end{aligned} \tag{8.61}$$

TABLE 8.2 Coefficients for Ammon approximation of coherence γ_h (for federal roads-highways in asphalt [51, 62]).

Coefficient	Value	Unit
$\Omega_{p\gamma}$	0.73	1/m
a_γ	0.6	-
w	2	-
p_γ	0.45	-

FIGURE 8.9 Ammon approximation of coherence γ_h (track width $2s$ = 1.46 m).

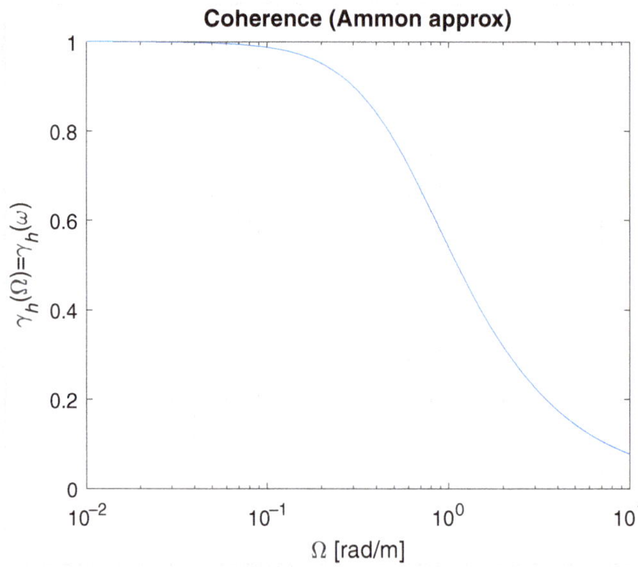

$$\begin{aligned}\Phi_{h\Delta\Delta}(\omega) &= \frac{1}{4}(\Phi_{hLL} + \Phi_{hRR} + \Phi_{hLR} + \Phi_{hRL}) \\ &\approx \frac{1}{2s^2}(1-\gamma_h)\Phi_h(\omega)\end{aligned} \tag{8.62}$$

8.4.2 Variance Computation

From Equations 8.55, 8.61, and 8.62, the variance of the generic variable q can be computed as

$$\sigma_q^2 \approx \int_0^{+\infty} \left|K_{q,h\Sigma}\right|^2 \left[\frac{1}{2}(1+\gamma_h)\Phi_h\right] d\omega + \int_0^{+\infty} \left|K_{q,h\Delta}\right|^2 \left[\frac{1}{2s^2}(1-\gamma_h)\Phi_h\right] d\omega \tag{8.63}$$

The heave and roll excitation power spectral densities are represented in **Figure 8.10**.

For the wheels' normal load, the frequency response functions of Equations 8.30–8.31 (heave) and Equations 8.40–8.47 (roll) can be used in Equation 8.63 giving

$$\begin{aligned}\sigma_{Fz,iL}^2 &\approx \int_0^{+\infty} \left|K_{Fzi,h\Sigma}\right|^2 \left[\frac{1}{2}(1+\gamma_h)\Phi_h\right] d\omega + \int_0^{+\infty} \left|K_{FziL,h\Delta}\right|^2 \left[\frac{1}{2s^2}(1-\gamma_h)\Phi_h\right] d\omega \\ \sigma_{Fz,iR}^2 &\approx \int_0^{+\infty} \left|K_{Fzi,h\Sigma}\right|^2 \left[\frac{1}{2}(1+\gamma_h)\Phi_h\right] d\omega + \int_0^{+\infty} \left|K_{FziR,h\Delta}\right|^2 \left[\frac{1}{2s^2}(1-\gamma_h)\Phi_h\right] d\omega\end{aligned} \tag{8.64}$$

where i = 1 indicates front axle and i = 2 indicates rear axle, and

$$K_{Fzi,h\Sigma} = K_{Fzi,hi\Sigma} \tag{8.65}$$

FIGURE 8.10 Heave and roll excitation power spectral densities (respectively $\Phi_{h\Sigma\Sigma}$ and $\Phi_{h\Delta\Delta}$).

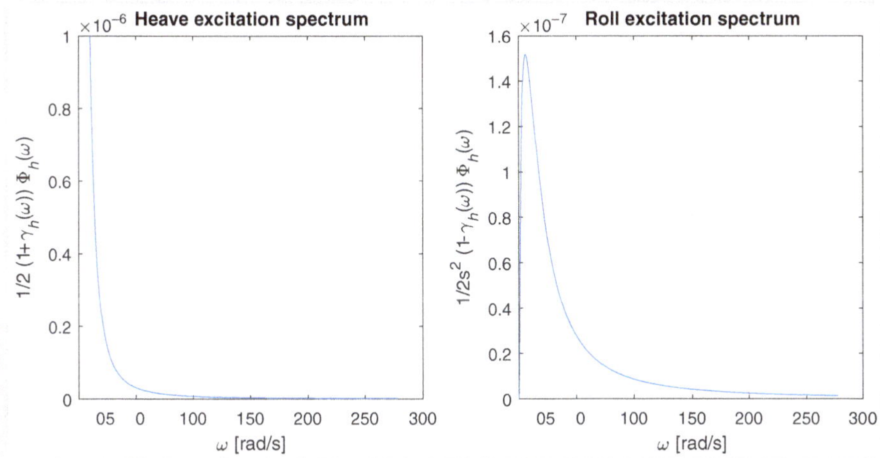

$$
\begin{aligned}
K_{FziL,h\Delta} &= K_{FziL,h1\Delta} + K_{FziL,h2\Delta} e^{-j\omega\Delta t} \\
K_{FziR,h\Delta} &= K_{FziR,h1\Delta} + K_{FziR,h2\Delta} e^{-j\omega\Delta t}
\end{aligned}
\tag{8.66}
$$

The magnitude of frequency response functions of wheels' load variation are represented in Figures 8.11 and 8.12.

FIGURE 8.11 Magnitude of frequency response functions of wheel load variation (heave).

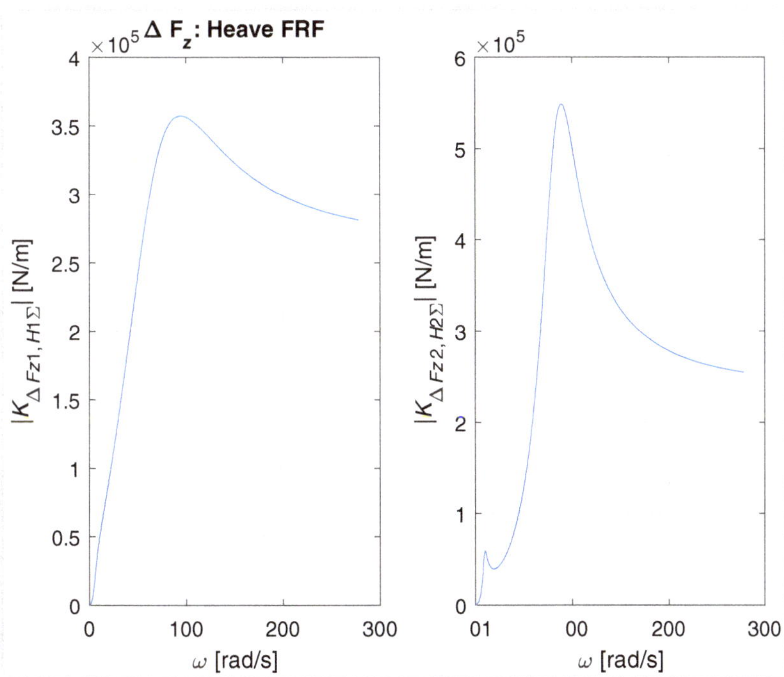

FIGURE 8.12 Magnitude of frequency response functions of wheel load variation (roll).

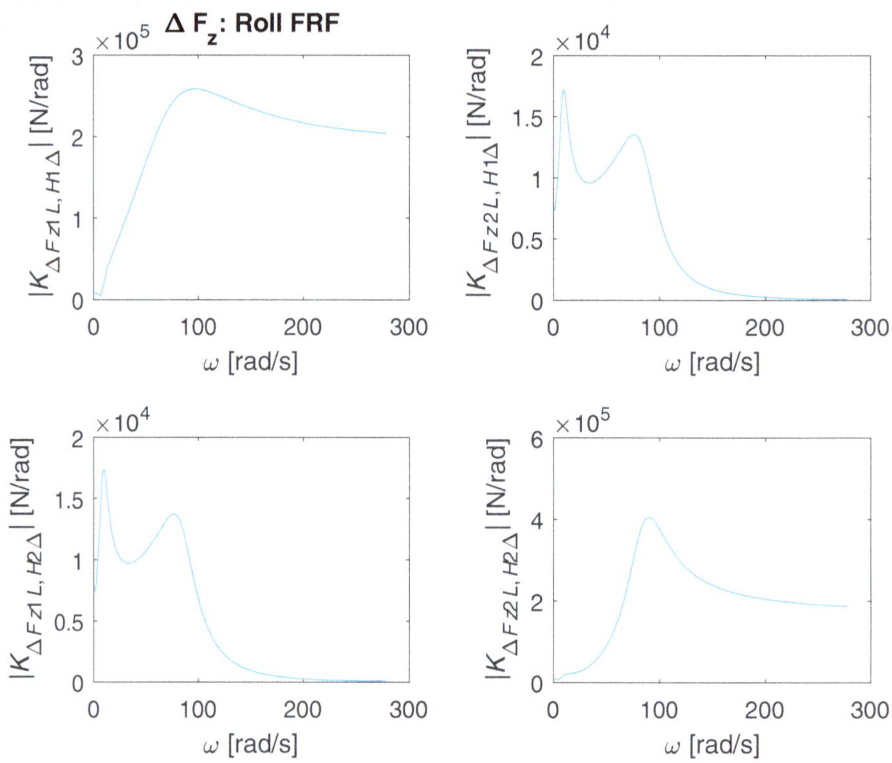

The results obtained for the parameters' values previously stated are

$$\sigma_{Fz,1L} = \sigma_{Fz,1R} = 563.2 \text{ N}$$
$$\sigma_{Fz,2L} = \sigma_{Fz,2R} = 539.2 \text{ N}$$

8.5 Variance of Slip Angles for Straight Motion

As it has been done for the wheels' normal load, it is assumed that the slip angles for the straight motion condition can also be divided into a static component and a dynamic component, related to road irregularity effect

$$\begin{aligned} \alpha_1(t) &= \alpha_{1,stat} + \alpha_{1,dyn}(t) \\ \alpha_2(t) &= \alpha_{2,stat} + \alpha_{2,dyn}(t) \end{aligned} \tag{8.67}$$

Their variance can be estimated from the variance of wheels' normal load.

The first order approximation of variance for a function $g(\underline{X})$ of a vector of random variables $\underline{X}$ reads

$$\sigma_g^2 = \sum_k^n \sum_l^n \left(\frac{\partial g(\underline{X})}{\partial X_k} \bigg|_{\underline{X}_{stat}} \right) \cdot \left(\frac{\partial g(\underline{X})}{\partial X_l} \bigg|_{\underline{X}_{stat}} \right) \cdot Cov(X_k, X_l) \tag{8.68}$$

The following relations between covariance and variance are then used

$$Cov(X_k,X_k)=\sigma^2_{Xk}$$
$$Cov(X_k,X_l)\le\sqrt{\sigma^2_{Xk}\cdot\sigma^2_{Xl}}\quad(\text{Cauchy}-\text{Schwarz inequality})$$

Therefore, the variances of front and rear slip angles can be overestimated as

$$\sigma^2_{\alpha_1}=\left(\frac{\partial\alpha_1}{\partial F_{z,1L}}\bigg|_{\substack{Fz,1L,stat\\Fz,1R,stat}}\right)^2\sigma^2_{Fz,1L}+\left(\frac{\partial\alpha_1}{\partial F_{z,1R}}\bigg|_{\substack{Fz,1L,stat\\Fz,1R,stat}}\right)^2\sigma^2_{Fz,1R}+$$
$$+2\left(\frac{\partial\alpha_1}{\partial F_{z,1L}}\bigg|_{\substack{Fz,1L,stat\\Fz,1R,stat}}\right)\left(\frac{\partial\alpha_1}{\partial F_{z,1R}}\bigg|_{\substack{Fz,1L,stat\\Fz,1R,stat}}\right)Cov(F_{z,1L},F_{z,1R})$$
$$\sigma^2_{\alpha_1}\le\left[\left(\frac{\partial\alpha_1}{\partial F_{z,1L}}\bigg|_{\substack{Fz,1L,stat\\Fz,1R,stat}}\right)^2+\left(\frac{\partial\alpha_1}{\partial F_{z,1R}}\bigg|_{\substack{Fz,1L,stat\\Fz,1R,stat}}\right)^2+\right.$$
$$\left.+2\left(\frac{\partial\alpha_1}{\partial F_{z,1L}}\bigg|_{\substack{Fz,1L,stat\\Fz,1R,stat}}\right)\left(\frac{\partial\alpha_1}{\partial F_{z,1R}}\bigg|_{\substack{Fz,1L,stat\\Fz,1R,stat}}\right)\right]\sigma^2_{Fz,1L/R}\tag{8.69}$$

$$\sigma^2_{\alpha_2}\le\left[\left(\frac{\partial\alpha_2}{\partial F_{z,2L}}\bigg|_{\substack{Fz,2L,stat\\Fz,2R,stat}}\right)^2+\left(\frac{\partial\alpha_2}{\partial F_{z,2R}}\bigg|_{\substack{Fz,2L,stat\\Fz,2R,stat}}\right)^2+\right.$$
$$\left.+2\left(\frac{\partial\alpha_2}{\partial F_{z,2L}}\bigg|_{\substack{Fz,2L,stat\\Fz,2R,stat}}\right)\left(\frac{\partial\alpha_2}{\partial F_{z,2R}}\bigg|_{\substack{Fz,2L,stat\\Fz,2R,stat}}\right)\right]\sigma^2_{Fz,2L/R}\tag{8.70}$$

where $\sigma^2_{Fz,1L}=\sigma^2_{Fz,1R}=\sigma^2_{Fz,1L/R}$ and $\sigma^2_{Fz,2L}=\sigma^2_{Fz,2R}=\sigma^2_{Fz,2L/R}$.

The relations between the straight-driving slip angles and the wheels' normal load can be expressed according to Equations 6.6–6.7 (neglecting elastokinematics contributions), rewritten here for convenience as

$$\alpha_1(F_{z,1L},F_{z,1R})=-\frac{(F_{z,1L}+F_{z,1R})\cdot\varphi_r}{C_{F\alpha,10L}+C_{F\alpha,10R}}-\frac{F_{y0,1L}+F_{y0,1R}}{C_{F\alpha,10L}+C_{F\alpha,10R}}$$
$$\alpha_2(F_{z,2L},F_{z,2R})=-\frac{(F_{z,2L}+F_{z,2R})\cdot\varphi_r}{C_{F\alpha,20L}+C_{F\alpha,20R}}-\frac{F_{y0,2L}+F_{y0,2R}}{C_{F\alpha,20L}+C_{F\alpha,20R}}$$

By introducing the piecewise polynomial approximations expressed in Equations 9.1 and 9.3, pertaining to three data points for vertical load F_z, the relationships read

$$\alpha_1(F_{z,1L},F_{z,1R})=\frac{-(F_{z,1L}+F_{z,1R})\cdot\varphi_r}{2p_{Cf,0}+p_{Cf,1}\cdot(F_{z,1L}+F_{z,1R})+p_{Cf,2}\cdot(F^2_{z,1L}+F^2_{z,1R})}+$$
$$+\frac{-\left[2p_{Fy,0}+p_{Fy,1}\cdot(F_{z,1L}+F_{z,1R})+p_{Fy,2}\cdot(F^2_{z,1L}+F^2_{z,1R})\right]}{2p_{Cf,0}+p_{Cf,1}\cdot(F_{z,1L}+F_{z,1R})+p_{Cf,2}\cdot(F^2_{z,1L}+F^2_{z,1R})}\tag{8.71}$$

$$\alpha_2(F_{z,2L},F_{z,2R})=\frac{-(F_{z,2L}+F_{z,2R})\cdot\varphi_r}{2p_{Cf,0}+p_{Cf,1}\cdot(F_{z,2L}+F_{z,2R})+p_{Cf,2}\cdot(F^2_{z,2L}+F^2_{z,2R})}+$$
$$+\frac{-\left[2p_{Fy,0}+p_{Fy,1}\cdot(F_{z,2L}+F_{z,2R})+p_{Fy,2}\cdot(F^2_{z,2L}+F^2_{z,2R})\right]}{2p_{Cf,0}+p_{Cf,1}\cdot(F_{z,2L}+F_{z,2R})+p_{Cf,2}\cdot(F^2_{z,2L}+F^2_{z,2R})}\tag{8.72}$$

where the piecewise polynomial interpolation coefficients $p_{Fy,0}$, $p_{Fy,1}$, $p_{Fy,2}$ and $p_{Cf,0}$, $p_{Cf,1}$, $p_{Cf,2}$ are computed as discussed in Chapter 9.

The partial derivatives of the previous expressions are computed as

$$\frac{\partial \alpha_i}{\partial F_{z,iL}} = \frac{-\left(\varphi_r + p_{Fy,1} + 2p_{Fy,2} \cdot F_{z,iL}\right) \cdot Den_{\alpha i} - \left(p_{Cf,1} + 2p_{Cf,2} \cdot F_{z,iL}\right) \cdot Num_{\alpha i}}{\left[Den_{\alpha i}\right]^2} \tag{8.73}$$

$$\frac{\partial \alpha_i}{\partial F_{z,iR}} = \frac{-\left(\varphi_r + p_{Fy,1} + 2p_{Fy,2} \cdot F_{z,iR}\right) \cdot Den_{\alpha i} - \left(p_{Cf,1} + 2p_{Cf,2} \cdot F_{z,iR}\right) \cdot Num_{\alpha i}}{\left[Den_{\alpha i}\right]^2} \tag{8.74}$$

8.6 Influence of Wheel Suspension with Road Irregularity Excitation

Several effects of wheel suspension are addressed by Mitschke [51] for a vehicle's vibration behavior under the influence of road irregularity. Apart from forces transmitted through the suspensions joints – that couple the accelerations of a vehicle's body and axles and cause noise at high frequencies – wheel travel causes track side shifts. These shifts (or displacements) induce side forces that affect the lateral vibration behavior of the vehicle. This influence is negligible in case of equal excitation for left and right track (the so-called *stroke excitation* [51]) since the lateral forces induced on left and right wheels are equal and opposite (if the vehicle is symmetrical). On the contrary, with unequal left and right track excitation (*roll excitation* [51]), roll and lateral motions are coupled, and the former has considerable influence on the latter. Therefore, lateral acceleration and forces produced by wheel suspension have to be considered in order to fully describe the vibration behavior of the vehicle [51].

FIGURE 8.13 Tire side force F_y produced by track shift y_0 (scheme adapted from [51]).

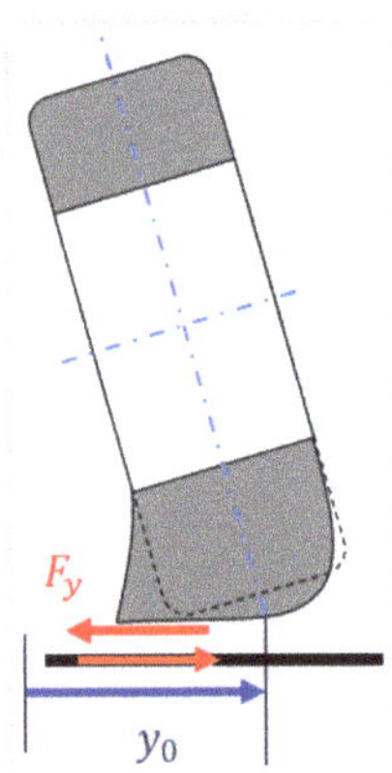

8.6.1 Tire Lateral Force Dependence on Track Shift

The relation between tire lateral force F_y and side shift of track y_0 (represented in Figure 8.13) can be expressed with a first order differential equation, as done by *Schlippe-Dietrich* [51]

$$\dot{F}_y + \frac{C_{Fy} \cdot V}{C_{F\alpha}} \cdot F_y = C_{Fy} \cdot \dot{y}_0 \tag{8.75}$$

where the sign convention of F_y in this case is opposite to the one adopted for y_0, as depicted in Figure 8.13. The parameter C_{Fy} [N/m] is the tire lateral spring constant, and the ratio $\frac{C_{F\alpha}}{C_{Fy}}$. is the *relaxation length* [7], governing the force buildup lag. It is noted that at steady state, Equation 8.75 assumes the well-known form (the absolute values are due to the different sign conventions used)

$$\left|F_y\right| = C_{F\alpha} \cdot \frac{\left|\dot{y}_0\right|}{V} = C_{F\alpha} \cdot \left|\alpha\right|$$

8.6.2 Motion Ratio of Wheel Suspensions

As for the wheel suspensions, the motion ratio concept has to be introduced in order to obtain the expressions for roll stiffness and damping coefficients, respectively, k_φ and c_φ. The motion ratio of a suspension mechanism is defined as the ratio between spring displacement and wheel displacement. Two types of suspension systems are considered in this part, namely *McPherson* and *Double Wishbone*, represented in **Figure 8.14**. Henceforward, the *McPherson* system will appertain to the front axle (identified with $i = 1$), and the *Double Wishbone* system to the rear axle (identified with $i = 2$).

The total roll stiffness $k_{\varphi 1}$ of the *McPherson* front suspension can be computed as the sum of a contribution from the anti-roll bar ($k_{\varphi ARB,1}$) and another contribution from the suspension's spring stiffness k_1, corrected via the motion ratio

$$k_{\varphi 1} = k_{\varphi ARB,1} + 2k_1 \cdot \left(\frac{b_{MP} \cdot d_{MP}}{a_{MP}} \right)^2 \tag{8.76}$$

Analogously, the roll damping $c_{\varphi 1}$ can be computed as function of the suspension's damping coefficient c_1, corrected with the motion ratio

$$c_{\varphi 1} = 2c_1 \cdot \left(\frac{b_{MP} \cdot d_{MP}}{a_{MP}} \right)^2 \tag{8.77}$$

The involved parameters can be identified by inspection of Figure 8.14.

For the *Double Wishbone* rear suspension, the total roll stiffness $k_{\varphi 2}$ and the roll damping $c_{\varphi 2}$ can be similarly computed as

$$k_{\varphi 2} = k_{\varphi ARB,2} + 2k_2 \cdot \left(\frac{b_{DW} \cdot d_{DW}}{a_{DW}} \cdot \frac{f_{DW}}{e_{DW}} \right)^2 \tag{8.78}$$

$$c_{\varphi 1} = 2c_2 \cdot \left(\frac{b_{DW} \cdot d_{DW}}{a_{DW}} \cdot \frac{f_{DW}}{e_{DW}} \right)^2 \tag{8.79}$$

FIGURE 8.14 *McPherson* (a) and *Double Wishbone* (b) suspension systems, with geometrical identification of roll center RC_r

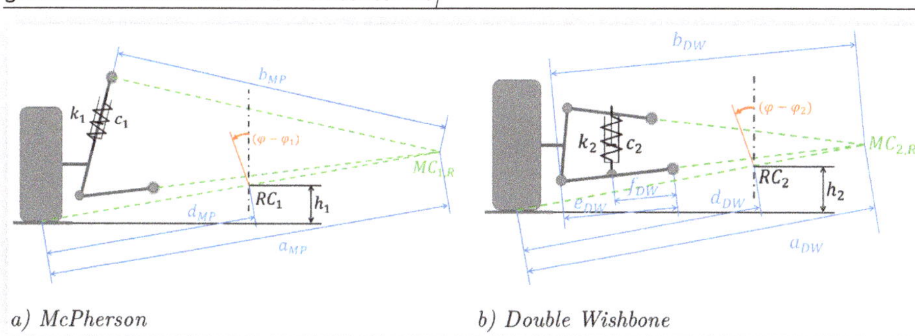

8.6.3 Extended Vehicle Model for Roll Excitation

The vehicle model for *roll excitation* analysis, represented in Figure 8.7, can be extended in order to incorporate also the description of side movements together with roll. This extended model is represented in **Figure 8.15**, where road irregularity excitations $h_{i\Delta}$ are defined according to Equation 8.33. With respect to the previously introduced model, this extended version includes the coordinate of lateral displacement y of a vehicle's body. Moreover, it includes the lateral tire forces $F_{yi,L/R}$ caused by track shifts $y_{0i,L/R}$, which are induced by road irregularity.

In order to write the force and moment equilibria to find the equations of motion, the vehicle model of Figure 8.15 can be decomposed into three subsystems, whose free body diagrams are represented in **Figure 8.16**.

The description of axles' movements is deliberately simplified in this study, by considering fictitious rigid axles. They are modeled with mass $2m_{ti}$ concentrated at the center, at a height equal to the wheel radius R_r and moment of inertia $I_{xi} = 2m_{ti} \cdot s^2$. With the equivalent rigid axles assumption, their fictitious roll movements (φ_1 for front axle, φ_2 for rear axle) are defined as in Equation 8.32, while axles' lateral movements y_i are functions of a vehicle's body roll φ, side motion y, and axle's roll φ_i

$$
\begin{aligned}
y_1 &= y - h' \cdot \varphi - (h_1 - R_r) \cdot \varphi_1 \\
y_2 &= y - h' \cdot \varphi - (h_2 - R_r) \cdot \varphi_2
\end{aligned}
\tag{8.80}
$$

FIGURE 8.15 Vehicle model for roll and side motion analysis, with different road irregularity excitations for left and right track ("roll excitation").

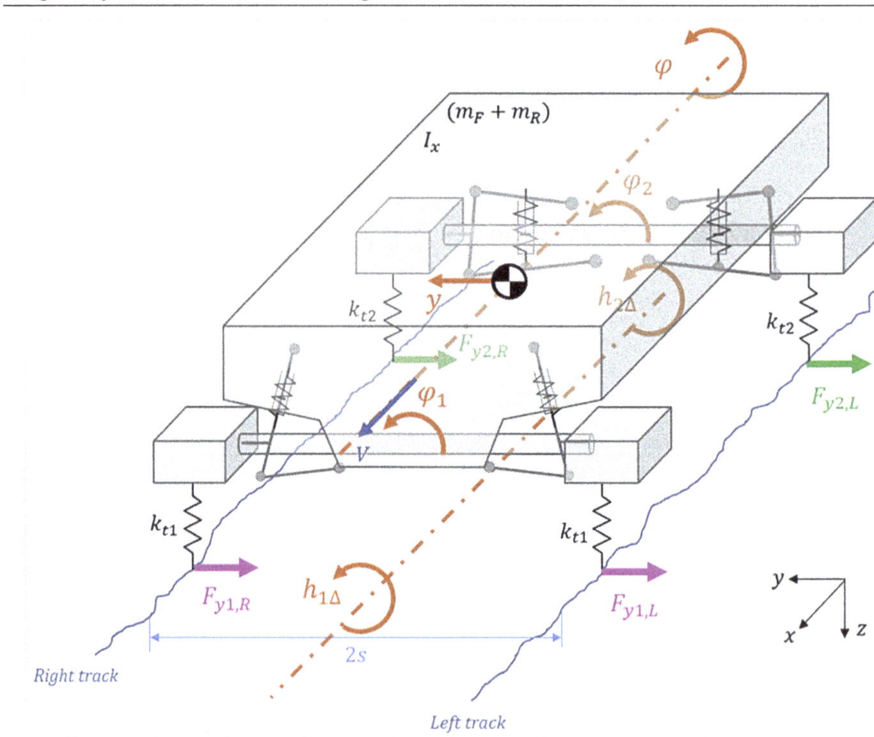

FIGURE 8.16 Free body diagrams of the three subsystems of vehicle model: a) vehicle's body, b) front axle, c) rear axle.

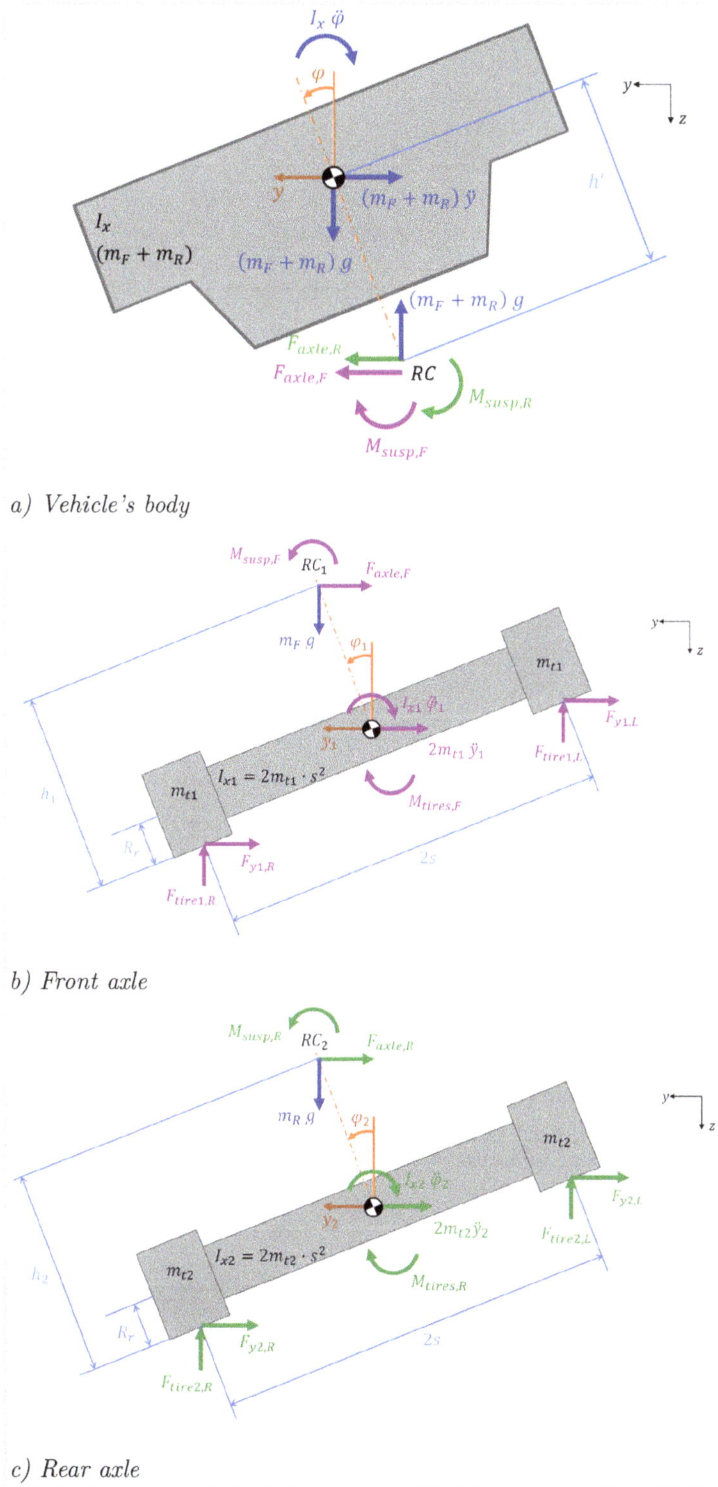

Analogously, the lateral track shifts $y_{0i,L/R}$, which are responsible for the lateral tire forces $F_{yi,L/R}$, can be defined as

$$\begin{aligned} y_{01,L/R} &= y - h' \cdot \varphi - h_1 \cdot \varphi_1 \\ y_{02,L/R} &= y - h' \cdot \varphi - h_2 \cdot \varphi_2 \end{aligned} \tag{8.81}$$

A more refined approach would require considering independent suspensions, with decoupled left and right wheel travels, with the drawbacks of additional degrees of freedom and increased model complexity.

The differential equations for tire lateral forces are obtained by merging the lateral force dependence on track shift, expressed by Equation 8.75 and the track shifts definitions in Equation 8.81. By considering $F_{yi,T} = F_{yi,L} + F_{yi,R}$, one draws

$$\dot{F}_{y1,T} + \left(\frac{C_{Fy,1}}{C_{F\alpha,10}} \cdot V \right) \cdot F_{y1,T} = 2C_{Fy,1} \cdot \left(\dot{y} - h' \cdot \dot{\varphi} - h_1 \cdot \dot{\varphi}_1 \right) \tag{8.82}$$

$$\dot{F}_{y2,T} + \left(\frac{C_{Fy,2}}{C_{F\alpha,20}} \cdot V \right) \cdot F_{y2,T} = 2C_{Fy,2} \cdot \left(\dot{y} - h' \cdot \dot{\varphi} - h_2 \cdot \dot{\varphi}_2 \right) \tag{8.83}$$

It is assumed that lateral forces (due to inertia and tires) are transmitted between the three subsystems of Figure 8.16 through the roll axis (see Figure 2.1). The moment contributions, produced by the minimal difference of heights between the roll centers of front axle, rear axle, and the vehicle's body, are neglected. Furthermore, it is assumed that inertial lateral forces can be split up between front and rear axle according to

$$\begin{aligned} m_F \cdot \ddot{y} &= F_{axle,F} \\ m_R \cdot \ddot{y} &= F_{axle,R} \end{aligned} \tag{8.84}$$

where front and rear sprung masses have been defined in Equation 8.24.

Hence, force and moment equilibria are written for the three subsystems of Figure 8.16, in order to obtain the equations of motion, with small angles approximation

- moment equilibrium for the vehicle's body, with respect to its roll center

$$I_x \cdot \dot{\varphi} + M_{susp,F} + M_{susp,R} + \left(m_F + m_R\right) \cdot h' \cdot \left(\ddot{y} - g \cdot \varphi\right) = 0 \tag{8.85}$$

The terms $M_{susp,F}$ and $M_{susp,R}$ are the moment contributions from roll stiffness and damping, respectively, of front *McPherson* suspension and rear *Double Wishbone* suspension. They can be computed considering the difference of roll angle between vehicle's body and axles, and the roll stiffness and damping coefficients defined in Equations 8.76–8.79

$$\begin{aligned} M_{susp,F} &= k_{\varphi 1} \cdot \left(\varphi - \varphi_1\right) + c_{\varphi 1} \cdot \left(\dot{\varphi} - \dot{\varphi}_1\right) \\ M_{susp,R} &= k_{\varphi 2} \cdot \left(\varphi - \varphi_2\right) + c_{\varphi 2} \cdot \left(\dot{\varphi} - \dot{\varphi}_2\right) \end{aligned} \tag{8.86}$$

- moment equilibrium for front axle, with respect to its center of gravity

$$I_{x1} \cdot \ddot{\varphi}_1 + \left(F_{axle,F} - m_F \cdot g \cdot \varphi_1\right) \cdot \left(h_1 - R_r\right) - M_{susp,F} - F_{y1,T} \cdot R_r + M_{tires,F} = 0 \tag{8.87}$$

where the moment contribution coming from tire stiffness and road irregularity is

$$M_{tires,F} = 2k_{t1} \cdot s^2 \cdot \left(\varphi_1 - h_{1\Delta}\right) \tag{8.88}$$

- moment equilibrium for rear axle, with respect to its center of gravity

$$I_{x2}\cdot\ddot{\varphi}_2+\left(F_{axle,R}-m_R\cdot g\cdot\varphi_2\right)\cdot\left(h_2-R_r\right)-M_{susp,R}-F_{y2,T}\cdot R_r+M_{tires,R}=0 \quad (8.89)$$

where the moment contribution coming from tire stiffness and road irregularity is akin to front axle's term

$$M_{tires,R}=2k_{t2}\cdot s^2\cdot\left(\varphi_2-h_{2\Delta}\right) \quad (8.90)$$

- lateral force equilibrium for the whole system

$$\begin{aligned}&\left(m_F+m_R\right)\cdot\ddot{y}+2m_{t1}\cdot\ddot{y}_1+2m_{t2}\cdot\ddot{y}_2+F_{y1,T}+F_{y2,T}=0\\ \Rightarrow&\left(m_F+m_R\right)\cdot\ddot{y}+2m_{t1}\cdot\left[\ddot{y}-h'\cdot\ddot{\varphi}-\left(h_1-R_r\right)\cdot\ddot{\varphi}_1\right]+\\ &+2m_{t2}\cdot\left[\ddot{y}-h'\cdot\ddot{\varphi}-\left(h_2-R_r\right)\cdot\ddot{\varphi}_2\right]+F_{y1,T}+F_{y2,T}=0\end{aligned} \quad (8.91)$$

The equations of motion, resulting from force and moment equilibria (Equations 8.85–8.91) and tire lateral force expressions (Equations 8.82–8.83), are reported in matrix form

$$M_\varphi\ \underline{\ddot{x}}+C_\varphi\ \underline{\dot{x}}+K_\varphi\ \underline{x}=R_\varphi\ \underline{h}_\Delta \quad (8.92)$$

$$M_\varphi=\begin{bmatrix}M_{11}&0&0&M_{14}&0&0\\0&M_{22}&0&M_{24}&0&0\\0&0&M_{33}&M_{34}&0&0\\M_{41}&M_{42}&M_{43}&M_{44}&0&0\\0&0&0&0&0&0\\0&0&0&0&0&0\end{bmatrix}$$

$$C_\varphi=\begin{bmatrix}C_{11}&C_{12}&C_{13}&0&0&0\\C_{21}&C_{22}&0&0&0&0\\C_{31}&0&C_{33}&0&0&0\\0&0&0&0&0&0\\C_{51}&C_{52}&0&C_{54}&1&0\\C_{61}&0&C_{63}&C_{64}&0&1\end{bmatrix}$$

$$K_\varphi=\begin{bmatrix}K_{11}&K_{12}&K_{13}&0&0&0\\K_{21}&K_{22}&0&0&K_{25}&0\\K_{31}&0&K_{33}&0&0&K_{36}\\0&0&0&0&1&1\\0&0&0&0&K_{55}&0\\0&0&0&0&0&K_{66}\end{bmatrix}$$

$$R_\varphi\underline{h}_\Delta=\begin{bmatrix}0&0\\2k_{t1}\cdot s^2&0\\0&2k_{t2}\cdot s^2\\0&0\\0&0\\0&0\end{bmatrix}\begin{Bmatrix}h_{1\Delta}\\h_{2\Delta}\end{Bmatrix},\quad \underline{x}\begin{Bmatrix}\varphi\\\varphi_1\\\varphi_2\\y\\F_{y1,T}\\F_{y2,T}\end{Bmatrix}$$

with

$$\begin{cases} M_{11} = I_x \\ M_{14} = (m_F + m_R)\cdot h' \\ M_{22} = 2m_{t1}\cdot s^2 \\ M_{24} = m_F\cdot(h_1 - R_r) \\ M_{33} = 2m_{t2}\cdot s^2 \\ M_{34} = m_R\cdot(h_2 - R_r) \\ M_{41} = -2\cdot(m_{t1} + m_{t2})\cdot h' \\ M_{42} = -2m_{t1}\cdot(h_1 - R_r) \\ M_{43} = -2m_{t2}\cdot(h_2 - R_r) \\ M_{44} = m_F + m_R + 2m_{t1} + 2m_{t2} \end{cases} \qquad \begin{cases} C_{11} = c_{\varphi1} + c_{\varphi2} \\ C_{12} = -c_{\varphi1} \\ C_{13} = -c_{\varphi2} \\ C_{21} = -c_{\varphi1} \\ C_{22} = c_{\varphi1} \\ C_{31} = -c_{\varphi2} \\ C_{33} = c_{\varphi2} \\ C_{51} = 2h'\cdot C_{Fy,1} \\ C_{52} = 2h_1\cdot C_{Fy,1} \\ C_{54} = -2\cdot C_{Fy,1} \\ C_{61} = 2h'\cdot C_{Fy,2} \\ C_{63} = 2h_2\cdot C_{Fy,2} \\ C_{64} = -2\cdot C_{Fy,2} \end{cases}$$

$$\begin{cases} K_{11} = k_{\varphi1} + k_{\varphi2} - (m_F + m_R)\cdot g\cdot h' \\ K_{12} = -k_{\varphi1} \\ K_{13} = -k_{\varphi2} \\ K_{21} = -k_{\varphi1} \\ K_{22} = k_{\varphi1} + 2k_{t1}\cdot s^2 - m_F\cdot g\cdot(h_1 - R_r) \\ K_{25} = -R_r \\ K_{31} = -k_{\varphi2} \\ K_{33} = k_{\varphi2} + 2k_{t2}\cdot s^2 - m_R\cdot g\cdot(h_2 - R_r) \\ K_{36} = -R_r \\ K_{55} = \frac{C_{Fy,1}}{C_{F\alpha,10}}\cdot V \\ K_{66} = \frac{C_{Fy,2}}{C_{F\alpha,20}}\cdot V \end{cases}$$

In order to solve the system in Equation 8.92, in the work of Mitschke [51] the flat vehicle assumption is made, namely $h_1 = h_2 = h' = r = 0$. Another approach can be adopted that does not involve such a restrictive assumption: moving to the Fourier domain. Equations 8.82–8.83 for the lateral forces can then be written as

$$\begin{aligned} F_{y1,T}(\omega) &= \frac{2C_{Fy,1}\cdot j\omega}{j\omega + \frac{C_{Fy,1}}{C_{F\alpha,10}}\cdot V}\cdot\left[y(\omega) - h'\cdot\varphi(\omega) - h_1\cdot\varphi_1(\omega)\right] \\ &= \frac{-C_{54}\cdot j\omega}{j\omega + K_{55}}\cdot\left[y(\omega) - h'\cdot\varphi(\omega) - h_1\cdot\varphi_1(\omega)\right] \end{aligned} \tag{8.93}$$

$$\begin{aligned} F_{y2,T}(\omega) &= \frac{2C_{Fy,2}\cdot j\omega}{j\omega + \frac{C_{Fy,2}}{C_{F\alpha,20}}\cdot V}\cdot\left[y(\omega) - h'\cdot\varphi(\omega) - h_2\cdot\varphi_2(\omega)\right] \\ &= \frac{-C_{64}\cdot j\omega}{j\omega + K_{66}}\cdot\left[y(\omega) - h'\cdot\varphi(\omega) - h_2\cdot\varphi_2(\omega)\right] \end{aligned} \tag{8.94}$$

In this way, the system in Equation 8.92 can be reduced to a system of four equations in the Fourier domain, according to

$$\left[-\omega^2 M_\varphi + j\omega C_\varphi + K_\varphi + K_{Fy}\right]\underline{x}(\omega) = R_\varphi \underline{h}_\Delta(\omega) \tag{8.95}$$

$$M_\varphi = \begin{bmatrix} M_{11} & 0 & 0 & M_{14} \\ 0 & M_{22} & 0 & M_{24} \\ 0 & 0 & M_{33} & M_{34} \\ M_{41} & M_{42} & M_{43} & M_{44} \end{bmatrix}$$

$$C_\varphi = \begin{bmatrix} C_{11} & C_{12} & C_{13} & 0 \\ C_{21} & C_{22} & 0 & 0 \\ C_{31} & 0 & C_{33} & 0 \\ 0 & 0 & 0 & 0 \end{bmatrix}$$

$$K_\varphi = \begin{bmatrix} K_{11} & K_{12} & K_{13} & 0 \\ K_{21} & K_{22} & 0 & 0 \\ K_{31} & 0 & K_{33} & 0 \\ 0 & 0 & 0 & 0 \end{bmatrix}$$

$$K_{Fy} = \begin{bmatrix} 0 & 0 & 0 & 0 \\ -K_{25}\cdot h'\cdot\frac{-C_{54}\cdot j\omega}{j\omega+K_{55}} & -K_{25}\cdot h_1\cdot\frac{-C_{54}\cdot j\omega}{j\omega+K_{55}} & 0 & K_{25}\cdot\frac{-C_{54}\cdot j\omega}{j\omega+K_{55}} \\ -K_{36}\cdot h'\cdot\frac{-C_{64}\cdot j\omega}{j\omega+K_{66}} & 0 & -K_{36}\cdot h_2\cdot\frac{-C_{64}\cdot j\omega}{j\omega+K_{66}} & K_{36}\cdot\frac{-C_{64}\cdot j\omega}{j\omega+K_{66}} \\ -h'\cdot\left(\frac{-C_{54}\cdot j\omega}{j\omega+K_{55}}+\frac{-C_{64}\cdot j\omega}{j\omega+K_{66}}\right) & -h_1\cdot\frac{-C_{54}\cdot j\omega}{j\omega+K_{55}} & -h_2\cdot\frac{-C_{64}\cdot j\omega}{j\omega+K_{66}} & \left(\frac{-C_{54}\cdot j\omega}{j\omega+K_{55}}+\frac{-C_{64}\cdot j\omega}{j\omega+K_{66}}\right) \end{bmatrix}$$

$$R_\varphi \underline{h}_\Delta = \begin{bmatrix} 0 & 0 \\ 2k_{t1}\cdot s^2 & 0 \\ 0 & 2k_{t2}\cdot s^2 \\ 0 & 0 \end{bmatrix} \begin{Bmatrix} h_{1\Delta} \\ h_{2\Delta} \end{Bmatrix}, \quad \underline{x} = \begin{Bmatrix} \varphi \\ \varphi_1 \\ \varphi_2 \\ y \end{Bmatrix}$$

The system expressed in Equation 8.95 can be solved to obtain the frequency response functions between the vehicle's body lateral motion y and the road excitations $h_{1\Delta}$ and $h_{2\Delta}$, respectively, $K_{y,h1\Delta} = \dfrac{y(\omega)}{h_{1\Delta}(\omega)}$ and $K_{y,h2\Delta} = \dfrac{y(\omega)}{h_{2\Delta}(\omega)}$. In this chapter, these frequency response functions have been obtained using the *Symbolic Math Toolbox*™ of *MATLAB*®.

The frequency response functions can be employed to evaluate the variance of the vehicle's body lateral displacement, according to Equation 8.63

$$\sigma_y^2 \approx \int_0^{+\infty} \left|K_{y,h1\Delta} + K_{y,h2\Delta}e^{-j\omega\Delta t}\right|^2 \left[\frac{1}{2s^2}(1-\gamma_h)\Phi_h\right] d\omega \tag{8.96}$$

8.7 Synthesis of the Response to the Random Process

In this section, a hint is provided on methods that can be used for the synthesis of road height profile. Two simple approaches can be considered, with basically equal quality of results obtainable [58]

- synthesis with harmonic functions with randomly distributed phases
- synthesis with linear filters

The former approach is implemented in this section, as presented in [63]. If the road irregularity $h(t)$ is a Gaussian random process with null mean $\bar{h}=0$, a realization of it can be written as the sum of a large number N of harmonic functions [62, 63]

$$h(t)=\sigma_h\sqrt{\frac{2}{N}}\cdot\sum_{k=1}^{N}cos(\omega_k t+\theta_k) \tag{8.97}$$

The standard deviation σ_h can be computed from the single-sided power spectral density $\Phi_h(\omega)$ or from the double-sided power spectral density $S_h(\omega)$ as already discussed

$$\sigma_h^2=\int_{-\infty}^{+\infty}S_h(\omega)d\omega=\int_{0}^{+\infty}\Phi_h(\omega)d\omega$$

The angular frequencies ω_k of Equation 8.97 are independent random variables, identically distributed with density function given by the normalized double-sided power spectral density $g(\omega)$

$$g_h(\omega)=\frac{S_h(\omega)}{\sigma_h^2} \tag{8.98}$$

They can be sampled with a Monte Carlo method, for instance. The phases θ_k are independent random variables, identically distributed in the interval $[0, 2\pi]$ with uniform density $\frac{1}{2\pi}$.

A realization of the generic output of the system $q(t)$ can be expressed by means of the frequency response function $K_{q,h}$ [63]

$$q(t)=\sigma_h\sqrt{\frac{2}{N}}\cdot\sum_{k=1}^{N}\left|K_{q,h}(\omega_k)\right|\cdot cos\left(\omega_k t+\theta_k+\delta_{q,h}(\omega_k)\right) \tag{8.99}$$

where

$$\delta_{q,h}(\omega_k)=arctan\left(\frac{Im\left[K_{q,h}(\omega)\right]}{Re\left[K_{q,h}(\omega)\right]}\right) \tag{8.100}$$

More refined road modeling approaches can be found in literature. An example is provided by [56, 57]: a Laplace process (namely a Gaussian process with randomly varying variance) is adopted in order to obtain more realistic rough road models, including shorter road sections with high irregularity that cannot be described with a stationary Gaussian model. Nevertheless, this is beyond the intent of this analysis and is of more interest for fatigue studies for vehicle and infrastructure [56, 64].

8.7.1 Synthesis of Two-Track Excitation

A realization of the height profiles for left and right track can be obtained, with different random phases $\theta_{L,k} \neq \theta_{R,k}$, from Equation 8.97

$$h_L(t)=\sigma_h\sqrt{\frac{2}{N}}\cdot\sum_{k=1}^{N}cos(\omega_k t+\theta_{L,k})$$
$$h_R(t)=\sigma_h\sqrt{\frac{2}{N}}\cdot\sum_{k=1}^{N}cos(\omega_k t+\theta_{R,k}) \tag{8.101}$$

The realization (time history) of the corresponding wheel normal load variations can be obtained according to Equation 8.99, with the heave and roll excitations from Equations 8.27, 8.30, 8.31, 8.33, 8.40–8.47

$$\begin{aligned}\Delta F_{z,1L}(t)=\sigma_h\sqrt{\frac{2}{N}}\cdot\Bigg\{&\sum_{k=1}^{N}[|K_{Fz1,h1\Sigma}(\omega_k)|\cdot cos(\omega_k t+\theta_{L,k}+\delta_{Fz1,h1\Sigma})]+\\&+\sum_{k=1}^{N}\Bigg[|K_{Fz1L,h1\Delta}(\omega_k)|\cdot\frac{1}{2s}\cdot[cos(\omega_k t+\theta_{L,k}+\delta_{Fz1L,h1\Delta})-\\&-cos(\omega_k t+\theta_{R,k}+\delta_{Fz1L,h1\Delta})]\Bigg]+\\&+\sum_{k=1}^{N}\Bigg[|K_{Fz1L,h2\Delta}(\omega_k)|\cdot\frac{1}{2s}\cdot[cos(\omega_k(t-\Delta t)+\theta_{L,k}+\delta_{Fz1L,h2\Delta})-\\&-cos(\omega_k(t-\Delta t)+\theta_{R,k}+\delta_{Fz1L,h2\Delta})]\Bigg]\Bigg\}\end{aligned} \tag{8.102}$$

$$\begin{aligned}\Delta F_{z,1R}(t)=\sigma_h\sqrt{\frac{2}{N}}\cdot\Bigg\{&\sum_{k=1}^{N}[|K_{Fz1,h1\Sigma}(\omega_k)|\cdot cos(\omega_k t+\theta_{R,k}+\delta_{Fz1,h1\Sigma})]+\\&+\sum_{k=1}^{N}\Bigg[-|K_{Fz1L,h1\Delta}(\omega_k)|\cdot\frac{1}{2s}\cdot[cos(\omega_k t+\theta_{L,k}+\delta_{Fz1L,h1\Delta})-\\&-cos(\omega_k t+\theta_{R,k}+\delta_{Fz1L,h1\Delta})]\Bigg]+\\&+\sum_{k=1}^{N}\Bigg[-|K_{Fz1L,h2\Delta}(\omega_k)|\cdot\frac{1}{2s}\cdot[cos(\omega_k(t-\Delta t)+\theta_{L,k}+\delta_{Fz1L,h2\Delta})-\\&-cos(\omega_k(t-\Delta t)+\theta_{R,k}+\delta_{Fz1L,h2\Delta})]\Bigg]\Bigg\}\end{aligned} \tag{8.103}$$

$$\begin{aligned}\Delta F_{z,2L}(t)=\sigma_h\sqrt{\frac{2}{N}}\cdot\Bigg\{&\sum_{k=1}^{N}[|K_{Fz2,h2\Sigma}(\omega_k)|\cdot cos(\omega_k(t-\Delta t)+\theta_{L,k}+\delta_{Fz2,h2\Sigma})]+\\&+\sum_{k=1}^{N}\Bigg[|K_{Fz2L,h1\Delta}(\omega_k)|\cdot\frac{1}{2s}\cdot[cos(\omega_k t+\theta_{L,k}+\delta_{Fz2L,h1\Delta})-\\&-cos(\omega_k t+\theta_{R,k}+\delta_{Fz2L,h1\Delta})]\Bigg]+\\&+\sum_{k=1}^{N}\Bigg[|K_{Fz2L,h2\Delta}(\omega_k)|\cdot\frac{1}{2s}\cdot[cos(\omega_k(t-\Delta t)+\theta_{L,k}+\delta_{Fz2L,h2\Delta})-\\&-cos(\omega_k(t-\Delta t)+\theta_{R,k}+\delta_{Fz2L,h2\Delta})]\Bigg]\Bigg\}\end{aligned} \tag{8.104}$$

$$\Delta F_{z,2R}(t)=\sigma_h\sqrt{\frac{2}{N}}\cdot\left\{\sum_{k=1}^{N}[|K_{Fz2,h2\Sigma}(\omega_k)|\cdot cos(\omega_k(t-\Delta t)+\theta_{R,k}+\delta_{Fz2,h2\Sigma})]+\right.$$
$$+\sum_{k=1}^{N}\left[-|K_{Fz2L,h1\Delta}(\omega_k)|\cdot\frac{1}{2s}\cdot[cos(\omega_k t+\theta_{L,k}+\delta_{Fz2L,h1\Delta})-\right.$$
$$\left.-cos(\omega_k t+\theta_{R,k}+\delta_{Fz2L,h1\Delta})]\right]+$$
$$+\sum_{k=1}^{N}\left[-|K_{Fz2L,h2\Delta}(\omega_k)|\cdot\frac{1}{2s}\cdot[cos(\omega_k(t-\Delta t)+\theta_{L,k}+\delta_{Fz2L,h2\Delta})-\right.$$
$$\left.\left.-cos(\omega_k(t-\Delta t)+\theta_{R,k}+\delta_{Fz2L,h2\Delta})]\right]\right\} \qquad (8.105)$$

Part II

Numerical Simulations

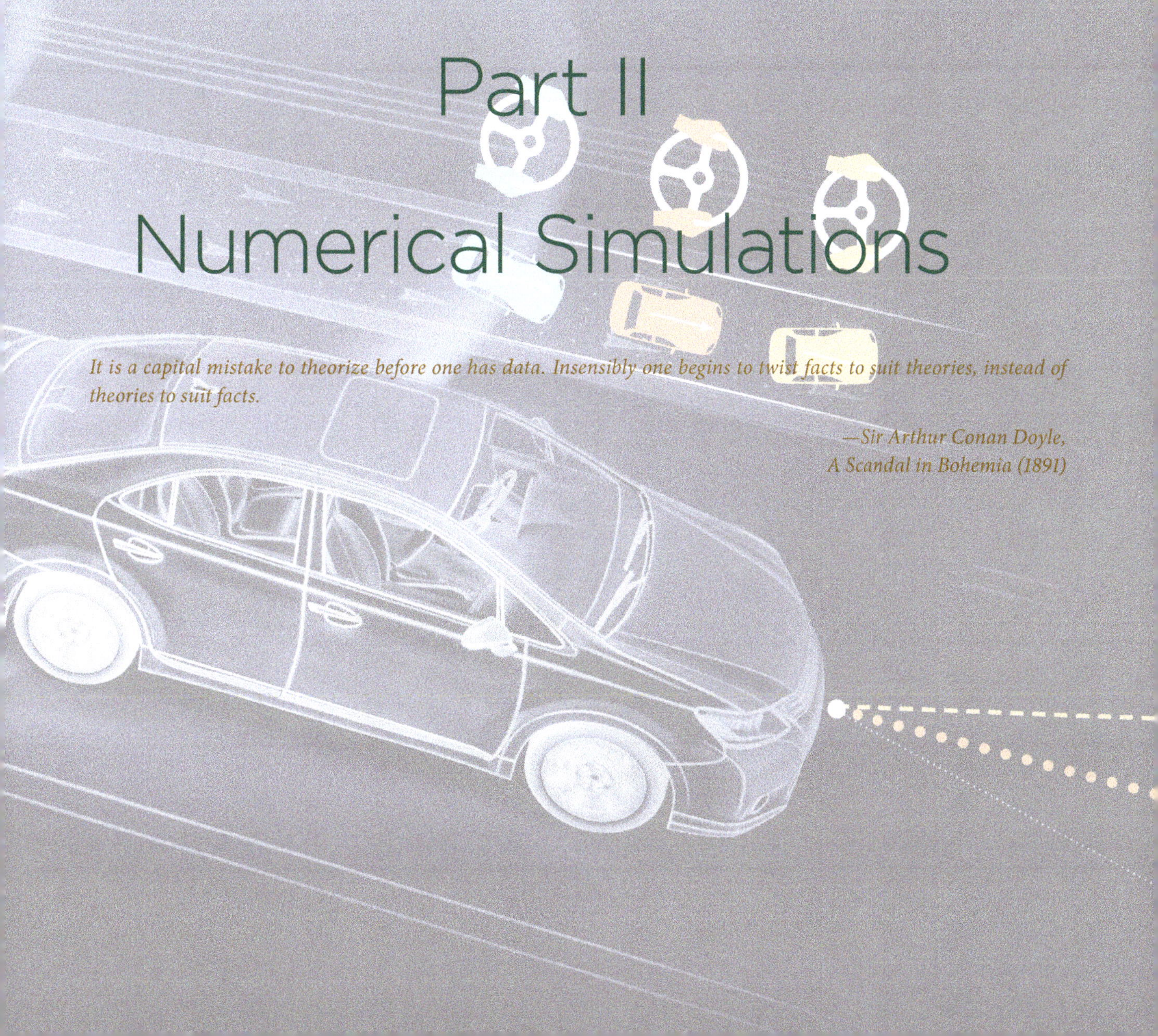

It is a capital mistake to theorize before one has data. Insensibly one begins to twist facts to suit theories, instead of theories to suit facts.

—*Sir Arthur Conan Doyle,*
A Scandal in Bohemia (1891)

CHAPTER 9

Tire Data

The data are referred to factory-new prototype tires of size *205/50R17*. Measurements have been obtained by *Smithers RAPRA Inc.* that performed tire tests with the *MTS Flat-Trac*® system, reversing the rotation direction in order to deduce ply-steer and conicity characteristics as discussed in Chapter 1.

9.1 Tire Types

The data, reported in Tables 9.1 and 9.2, are represented graphically in Figures 9.1 and 9.2.

They refer to five different types of tires

- **Normal-1**. Tire without any modification, with inflation pressure $p = 250$ kPa (recommended for front tires of the vehicle).
- **Normal-2**. Tire without any modification, with inflation pressure $p = 220$ kPa (recommended for rear tires of the vehicle).
- **Mod-A**. As *Normal-1* ($p = 250$ kPa), but with tire tread design aimed at obtaining positive PRAT.
- **Mod-B**. As *Normal-1* ($p = 250$ kPa), but with tire tread design aimed at obtaining **negative** PRAT.
- **Buffed**. As *Normal-1* ($p = 250$ kPa), but with tread part completely **removed**.

Data for $p = 220$ kPa for tire types *Mod-A*, *Mod-B*, and *Buffed* were estimated comparing the difference between *Normal-1* ($p = 250$ kPa) and *Normal-2* ($p = 220$ kPa) and applying the same difference to data of *Mod-A*, *Mod-B* and *Buffed* (which are all with $p = 250$ kPa).

TABLE 9.1 Tire data *205/50R17* (inflation pressure p = 250 kPa).

Tire ID	P [kPa]	F_z [N]	$C_{F\alpha}$ [N/rad]	F_{ply} [N]	F_{con} [N]	$C_{M\alpha}$ [Nm/rad]	M_{ply} [Nm]	M_{con} [Nm]
Normal-1	250	892.125	22055.825	1.5934	7.9318	217.394	−0.2404	0.0558
Normal-1	250	2796.151	60758.404	59.4408	29.2221	1182.926	−1.1660	0.4700
Normal-1	250	4503.307	91867.558	138.4634	45.8378	2466.667	−3.0850	1.0148
Normal-1	250	6401.682	108041.740	177.0719	48.1737	4142.379	−6.1438	1.2379
Mod-A	250	895.803	22944.035	0.8662	7.1004	242.898	−0.0329	0.2755
Mod-A	250	2794.812	61957.081	58.9329	11.1722	1247.269	−0.1195	−0.6851
Mod-A	250	4503.170	92767.048	141.6566	−1.9682	2613.895	−0.8844	−1.3358
Mod-A	250	6402.350	108440.506	192.4238	−47.3681	4222.576	−3.0410	−2.1095
Mod-B	250	895.681	22937.170	2.4705	6.3180	221.228	−0.3487	0.3015
Mod-B	250	2794.450	64102.033	60.3375	21.9089	1288.641	−2.1241	−0.0548
Mod-B	250	4501.326	96433.376	123.4718	30.5782	2734.062	−5.0122	−0.3012
Mod-B	250	6403.458	112014.582	151.6593	20.4326	4486.210	−9.0292	−0.2418
Buffed	250	896.254	34470.526	9.7575	−1.8077	310.366	−0.0878	0.0405
Buffed	250	2798.001	107505.466	145.5975	−27.0804	3084.862	−5.0270	−0.2909
Buffed	250	4501.987	124444.716	190.9446	−64.3202	6186.168	−11.0235	−2.1323
Buffed	250	6402.473	118048.901	161.6557	−64.2132	8357.596	−14.2461	−0.9333

TABLE 9.2 Tire data *205/50R17* (inflation pressure p = 220 kPa).

Tire ID	P [kPa]	F_z [N]	$C_{F\alpha}$ [N/rad]	F_{ply} [N]	F_{con} [N]	$C_{M\alpha}$ [Nm/rad]	M_{ply} [Nm]	M_{con} [Nm]
Normal-2	220	893.712	23483.282	5.6124	8.1455	243.987	−0.3767	0.0765
Normal-2	220	2797.927	64950.011	75.8755	24.5375	1392.026	−1.5745	0.5765
Normal-2	220	4504.486	93467.843	153.3899	45.8147	2819.091	−4.0373	0.9089
Normal-2	220	6401.824	105479.354	186.8229	43.5425	4598.447	−7.6909	1.6770
Mod-A	220	895.803	24371.492	4.8851	7.3141	269.491	−0.1691	0.2962
Mod-A	220	2794.812	66148.689	75.3676	6.4876	1456.370	−0.5280	−0.5786
Mod-A	220	4503.170	94367.334	156.5831	-1.9914	2966.319	−1.8367	−1.4417
Mod-A	220	6402.350	105878.120	202.1747	-51.9994	4678.644	−4.5881	−1.6705
Mod-B	220	895.681	24364.627	6.4895	6.5316	247.822	−0.4850	0.3223
Mod-B	220	2794.450	68293.641	76.7722	17.2242	1497.741	−2.5326	0.0517
Mod-B	220	4501.326	98033.662	138.3983	30.5551	3086.486	−5.9645	−0.4071
Mod-B	220	6403.458	109452.196	161.4103	15.8013	4942.278	−10.5762	0.1972
Buffed	220	896.254	35897.983	13.7765	−1.5941	336.960	−0.2241	0.0612
Buffed	220	2798.001	111697.074	162.0321	−31.7651	3293.962	−5.4355	−0.1844
Buffed	220	4501.987	126045.001	205.8712	−64.3434	6538.593	−11.9758	−2.2382
Buffed	220	6402.473	115486.516	171.4066	−68.8444	8813.663	−15.7931	−0.4942

FIGURE 9.1 Tire property variation with load, front tires (p = 250 kPa). The piecewise interpolation (dotted line) is used to compute tire characteristics in the neighborhood of F_z = 4500 N.

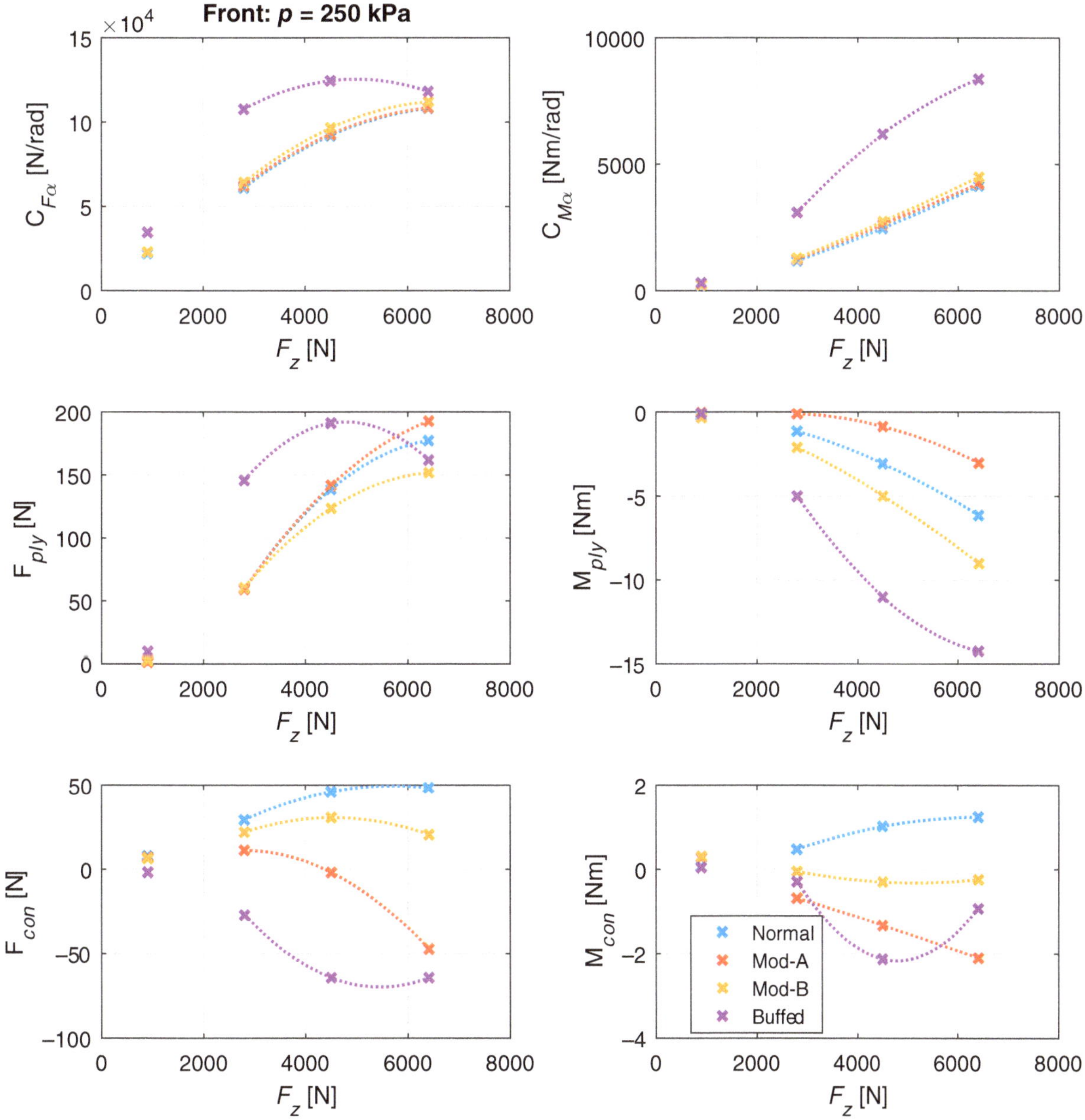

FIGURE 9.2 Tire property variation with load, rear tires (p = 220 kPa). The piecewise interpolation (dotted line) is used to compute tire characteristics in the neighborhood of F_z = 2800 N.

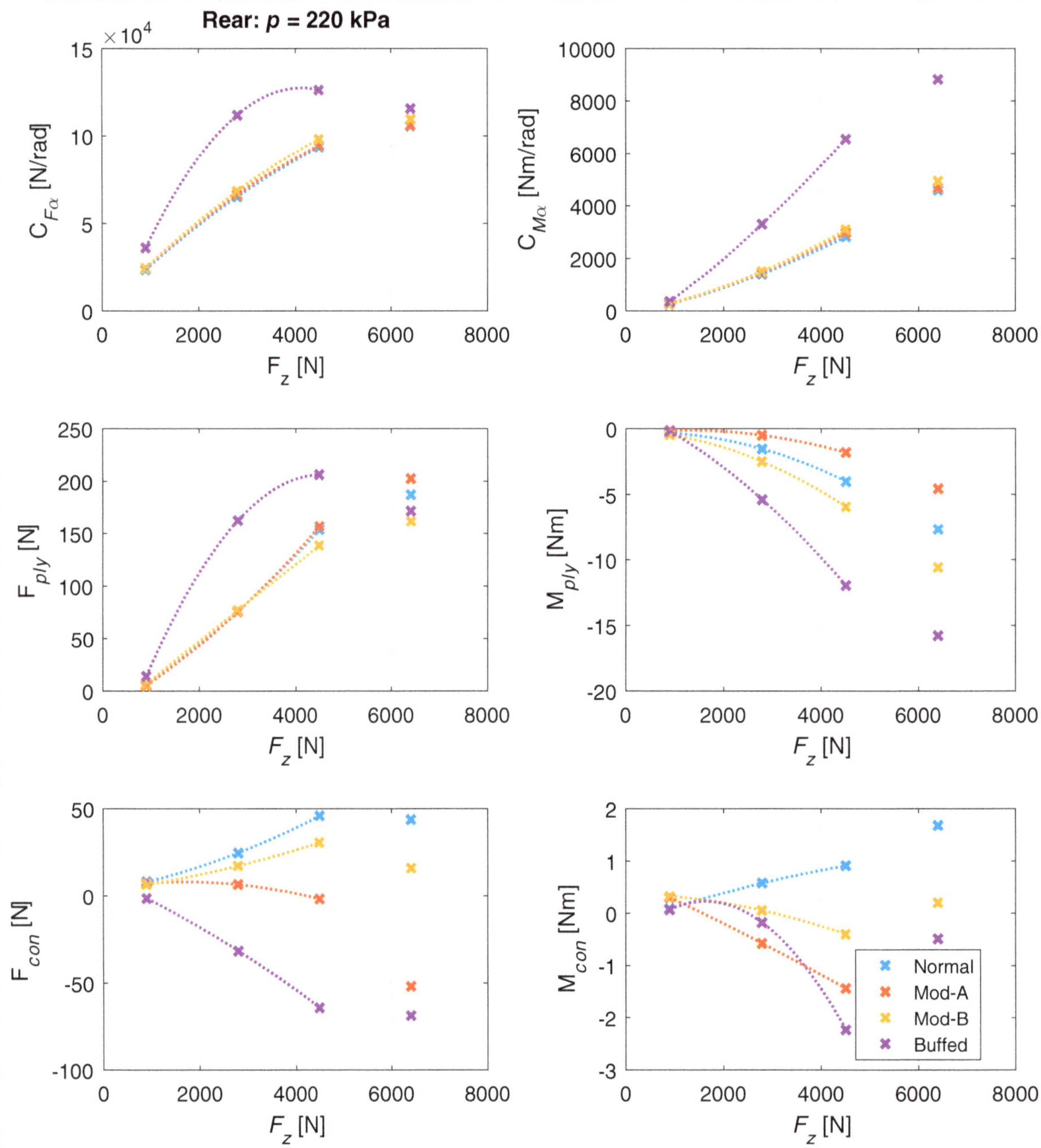

9.2 Load Conditions

The different types of tires were tested under four different loading conditions

- $F_z = 4500$ N. Front axle tire, nominal load
- $F_z = 2800$ N. Rear axle tire, nominal load
- $F_z = 6400$ N. Front axle tire, with load transfer during a braking event
- $F_z = 900$ N. Rear axle tire, with load transfer during a braking event

9.2.1 Polynomial Interpolation of Data with Small Load Variations

The characteristics are approximated in the neighborhood of the front and rear nominal loads, with a piecewise interpolation of the data, with 2nd order polynomials passing through three points

- For front axle tire (in the neighborhood of nominal load $F_z = 4500\ N$) polynomial passing through
  - data at $F_z = 2800$ N
  - data at $F_z = 4500$ N
  - data at $F_z = 6400$ N
- For rear axle tire (in the neighborhood of nominal load $F_z = 2800\ N$) polynomial passing through
  - data at $F_z = 900$ N
  - data at $F_z = 2800$ N
  - data at $F_z = 4500$ N

The 2nd order approximation formulae for the different tire characteristics are reported in Equations 9.1–9.4.

Actually, as reported in [65], cornering stiffness $C_{F\alpha}$ and aligning moment stiffness $C_{M\alpha}$ should be approximated by a parabola passing through the origin: indeed, with null vertical load, no lateral force is expected. Nevertheless, the polynomial approximation is intended to be used to compute the coefficient values for small load variations around the data points. Therefore, the best fit of the available data is pursued rather than a realistic description of the behavior.

The 2nd order polynomial is selected, since it ensures the passage through the three data points for the piecewise interpolation, without increasing unnecessarily the complexity of the problem.

$$C_{F\alpha,i0\,L/R} = p_{Cf,i0} + p_{Cf,i1} \cdot \left(F_{z,iL/R}\right) + p_{Cf,i2} \cdot \left(F_{z,iL/R}\right)^2 \tag{9.1}$$

$$C_{M\alpha,i0\,L/R} = p_{Cm,i0} + p_{Cm,i1} \cdot \left(F_{z,iL/R}\right) + p_{Cm,i2} \cdot \left(F_{z,iL/R}\right)^2 \tag{9.2}$$

$$F_{y0,iL/R} = p_{Fy,i0} + p_{Fy,i1} \cdot \left(F_{z,iL/R}\right) + p_{Fy,i2} \cdot \left(F_{z,iL/R}\right)^2 \tag{9.3}$$

$$M_{z0,iL/R} = p_{Mz,i0} + p_{Mz,i1} \cdot \left(F_{z,iL/R}\right) + p_{Mz,i2} \cdot \left(F_{z,iL/R}\right)^2 \tag{9.4}$$

9.3 Subjective Feedback

Different sets of tires have been tested by an automotive OEM on a vehicle having load distribution according to the nominal loads of the tire data, on a rather flat road test surface. The professional test driver reported in a subjective evaluation that among the different sets

- **Mod-A** was the worst performing one, both in terms of
  - *steering pull*: the driver noticed the largest rightwards deviation from straight path in a free control test.
  - *on-center handling*: the driver experienced markedly different steering "feels" (feedback) when turning left and right.
- **Mod-B** was the best performing one, both in terms of
  - *steering pull*: the driver noticed the smallest deviation from straight path in a free control test.
  - *on-center handling*: the driver experienced similar steering wheel feedback when turning left and right.

CHAPTER 10

Single-Track Model Validation

The single-track (or bicycle) model, equipped with tires with linear characteristics introduced in the previous chapters, is validated by means of a full-vehicle model (represented in Figure 10.1) developed with the multibody dynamics simulation software *MSC Adams*™. The model parameters are reported in Appendix A.9 in Tables A.3 and A.4. In particular, the mass and the weight distribution are modeled to replicate the same nominal tire loads of the tire data available (see Chapter 9).

10.1 *MSC Adams*™ Tire Characteristics

The tire property files of the *MSC Adams*™ full-vehicle model for the eight different types of tires (*Normal, Mod-A, Mod-B* and *Buffed*; front and rear) have been created using the *PAC2002 Tire Data and Fitting Tool (TDFT)* of *MSC Adams*™*/Tire*. This tool allows to compute *PAC2002*[1] tire model parameters using data (from experimental measurements or virtual tests) [66].

In this case, the tool has been used to compute realistic tire parameters for steady-state pure lateral slip, i.e., lateral force F_y and aligning moment M_z with respect to side

[1] *PAC2002* is the latest version of a Magic-Formula tire model available in *MSC Adams*™*/Tire*, developed according to [7].

MATLAB® programs are provided online on the institutional repository of *Politecnico di Milano* (Technical University of Milan) https://re.public.polimi.it/handle/11311/1132512#.XmFR06hKjQC. They perform the comparison between the single-track model and the full-vehicle model results pertaining to the *null steer* test presented in this chapter. Details on such programs can be found at the link https://www.mathworks.com/matlabcentral/fileexchange/74425-straight-motion-of-road-vehicles.

FIGURE 10.1 Full-vehicle model in *MSC Adams*™.

slip angle α. The linear data for different tires (see Chapter 9) have been used as virtual test values, together with additional fictitious points outside the linear region from a default *PAC2002* tire model provided in *MSC Adams*™. This operation has been performed in order to obtain tire property files having both realistic coefficients and the desired linear characteristics and shifts, corresponding to the values of data in the corresponding range of slip angle.

PAC2002 TDFT uses the ISO convention for tire characteristics (see Appendix A.2); therefore the data for the linear characteristics have been provided to the tool, transformed according to this sign convention. Moreover, different data for left and right tires have been used to differentiate left and right shifts F_{y0} and M_{z0}, and the side (*'LEFT'* or *'RIGHT'*) has been specified in the resulting tire property files.

An example of the curves obtained for front *Normal* left tire is represented in **Figure 10.2**. The values for the different sets are reported in **Table 10.1**.

It is underlined that not just the shifts, but also left and right cornering stiffness $C_{F\alpha}$ are different for the same tire type. This feature, obtained as a consequence of the fitting of fictitious data, has been maintained in order to allow the study of the effect of unavoidable small manufacturing tolerances. Independent of manufacturing tolerances, different values for left and right tire's cornering stiffness are obtained when road cross slope φ_r is introduced. This happens also in case left and right tire's cornering stiffness are equal at nominal load, and is due to the induced static load transfer, as discussed in Chapter 6.

Longitudinal and combined slip parameters could not be obtained from available data. Thus, they have been left unmodified with respect to the ones of a default *PAC2002* tire model provided in *MSC Adams*™.

10.2 Null Steer Test: Trajectory Comparison with Null Road Cross Slope $\varphi_r = 0$

A null steer test (as described in Chapter 5) is performed both with the single-track model in *MATLAB*® and with the full-vehicle model in *MSC Adams*™, to compare the resulting trajectories. The initial velocity is set to u = 100 km/h and the steering input is fixed at $\delta = 0$. Initially, the road cross slope is set to $\varphi_r = 0$, simulating a test run on a flat and smooth track. The powertrain is not included: a free rolling analysis is conducted, with the aim of removing the influence of longitudinal forces.

The longitudinal velocity at the different time instants of the full-vehicle simulation is provided as an input to the single-track model.

All the subsystems of the full-vehicle model are set to *kinematic mode* instead of *compliant mode*, in order to exclude the compliance of bushings and other components, that are not taken into account in the single-track model. Moreover, static toe Ψ_{i0} and camber γ_{i0} angles are set to null value both for front and rear wheels.

FIGURE 10.2 *ADAMS* tire properties, *PAC2002 TDFT* resulting curves for front left (1*L*) tires with F_z = 4500 N. It is possible to compare values of lateral force and aligning moment at α = 0, respectively, with values in 'F_{y0} left' and 'M_{z0} left' columns of Table 10.1.

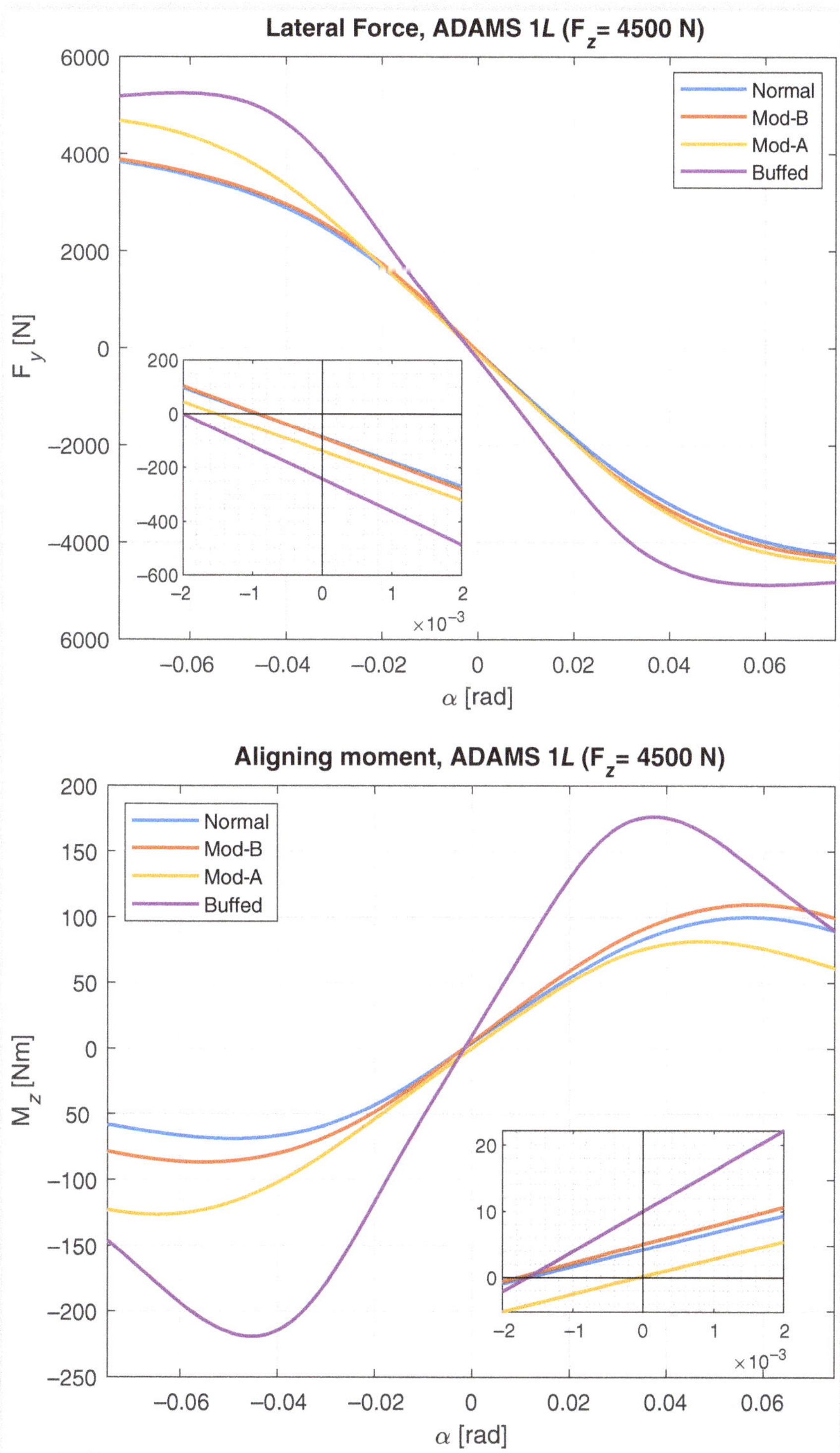

TABLE 10.1 *ADAMS* tire characteristics values, properties created with *PAC2002 TDFT*.

Tire ID	F_z [N]	$C_{F\alpha}$ left [N/rad]	$C_{F\alpha}$ right [N/rad]	$C_{M\alpha}$ left [Nm/rad]	$C_{M\alpha}$ right [Nm/rad]	F_{y0} left [N]	F_{y0} right [N]	M_{z0} left [Nm]	M_{z0} right [Nm]
Normal	4500	91658.7	92143.8	2617.2	2595.6	88.3	182.5	−4.17	−2.92
Mod-A	4500	91031.8	93331.5	2654.9	2736.7	140.6	137.4	−0.15	−2.99
Mod-B	4500	96147.8	95790.1	2829.2	2818.1	90.6	148.5	−5.02	−5.52
Buffed	4500	121000.0	119950.0	6081.9	6033.6	243.9	123.9	−9.99	−11.52
Normal	2800	68346.2	68480.5	1393.3	1414.9	51.1	107.9	−1.94	−1.15
Mod-A	2800	67910.0	69813.3	1463.2	1475.7	68.1	86.2	−0.08	−0.82
Mod-B	2800	70603.1	72209.6	1536.5	1528.7	59.8	95.5	−2.62	−2.31
Buffed	2800	116610.0	115610.0	3686.6	3752.5	166.1	135.7	−3.85	−7.15

10.2.1 Results: Full-Vehicle Decelerating and Single-Track Model not Accounting for Roll

In the first simulation, the single-track model does not account for any elastokinematics nor load transfer or body roll, namely $C^*_{eff,i} = \left(C_{F\alpha,i0L} + C_{F\alpha,i0R}\right)$. The resulting trajectories are represented in **Figure 10.3**. It can be observed that the lateral drifts of the trajectories simulated with *MATLAB*® single-track model are overestimated with respect to the ones obtained with the full-vehicle model, with all the tire sets.

FIGURE 10.3 Comparison between the trajectories simulated by single-track model and full-vehicle model for a null steer test. Full-vehicle decelerating, single-track model not accounting for roll. Data in Tables 10.1, A.1-A.4.

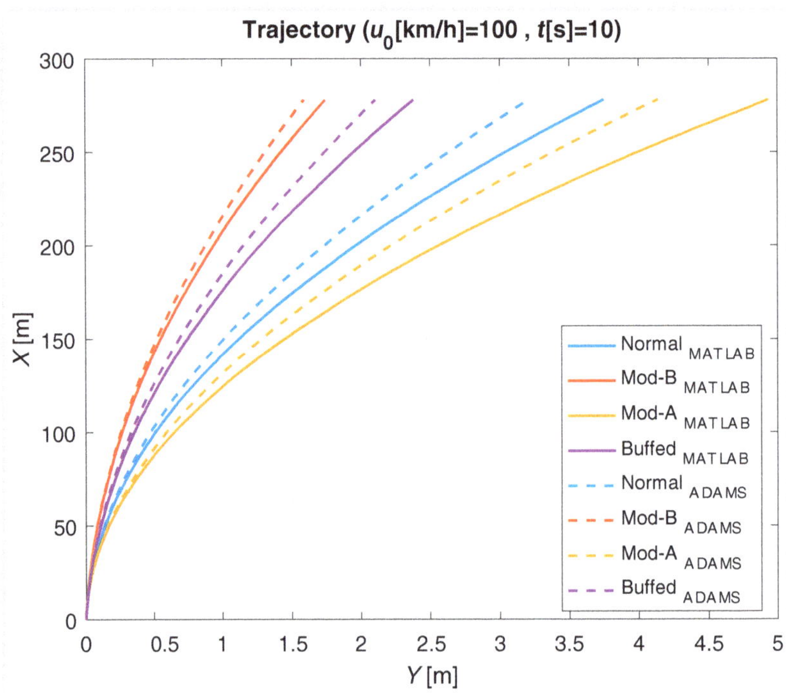

FIGURE 10.4 Longitudinal velocity time history, full-vehicle model simulation during a null steer test.

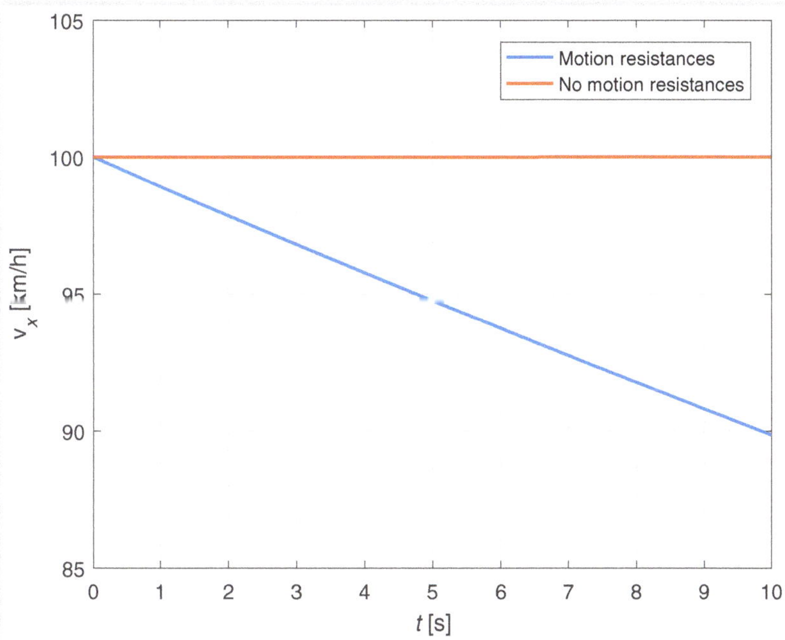

This discrepancy can be explained by two main differences between the single-track and the full-vehicle model

- **Deceleration of full-vehicle model**. The full-vehicle model decelerates, mainly due to rolling resistance and aerodynamic resistance, which is significant at 100 km/h. The longitudinal velocity time history is reported in Figure 10.4. It cannot be considered quasi-static for the model, meaning that the resulting deceleration causes a small load transfer from the rear toward the front axle in the full-vehicle model. This is not taken into account by the single-track model, which does not include pitch dynamics. The normal load variation modifies tire characteristics in the full-vehicle model, while the same variation is not described by the single-track model.
- **Body roll**. The single-track model – without effective axle cornering stiffness – does not account for suspension elastokinematics, load transfer, and body roll. Depending on vehicle and tire characteristics, a small roll angle can cause a reduction of the effective axle side force.

10.2.2 Results: Full-Vehicle not Decelerating and Single-Track Model Accounting for Roll

A second set of simulations is performed, implementing modifications aimed at reducing the aforementioned differences between single-track model and full-vehicle model

- **Deceleration of full-vehicle model**. All the motion resistances of the *MSC Adams*™ full-vehicle model are removed. Aerodynamic resistance is nullified by setting *air density* = 0. Rolling resistance is canceled by assigning null value to the scale factor of rolling resistance torque ($LMY = 0$) in all tire property files. With these changes, the longitudinal velocity is constant during the simulation, as it can be observed in Figure 10.4.

- **Body roll**. Roll contribution can be included in the single-track model as explained in Chapter 2, by considering the effective axle cornering stiffness, that is defined in Equation 2.30 and reported here as

$$C^*_{eff,i} = \frac{2C_{F\alpha,i0}}{1 + C_{roll,i} - C_{susp,i} + C_{steer,i} + C^*_{\Delta F_{z,i}}}$$

Since the compliances are deactivated in the full-vehicle model, the terms related to steering system and suspension elasticities are null ($C_{susp,i} = 0$, $C_{steer,i} = 0$). The remaining terms are related to roll motion and load transfer and have been introduced in Chapter 2

$$C_{roll,i} = \frac{2l}{l - a_i} \cdot \frac{h'}{k_{\varphi 1} + k_{\varphi 2} - mgh'} \left(\varepsilon_i C_{F\alpha,i0} + \tau_i C_{F\gamma,i0}\right)$$

$$C^*_{\Delta F_{z,i}} = \sigma_i \frac{l}{l - a_i} \left[\xi_{\alpha,i}\left(2\Psi_{i0} - \alpha_{ply,iL} + \alpha_{ply,iR}\right) + \xi_{\gamma,i}\left(2\gamma_{i0} - \gamma_{con,iL} + \gamma_{con,iR}\right)\right]$$

The equivalent angles for conicity and ply-steer are computed from tire data as discussed in Chapter 1, according to Equation 1.2.

Some vehicle parameters are estimated from the front and rear suspension assemblies in *MSC Adams*™*/Car*, represented, respectively, in **Figure 10.5** and **Figure 10.6**, using the *Roll & Vertical Force Analysis*

- Rolling stiffness ($k_{\varphi 1}$, $k_{\varphi 2}$) is estimated from the *total_ roll_ rate* result of suspension analysis, which is *"the torque, applied as vertical forces at the tire contact patches, per degree of roll, measured at the tire contact patches"* [67].
- Roll center height (h_1, h_2) is estimated from the result of suspension analysis *roll_ center_location_vertical*, as its value at null roll angle $\varphi = 0$.
- Roll-steering coefficient (ε_1, ε_2) is estimated from the result of suspension analysis *roll_steer*, as its value at null roll angle $\varphi = 0$. Since *MSC Adams*™ adopts the ISO sign convention (with *z-axis* upwards), ε_i value is opposite for the convention used in this book (with *z-axis* downwards).
- Roll-camber coefficient (τ_1, τ_2) is estimated from the result of suspension analysis *roll_camber_coefficient*, as its value at null roll angle $\varphi = 0$. Since the *x-axis* is directed frontwards for both the ISO convention (*MSC Adams*™) and the convention used in this book, the value is the same.

FIGURE 10.5 *McPherson* front suspension assembly.

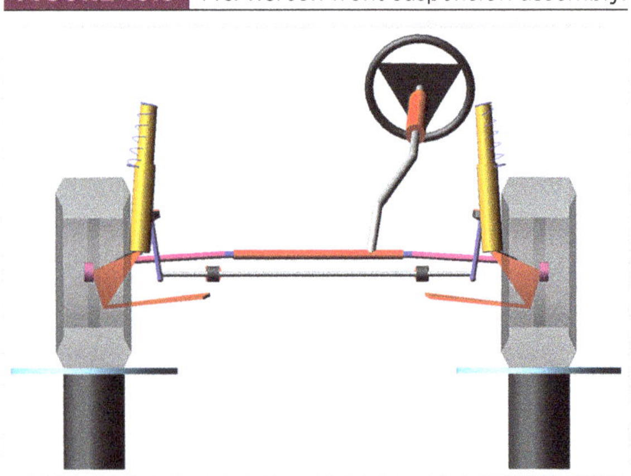

FIGURE 10.6 *Multi-link* rear suspension assembly.

The results of suspension analyses are presented in **Figures 10.7** and **10.8**, and the values of the listed parameters are reported in Appendix A.9 in Table A.4. The camber

FIGURE 10.7 *MSC Adams*™ front suspension *Roll & Vertical Force Analysis* results.

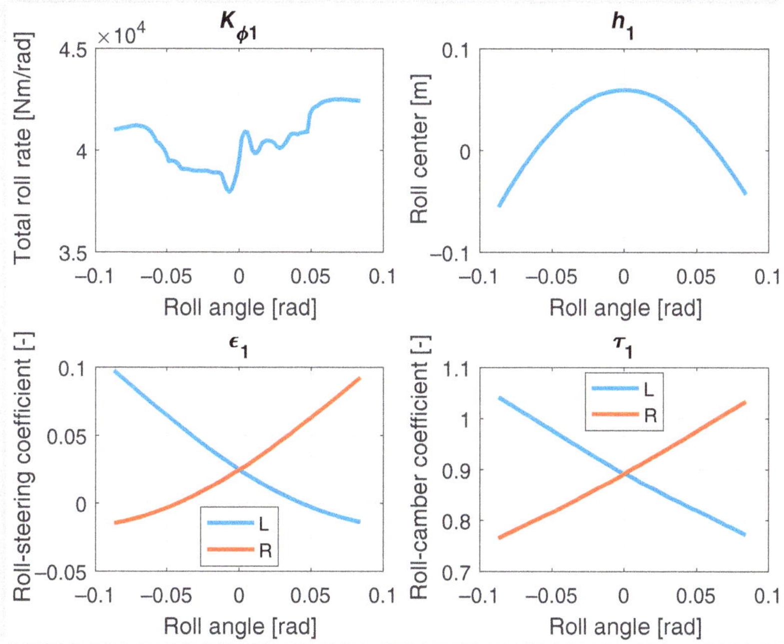

FIGURE 10.8 *MSC Adams*™ rear suspension *Roll & Vertical Force Analysis* results.

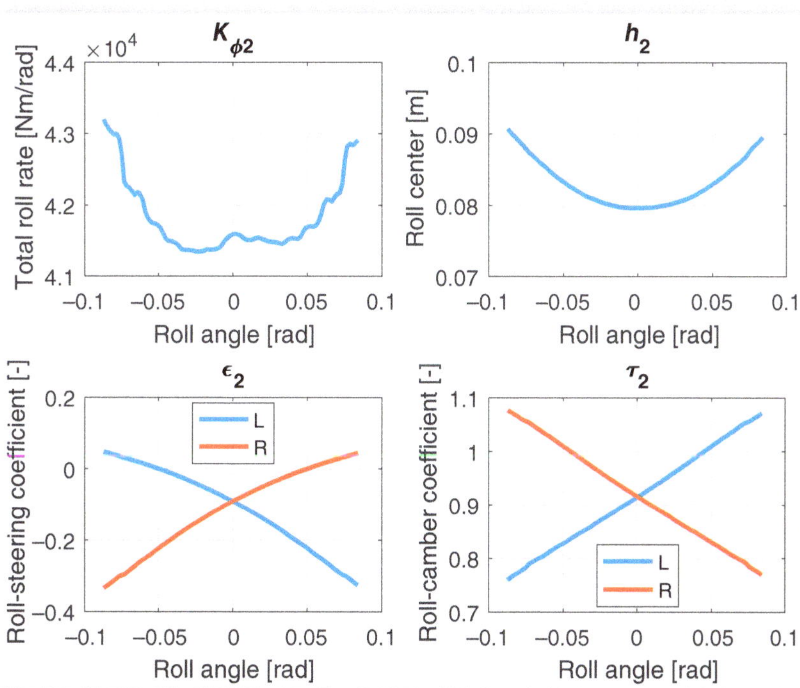

coefficient's linear dependence on load is set to $\xi_{y,i} = 1$. The cornering stiffness's linear dependence on load $\xi_{\alpha,i}$, on the other hand, is obtained from the piecewise interpolation of tire characteristics, discussed in Chapter 9, according to

$$\xi_{\alpha,i} = \left(\frac{\partial C_{F\alpha,i}}{\partial F_{z,i}} \bigg|_{F_{z0}} \right) = p_{Cf,i1} + 2 \cdot p_{Cf,i2} \cdot \frac{F_{z,iL,stat} + F_{z,iR,stat}}{2} \tag{10.1}$$

Moreover, the results of $\xi_{\alpha,i}$ – computed from data according to Equation 10.1 – are compared to the values obtained from *MSC Adams*™ tire property files according to the *PAC2002* cornering stiffness formula [66]

$$\xi_{\alpha,i} = \frac{P_{Ky1} \cdot F_{z0} \cdot \left\{ sin\left[2arctan\left(\frac{F_{z2}}{P_{Ky2} \cdot F_{z0}} \right) \right] - sin\left[2arctan\left(\frac{F_{z1}}{P_{Ky2} \cdot F_{z0}} \right) \right] \right\}}{F_{z2} - F_{z1}} \tag{10.2}$$

where F_{z1} and F_{z2} are [4500 ± 5 N] for front tires and [2800 ± 5 N] for rear tires, while the other parameters are the Magic Formula parameters defined in tire property files [7, 66]. The results obtained with the two approaches (Equations 10.1 and 10.2) are in agreement, showing consistency between the available experimental tire data and the created *MSC Adams*™ tire properties.

The results of the simulations are shown in **Figure 10.9** and reported in **Table 10.2**. The trajectories obtained with the simple single-track model match the results simulated with the full-vehicle model.

FIGURE 10.9 Comparison between the trajectories simulated by single-track model and full-vehicle model for a null steer test. Full-vehicle not decelerating, single-track model accounting for roll. Data in Tables 10.1, A.1–A.4.

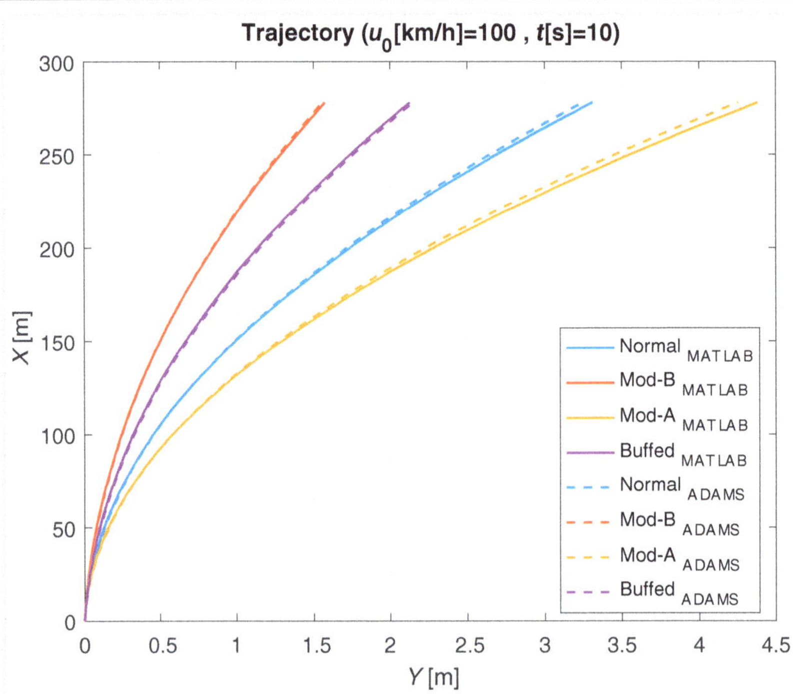

TABLE 10.2 Results of the null steer test trajectories simulated with single-track model with the four sets of tires. δ_{st} is the steering angle that would be necessary to run straight ahead, according to Equation 6.11. Data in Tables 10.1, A.1-A.4.

Tire set	Δy drift at $x = 100m$ [m]	δ_{st} [10^{-4} rad]	$M^{*}_{S,st}$ [Nm]
Normal	0.456	-3.151	-0.605
Mod-A	0.585	-4.259	-5.135
Mod-B	0.242	-1.379	3.606
Buffed	0.320	-2.129	3.138

As it can be inferred from results of Table 10.2, the tire set *Mod-B* is the best performing one in the null steer test, while *Mod-A* is the worst one, showing the largest rightwards path deviation. The straight-driving steering angle δ_{st} (computed according to Equation 6.11) is the steering angle that would be needed to run straight ahead. The straight-driving steering torque $M^{*}_{S,st}$, on the other hand, does not have any influence on the trajectory in a fixed or null steer test: it affects the torque M_S that the driver has to apply on the steering wheel, and becomes relevant in a free control test.

10.3 Null Steer Test: Trajectory Comparison with Road Cross Slope φ_r = 0.02 rad

The road cross slope is set to $\varphi_r = 0.02$ rad, in order to validate the model also with load transfer and roll induced by road cross section inclination.

10.3.1 *MSC Adams*™ Road Model

A 3D road model is created in *MSC Adams*™*/Car* with the specified lateral inclination parameter *BANK* = 0.02 rad. Moreover, since the single-track model has a constant road cross slope parameter as an input, the *MSC Adams*™ road is modeled so that its centerline is nearly following the vehicle's drift trajectory, as represented in **Figure 10.10**. By doing so, the full-vehicle model is subjected to an approximately constant road cross slope throughout its motion, like the *MATLAB*® single-track model, and does not experience any longitudinal slope. That would cause both a longitudinal velocity variation and a load transfer between rear and front axle. Even though the weight of these effects would be small, the aim of the road model described is to minimize the discrepancies between the inputs provided to the full-vehicle model and to the single-track model.

10.3.2 Single-Track Model Adaptations

Tire characteristics for the *MATLAB*® single-track model are computed at the actual wheel normal load, with road cross slope-induced static load transfer (computed as specified in Equation 6.4).

FIGURE 10.10 *MSC Adams*™ road model with centerline following vehicle's drift trajectory.

Moreover, an additional term is introduced in the single-track model for the axle side force $F_{y,i}$: static vehicle roll $\varphi_{rv} = 0.0016$ rad $\simeq 0.1$ deg, which is induced by road cross slope $\varphi_r = 0.02$ rad and is computed as in Equation 6.2, causes a static inclination of the tires, computed as

$$\gamma_{roll,i} = \tau_i \cdot \varphi_{rv} \tag{10.3}$$

where $\tau_i = \frac{\gamma_{r,i}}{\varphi}$ is the roll-camber coefficient. Thus, each *i-axle* can account for an additional lateral force contribution, given by the normal load projection (since the tire lateral force camber coefficient is $C_{F_y} = F_z$)

$$F_{y,i\ tot} = F_{y,i} +_{roll,i} \cdot F_{zi} \tag{10.4}$$

10.3.3 Results: Road Cross Slope Effect

The comparison between trajectories simulated with the full-vehicle model in *MSC Adams*™ and the ones obtained with the single-track model is shown in **Figure 10.11**, while results are reported in **Table 10.3**. The matching between the results of full-vehicle and single-track model holds also for the case with road cross slope introduction.

From results of Table 10.3, it can be argued that as in the case with null road cross slope $\varphi_r = 0$, the tire set *Mod-B* is the best performing one in the null steer test, while

FIGURE 10.11 Comparison between the trajectories simulated by single-track model and full-vehicle model for a null steer test, with road cross slope $\varphi_r = 0.02$ rad. Data in Tables 10.1, A.1-A.4.

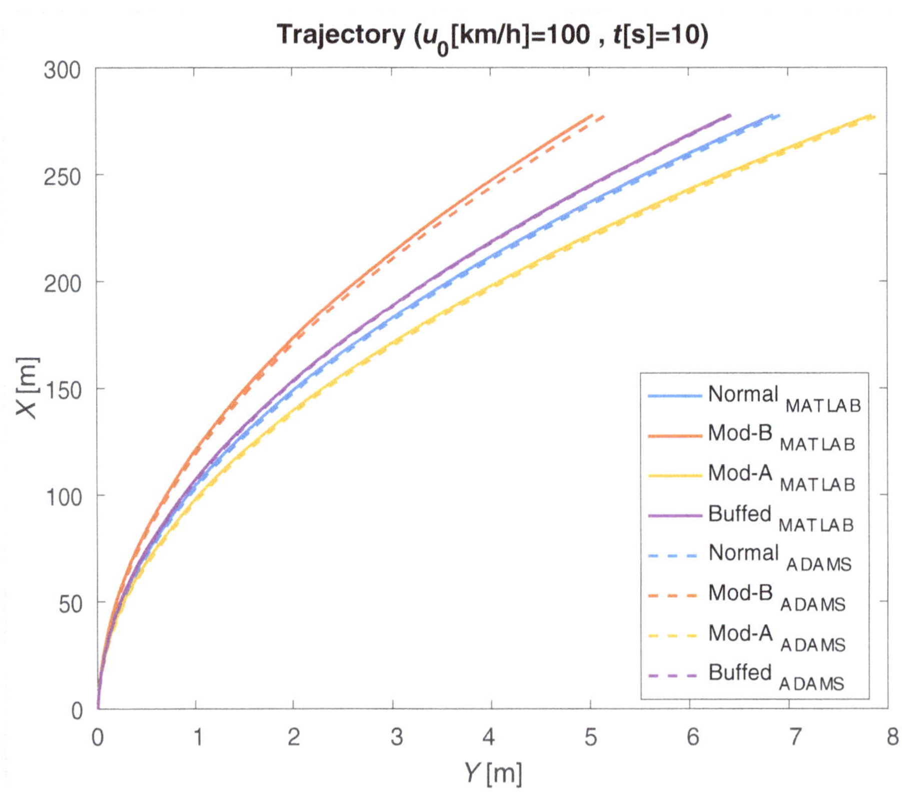

TABLE 10.3 Results of the null steer test trajectories simulated with single-track model with the four sets of tires, with road cross slope φ_r = 0.02 rad. δ_{st} is the steering angle that would be necessary to run straight ahead, according to Equation 6.11. Data in Tables 10.1, A.1–A.4.

Tire set	Δy drift at $x = 100m$ [m]	δ_{st} [10^{-4} rad]	$M^*_{S,st}$ [Nm]
Normal	0.923	-6.589	-9.958
Mod-A	1.042	-7.626	-14.695
Mod-B	0.699	-4.715	-6.069
Buffed	0.877	-6.838	-10.763

Mod-A is the worst one. The straight-driving steering angle δ_{st} (computed according to Equation 6.11) is the steering angle that would be needed to run straight ahead.

Among the four tire sets, the *Buffed* shows a different significance in the variation of the trajectory with the introduction of road cross slope, although the trend is coherent. This aspect can be explained considering that *Mod-A* and *Mod-B* tire sets are derived from the *Normal* one, therefore these three sets have characteristics with similar order of magnitude. The *Buffed* set, on the other hand, has characteristics deviating consistently from the ones of the other three sets.

10.4 Discussion

The results obtained in this chapter are consistent with the subjective evaluation of the different tire sets: both with and without road cross slope, the vehicle equipped with *Mod-A* tire set shows the largest lateral deviation from the straight path, while with the one with *Mod-B* tire set shows the smallest deviation.

Moreover, the direct link between the straight motion steering angle δ_{st} and the lateral drift is pointed out for a null steer test, while the steering torque offset $M^*_{S,st}$ does not have an influence on the trajectory for this kind of test. This aspect can be observed both in Tables 10.2 and 10.3.

Finally, the comparison between the results in Figures 10.3 and 10.9 highlights that the suspension and steering system elastokinematics has to be taken into account in order to have an agreement between the full-vehicle simulations and the single-track model and analytical results: indeed, elastokinematics modifies the effective axle side force, which is a key parameter for the trajectory. When the effective axle side force is considered, satisfying matching is obtained between the *MSC Adams*™ full-vehicle model simulations and the *MATLAB*® single-track model results, both for the case without road cross slope (see Figure 10.9) and with road cross slope (see Figure 10.11). In case the elastokinematics is not considered, the results of the single-track model show correct trend with respect to the full-vehicle simulations, but the description is not quantitatively accurate, as represented in Figure 10.3. In his analysis [17], Lee observed variations in the trajectory simulated with a full-vehicle model when modifying the suspension spring stiffness, while changes to the damping coefficients did not show meaningful effect. He did not proceed with the investigation on that aspect, but his results actually can be explained in light of the results presented in this chapter, considering the elastokinematics influence on the trajectory: spring stiffness is directly affecting the roll rate parameter $k_{\varphi i}$, which is involved in the computation of the coefficients $C_{roll,i}$ and σ_i for the effective axle characteristics, employed in this chapter.

CHAPTER 11

Straight-Driving Analytical Solutions Validation

In this chapter, the full-vehicle model in *MSC Adams*™ is employed to validate the analytical expressions for side slip angles (α_1: Equation 6.6, α_2: Equation 6.7), vehicle attitude (β: Equation 6.12), and steering torque (M_S^*: Equation 6.15) for the straight-driving condition, namely when $a_y = 0$.

The validation is carried out by means of simulations of *Straight-Line event* of the type *Maintain* in *MSC Adams*™, selecting the *steering input* option *straight line*. The simulations are performed on the full-vehicle model equipped with the *Normal* set of tires, running at a constant velocity u = 100 km/h, both with null and nonnull road cross slope φ_r.

The analysis is conducted by comparing the analytical results with the final value assumed by five outputs of the *MSC Adams*™ simulations, after the initial transient part is completed and the vehicle is running straight

- **Front slip angles**. The slip angles of left and right front tires resulting from *MSC Adams*™ simulation are compared with the front axle's straight-motion slip angle expressed in Equation 6.6

$$\alpha_1 = -K_{\alpha 1} \cdot \varphi_r + \alpha_{1,st}$$

- **Rear slip angles**. The slip angles of left and right rear tires resulting from *MSC Adams*™ simulation are compared with the rear axle's straight motion slip angle expressed in Equation 6.7

$$\alpha_2 = -K_{\alpha 2} \cdot \varphi_r + \alpha_{2,st}$$

- **Lateral velocity.** The lateral velocity assumed by the vehicle during straight-driving is analytically computed as the product between straight motion vehicle attitude and longitudinal velocity

$$v_y = \beta_{st} \cdot u$$

- **Yaw angle.** The yaw angle of the vehicle during straight-driving is compared with the rear axle's straight motion slip angle expressed in Equation 6.7, which is opposite to the straight motion vehicle attitude expressed in Equation 6.12

$$\Psi = -\beta_{st} = \alpha_2$$

- **Steering torque.** The torque on the steering linkage side of the steering gear M_S^*, computed from *MSC Adams*™ simulation results as expressed in Equation 4.1

$$M_S^* = \left(F_{y,1L} + F_{y,1R}\right) \cdot n - \left(M_{z,1L} + M_{z,1R}\right)$$

  is compared with the straight motion analytical expression of $M_{S,st}^*$ of Equation 6.15.

11.1 Results Without Road Cross Slope (φ_r = 0)

The results of the simulation of the full-vehicle model, equipped with *Normal* set of tires on a flat road, are reported in **Figure 11.1**, together with the results of the analytical expressions (dotted lines). After the end of the transient part (caused by the initial conditions on *MSC Adams*™, which dictate null side slip angles), the values from the full-vehicle simulation have a considerable correspondence with the analytical results, as it can be observed in **Table 11.1**.

11.2 Results with Road Cross Slope (φ_r = 0.02 rad)

The results of the simulation of the full-vehicle model, equipped with *Normal* set of tires on a road with cross slope $\varphi_r = 0.02$ rad, are reported in **Figure 11.2**, together with the results of the analytical expressions (dotted lines). After the end of the transient part, the values from the full-vehicle simulation are also in this case substantially consistent with the analytical results, as it can be observed in **Table 11.2**.

FIGURE 11.1 Comparison between straight line maintaining test results in *MSC Adams*™ and analytical results (dotted lines). Road cross slope $\varphi_r = 0$. Vehicle equipped with *Normal* set of tires. Data in Tables 10.1, A.1-A.4.

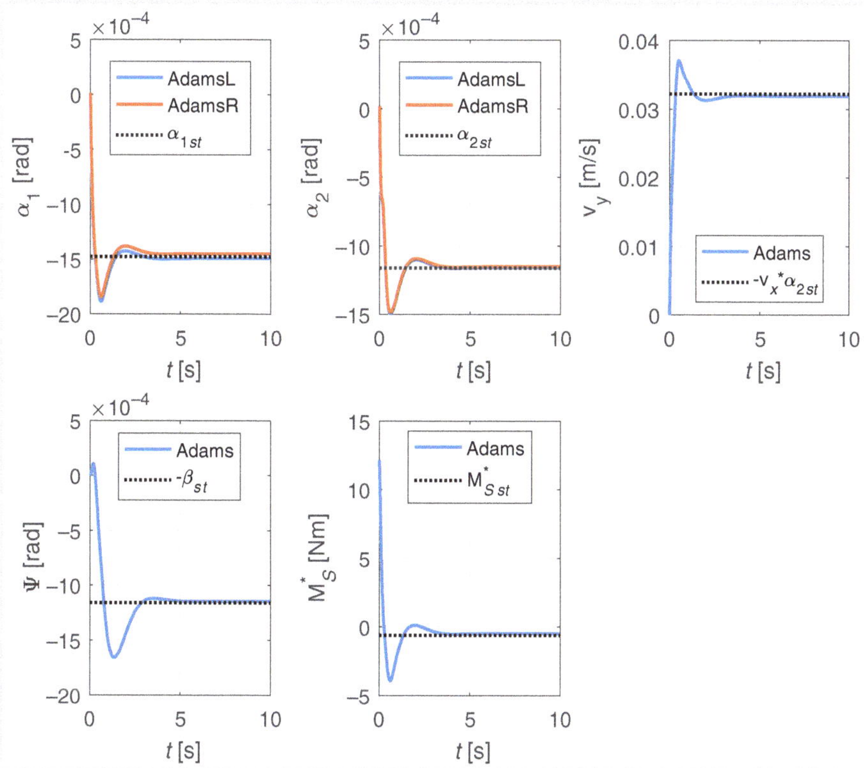

TABLE 11.1 Comparison between straight line maintaining test results in *MSC Adams*™ and analytical results. Road cross slope $\varphi_r = 0$. Vehicle equipped with *Normal* set of tires. Data in Tables 10.1, A.1-A.4.

Parameter	Analytical expression	*MSC Adams*™ result	Analytical result
α_1 [10^{-3} rad]	$-K_{\alpha 1} \cdot \varphi_r + \alpha_{1,st}$	Left: −1.49 Right: −1.45	−1.48
α_2 [10^{-3} rad]	$-K_{\alpha 2} \cdot \varphi_r + \alpha_{2,st}$	Left: −1.16 Right: −1.15	−1.16
v_y [m/s]	$\beta_{st} \cdot u$	0.032	0.032
Ψ [10^{-3} rad]	$-\beta_{st}$	−1.15	−1.16
M_S^*[Nm]	$M_{S,st}^*$	−0.51	−0.60

FIGURE 11.2 Comparison between straight line maintaining test results in *MSC Adams*™ and analytical results (dotted lines). Road cross slope φ_r = 0.02 rad. Vehicle equipped with *Normal* set of tires. Data in Tables 10.1, A.1–A.4.

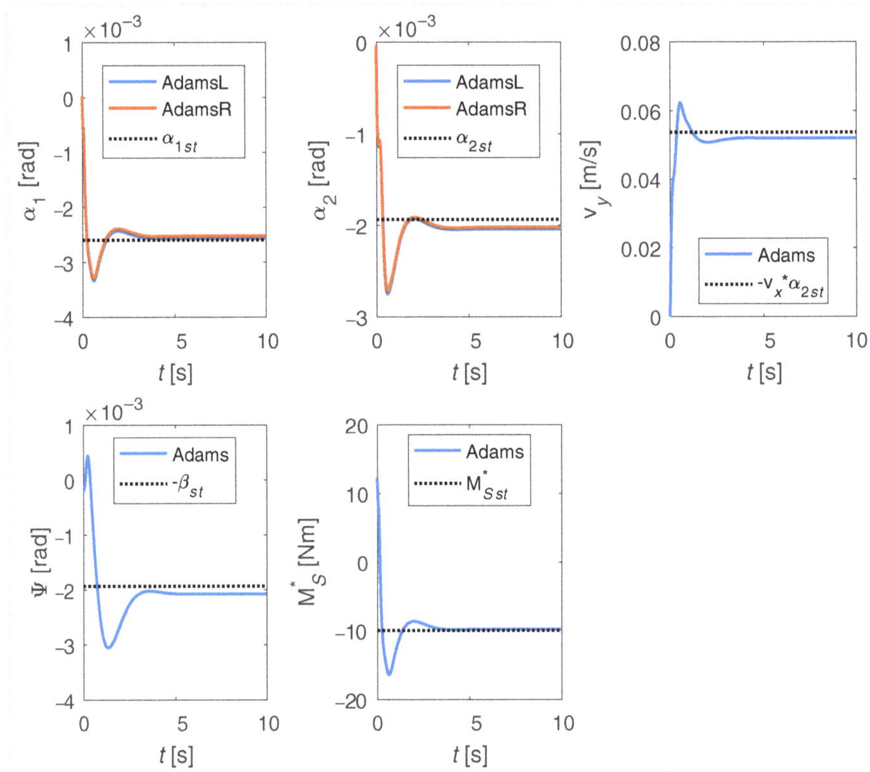

TABLE 11.2 Comparison between straight line maintaining test results in *MSC Adams*™ and analytical results. Road cross slope φ_r = 0.02 rad. Vehicle equipped with *Normal* set of tires. Data in Tables 10.1, A.1–A.4.

Parameter	Analytical expression	*MSC Adams*™ result	Analytical result
α_1 [10^{-3} rad]	$-K_{\alpha 1} \cdot \varphi_r + \alpha_{1,st}$	Left: −2.57 Right: −2.52	−2.59
α_2 [10^{-3} rad]	$-K_{\alpha 2} \cdot \varphi_r + \alpha_{2,st}$	Left: −2.04 Right: −2.02	−1.93
v_y [m/s]	$\beta_{st} \cdot u$	0.052	0.054
Ψ [10^{-3} rad]	$-\beta_{st}$	−2.08	−1.93
M_S^* [Nm]	$M_{S,st}^*$	−9.82	−9.96

11.3 Road Irregularity Effect

In this section, the impact of road irregularity on straight-driving slip angles and chassis lateral displacement y is analyzed.

11.3.1 Variance of Slip Angles for Straight Motion

In order to have an estimate of the robustness of the analytical solutions with respect to road irregularity effect, the standard deviations of straight-driving slip angles are analytically overestimated according to Equation 8.70. This equation refers to the simple vehicle model for road excitation analysis (see Figure 8.5), which is assumed to be suitable to estimate the standard deviation of slip angles: indeed, in the work of Mitschke [51], track shifts and lateral forces induced by road irregularity are claimed to have negligible effects on roll and consequently on load variation. Therefore, the variance of axle slip angles caused by load variation on wheels can be evaluated with the analytical model considering just heave and roll motions, without the lateral coordinate. The computed estimations are then compared to the standard deviations obtained from *MSC Adams*™ full-vehicle simulations. The same *straight line maintaining test* employed for the previous validation is simulated, with different sets of tires on three flat road types:

1. *smooth* road, with no irregularity, corresponding to the one used in the previously described validation, to be used as reference
2. *class B* road, with slight irregularity (see Table 8.1)
3. *class D* road, with sizable irregularity (see Table 8.1)

The full-vehicle simulation results, concerning the front and rear axles' slip angles, are reported, respectively, in **Figure 11.3** and **11.4**.

The comparisons between the results of *MSC Adams*™ full-vehicle simulations and analytical overestimations are presented in **Table 11.3**. As it can be observed, for a fairly smooth road (namely class B), the standard deviation is substantially smaller than the analytical solution of straight-driving slip angle. Moreover, the four-fold proportionality between standard deviations of road class B and class D, which is predicted from analytical formulation, is obtained also with *MSC Adams*™ simulation results. Furthermore, the analytical results always overestimate the standard deviations obtained from the simulations, although the order of magnitude is in most cases comparable. This is in line with the expectations, since the analytical expressions (see Equation 8.70) have been actually obtained as overestimations: the Cauchy-Schwarz inequality has been used, since the covariance of left and right wheel normal loads is analytically unknown. Other reasons for the gap between the results of standard deviations are attributable to the differences between the analytical estimation and the *MSC Adams*™ simulation; for instance, while the former takes into account the whole spectrum, the latter is limited by the time step used (in this case, $\Delta t = 0.005$ s): a trade-off between the time span of the simulation and the number of time steps must be considered. Moreover, the *MSC Adams*™ *Stochastic Uneven 2D road* model allows only to set a constant value for the left and right correlation function (set to 0.6 in this study), while the analytical model

FIGURE 11.3 Results for front axle's side slip α_1, *MSC Adams*™ straight line maintaining test performed on smooth, class B and class D roads.

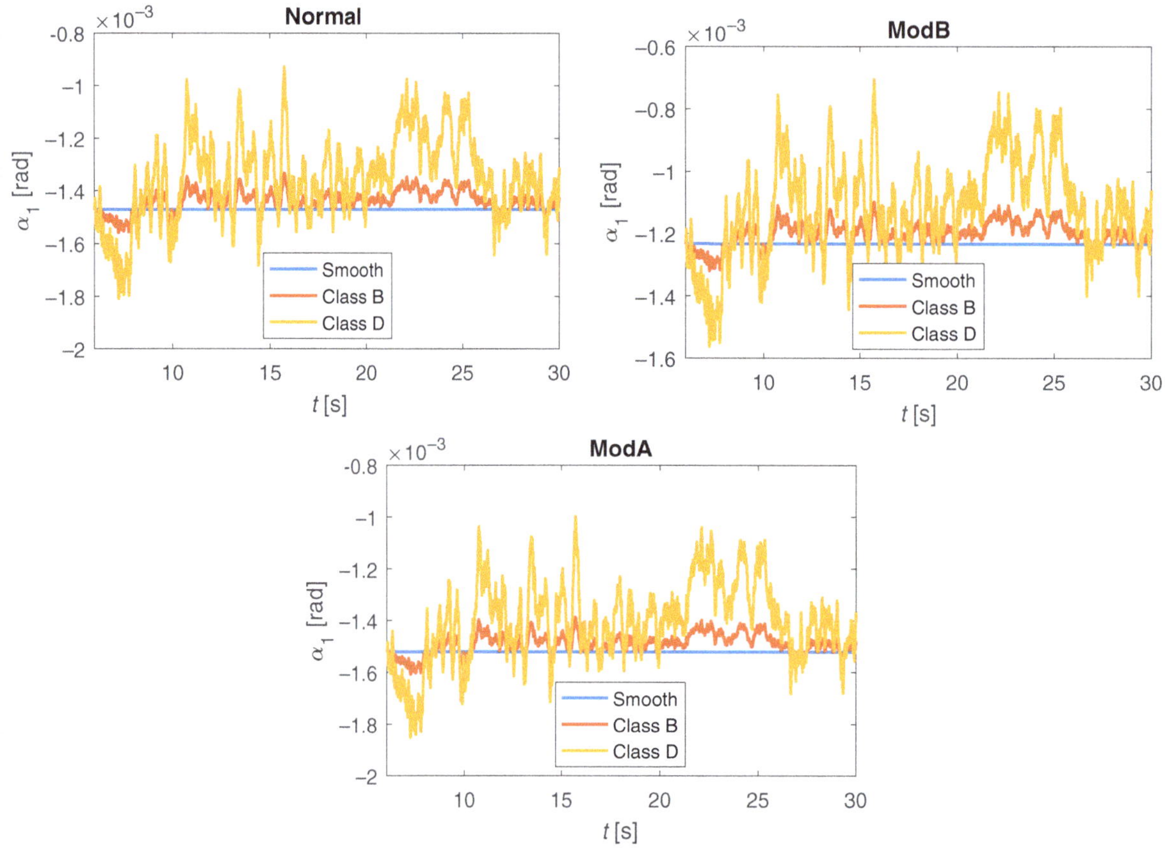

includes the Ammon approximation of coherence function γ_h, varying with ω. Finally, it has to be underlined that substantial simplifications are intrinsic to the analytical model considered in Chapter 8 that for instance neglects – among the others – the internal damping of tires, and considers rigid axles' movements. All these aspects may have severe impact on the estimation of variance.

11.3.2 Variance of Vehicle's Body Lateral Motion

The extended vehicle model depicted in Figure 8.15 is then considered in order to evaluate the effects of road irregularity on chassis lateral displacement y. This model, with respect to the simplified version previously considered, accommodates the description of track shifts induced on wheels by road irregularity. These displacements cause lateral forces, which have to be taken into account to describe the lateral vibration behavior of the vehicle [51].

FIGURE 11.4 Results for rear axle's side slip α_2, *MSC Adams*™ straight line maintaining test performed on smooth, class B and class D roads.

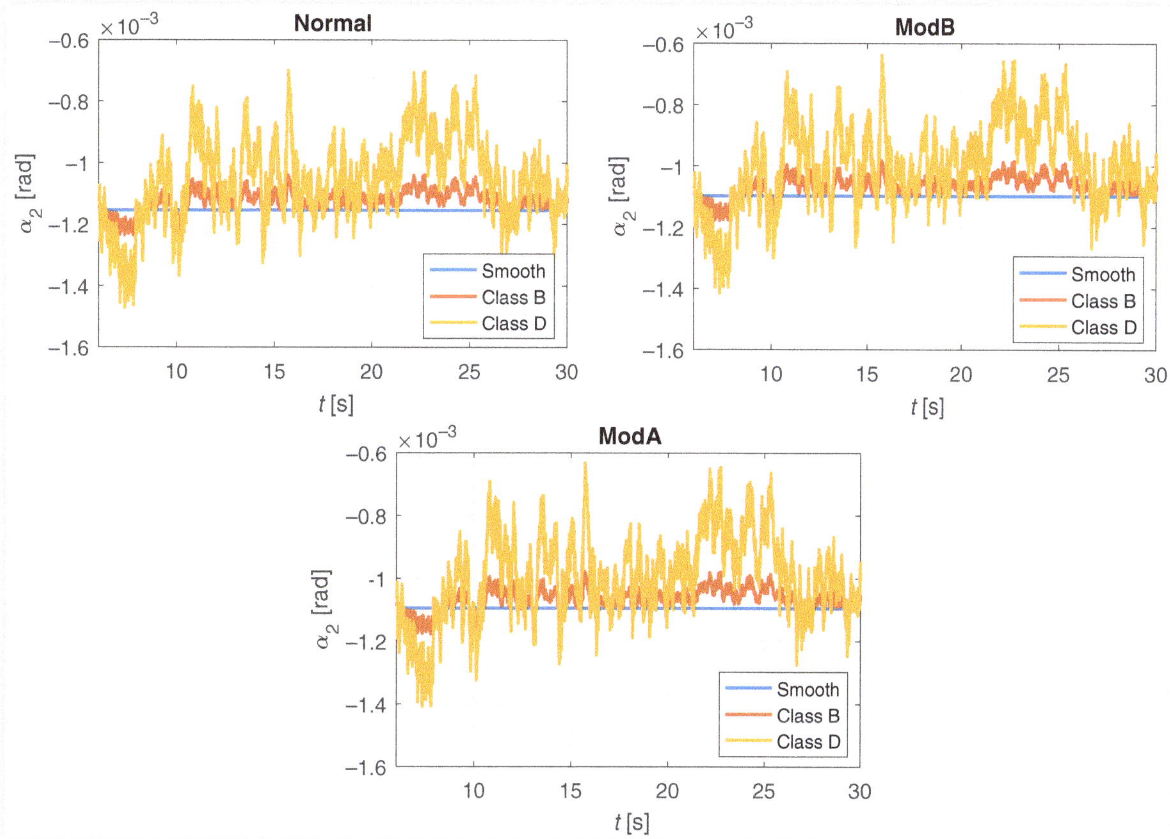

TABLE 11.3 Road irregularity effect on the straight-driving slip angles: comparison of their standard deviations obtained with analytical approach and with *MSC Adams*™ simulation, for different tire sets on smooth, class B and class D roads, without road cross slope. Data in Tables 10.1, A.1-A.4.

		$\sigma_{\alpha 1}$ [10^{-3} rad]				$\sigma_{\alpha 2}$ [10^{-3} rad]		
Tire set	**α_1 [10^{-3} rad]**	**Road class**	***MSC Adams*™**	**$\leq\sigma_{\alpha 1}$ analytic**	**α_2 [10^{-3} rad]**	**Road class**	***MSC Adams*™**	**$\leq\sigma_{\alpha 2}$ analytic**
Normal	-1.48	B	0.04	0.09	-1.16	B	0.03	0.13
		D	0.14	0.36		D	0.12	0.52
Mod-A	-1.52	B	0.03	0.19	-1.09	B	0.03	0.21
		D	0.13	0.76		D	0.12	0.84
Mod-B	-1.24	B	0.03	0.08	-1.09	B	0.03	0.09
		D	0.13	0.32		D	0.12	0.36

The standard deviations of y are estimated according to the analytical expression of Equation 8.96. These estimations are then compared to the standard deviations of vehicle's body displacement, obtained from the same previously employed *MSC Adams*™ simulations. The results of the full-vehicle simulations are reported in **Figure 11.5**.

In order to evaluate the integral of Equation 8.96, some parameters have to be estimated beforehand

- $c\varphi_1$: the front roll damping coefficient is estimated according to Equation 8.77. The dimensions of the *McPherson* suspension system (refer to Figure 8.14) are computed from the hardpoints of the *MSC Adams*™ full-vehicle model
    - $a_{MP} = 2.120$ m
    - $b_{MP} = 1.951$ m
    - $d_{MP} = 0.738$ m

FIGURE 11.5 Results for vehicle's body lateral motion *y*, *MSC Adams*™ straight line maintaining test performed on smooth, class B and class D roads.

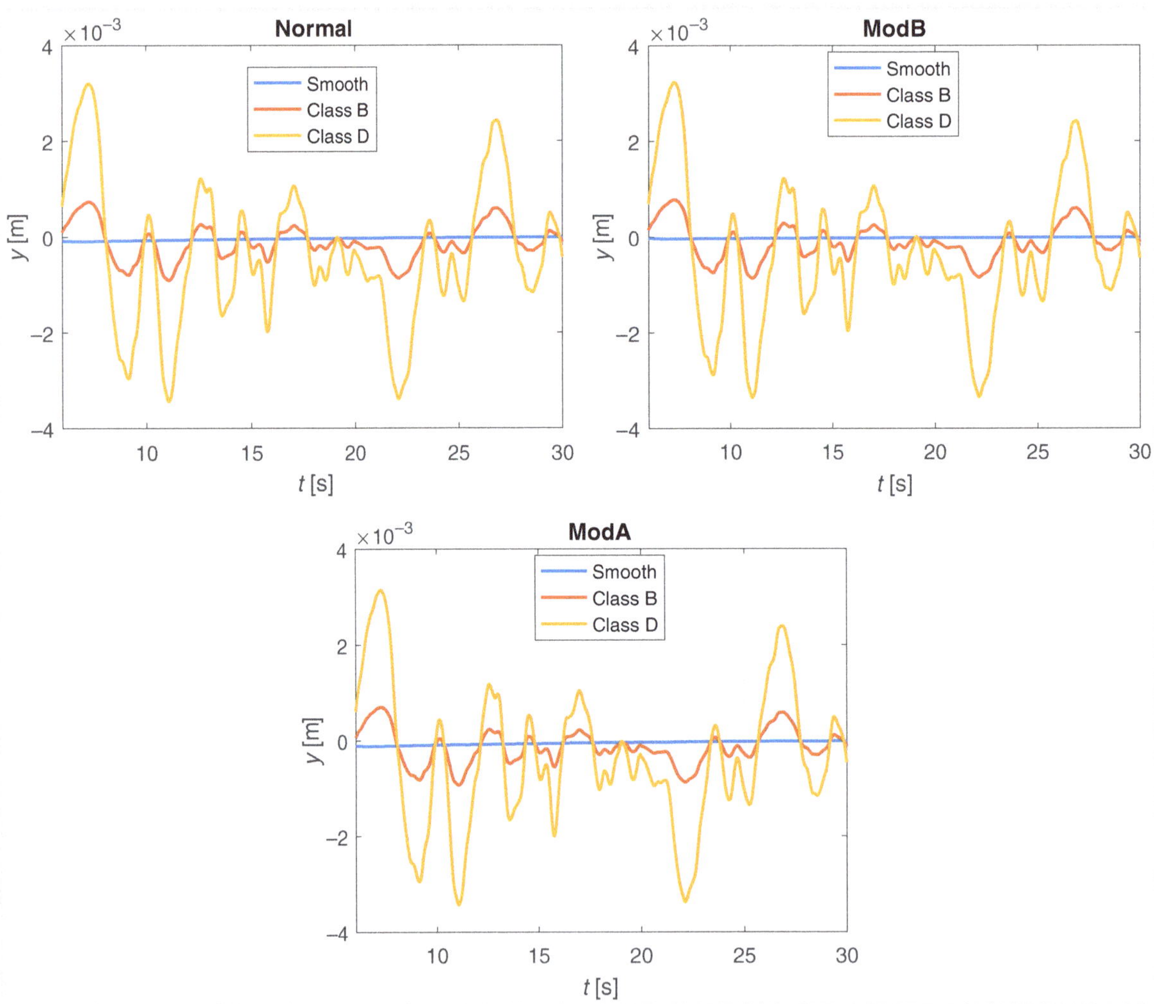

Therefore,

$$c_{\varphi 1} = 3508.9 \frac{\text{Nm s}}{\text{rad}}$$

- $c\varphi_2$: the rear roll damping coefficient is estimated in the same way. It is assumed that the *multi-link* suspension system can be described as an equivalent *double wishbone*, as far as the *y–z plane* is concerned. The dimensions of the suspension system (refer to Figure 8.14) are computed from the hardpoints of the *MSC Adams*™ full-vehicle model
  - $a_{DW} = 9.221$ m
  - $b_{DW} = 9.125$ m
  - $d_{DW} = 0.739$ m
  - $f_{DW} = 0.351$ m
  - $e_{DW} = 0.431$ m

  Thus, from Equation 8.79

$$c_{\varphi 2} = 922.2 \frac{\text{Nm s}}{\text{rad}}$$

- $C_{Fy,i}$: the value of tire lateral stiffness is set according to a realistic measure provided by Pacejka [7], and is assumed equal for front and rear tires

$$C_{Fy,1} = C_{Fy,2} = 125000 \frac{\text{N}}{\text{m}}$$

The comparisons between the standard deviations obtained with the analytical expression and the ones from the *MSC Adams*™ simulations are represented in **Table 11.4**. In this case, the values obtained using the two different methods have comparable order of magnitude. Additionally, the standard deviations of the analytical model are about twice the corresponding standard deviations of the *MSC Adams*™ model, for each tire set. Differences are expected due to the simplifying assumptions of the analytical model: for example, track shifts are computed as rigid axle's motion, in order to limit the complexity with respect to a more authentic case with independent suspensions. Although further studies may be needed, the indication that emerges from both the analytical and

TABLE 11.4 Road irregularity effect on the vehicle's body lateral motion *y*: comparison of standard deviations obtained with analytical approach and with *MSC Adams*™ simulation, for different tire sets on smooth, class B and class D roads, without road cross slope.

	σ_y [10^{-3} m]		
Tire set	**Road class**	***MSC Adams*™**	**Analytical**
Normal	B	0.33	0.65
	D	1.28	2.58
Mod-A	B	0.32	0.65
	D	1.26	2.50
Mod-B	B	0.32	0.66
	D	1.26	2.64

the *MSC Adams*™ results in Table 11.4 is that the specific set of tire does not influence vehicle's body lateral displacement y when the road irregularity is introduced.

11.4 Discussion

A straight-driving condition – similar to a fixed control test – has been analyzed in this chapter, by means of simulations with full-vehicle model equipped with the *Normal* set of tires.

It is shown that the analytical results are accurately predicting the steady-state condition (once the transient part is concluded) for the straight path motion, resulting from *MSC Adams*™ full-vehicle model simulations. The validation has been performed by monitoring five full-vehicle outcomes: the simulated straight-driving slip angles α_i, yaw angle Ψ, lateral velocity v_y, and steering torque M_S^* have been compared to the expected values of the corresponding dimensions computed with the analytical expressions. From Figures 11.1 and 11.2, it can be inferred that the analytical results are sufficiently consistent with the full-vehicle model simulation results, both in case of null or non-null road cross slope.

Furthermore, the values obtained in this chapter have a magnitude compatible with the results of an experimental study of *General Motors Co.* [2]. The test, performed with a midsize sedan on a road with cross slope $\varphi_r = 0.02$ rad, provided the following values, that can be compared with the results in Table 11.2

- $M_{S,st}^* = -10.2$ Nm
- $\beta_{st} = 1.57 \cdot 10^{-3}$ rad
- $\delta_{st} = -1.74 \cdot 10^{-3}$ rad $(=\alpha_1 - \alpha_2)$

When comparing these real vehicle test results with the results of the analysis object of this chapter, it must be taken into consideration that the two studies are disjointed: they are referred to vehicles and tires with different parameters. Nevertheless, the magnitude of the results obtained is in line in the two analyses.

Finally, the standard deviations of the straight-driving slip angles and of the lateral displacement of vehicle's body have been analyzed. The analytical estimations and *MSC Adams*™ simulation results have been compared in Tables 11.3–11.4 in order to study the effect of road irregularity on the analytical solutions and to have a figure of their robustness. The semi-qualitative outcome is that the steady-state solutions are approximately kept also when the road irregularity is introduced: a fluctuation is superimposed to the steady-state solutions, as it can be observed in Figures 11.3–11.5. The amplitude of these fluctuations does not appear to depend distinctly on the specific set of tire. Vehicle's body lateral displacement y is prone to depending mostly on vehicle and suspension characteristics rather than on the set of tires, which becomes evident from Table 11.4.

CHAPTER 12

On-Center Handling Test Simulation

Using the presented full-vehicle model developed in *MSC Adams*™*/Car*, a *weave test* [37, 40] is simulated in order to analyze the on-center handling, as discussed in chapter 5. In this way, the influence of tire characteristics on the on-center vehicle behavior is studied using objective metrics from the output results.

An adjusted *weave test*, similar to the one adopted in the study of Salaani et al. [37], is performed through an *open-loop sine steer event* in *MSC Adams*™ with

- low speed: $u = 50$ km/h
- small steering wheel displacement amplitude: $|\delta_{SW}| = 20$ deg
- low and constant steering input frequency: $f = 0.3$ Hz.

The steering wheel angle δ_{SW} and velocity u time histories from *MSC Adams*™ simulations are also provided as inputs to the single-track model (considered in Chapter 10), in order to compare the analysis results between the simple model and the full-vehicle model.

In order to do that, the steering ratio n_{st} for the single-track model is derived as follows. The steering wheel angle (δ_{SW}, *steering_ displacements. angle_front*) and the left and right front wheels' steering angles (δ_{1L} and δ_{1R}, respectively *steering_ system. steer_ angle. left* and

MATLAB® programs are provided online on the institutional repository of *Politecnico di Milano* (Technical University of Milan) https://re.public.polimi.it/handle/11311/1132512#.XmFR06hKjQC. They perform the comparison between the single-track model and the full-vehicle model results pertaining to the *weave test* presented in this chapter. Details on such programs can be found at the link https://www.mathworks.com/matlabcentral/fileexchange/74425-straight-motion-of-road-vehicles.

FIGURE 12.1 Steering ratio n_{st} derivation with steering analysis of *MSC Adams*™ front suspension assembly. The value around $\delta_{SW} = 0$ is omitted since it is not physically meaningful due to numerical error.

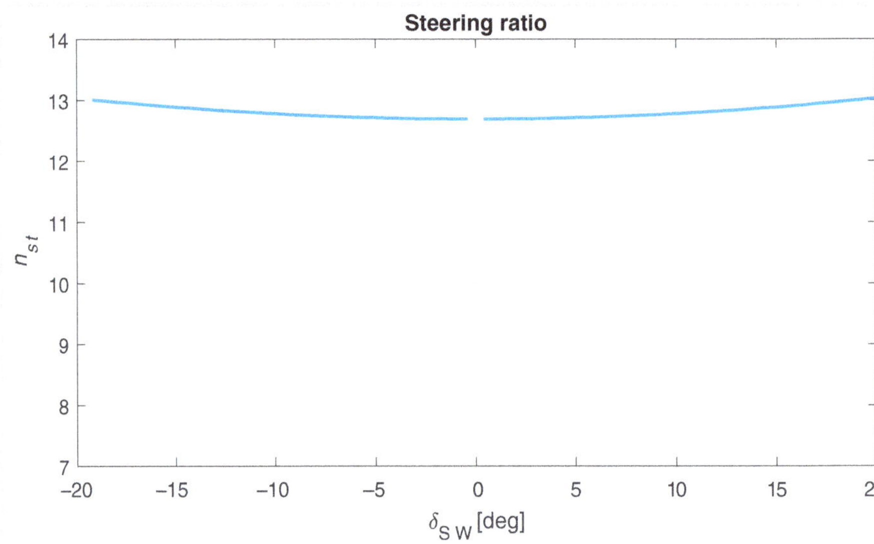

steering_system. steer_angle. right) are taken from the *MSC Adams*™ model to compute the steering ratio n_{st} as

$$n_{st} = \frac{\delta_{SW}}{\frac{1}{2}\left(\delta_{1L} + \delta_{1R}\right)} \tag{12.1}$$

The result of the derivation is represented in **Figure 12.1**. The steering ratio n_{st} is not constant throughout the steering wheel rotation, due to Hooke's joints (or universal or cardan joints) involved in the upper steering system mechanism. The constant value $n_{st} = 12.8$, namely the average over the range $\delta_{SW} = \pm 20$ deg, is considered for the *weave test* analysis.

12.1 Without Road Cross Slope ($\varphi_r = 0$)

A first set of simulations is performed without road cross slope ($\varphi_r = 0$), both with single-track model in *MATLAB*® and full-vehicle model in *MSC Adams*™. A selection of results is reported in **Figure 12.2** for *Normal* tire set, in order to show the agreement between full-vehicle model and single-track model results.

The steering torque M_S^* (at the steering linkage) is represented with respect to vehicle response (in terms of lateral acceleration a_y) in **Figure 12.3**, in the range $a_y = \pm 0.5$ m/s^2 for the sake of readability. In the ideal case, the values of M_S^* should be symmetric with respect to the origin, meaning that the steering effort for turning left and right is equal and opposite. Actually, due to tire characteristics, the upper and lower curves of the loop show an offset that matches with the value of $M_{S,st}^*$ (analytically computed according to Equation 6.15), as reported in **Table 12.1**.

FIGURE 12.2 Weave test simulation, set of results with *Normal* tires. Road slope $\varphi_r = 0$. Data in Tables 10.1, A.1–A.4.

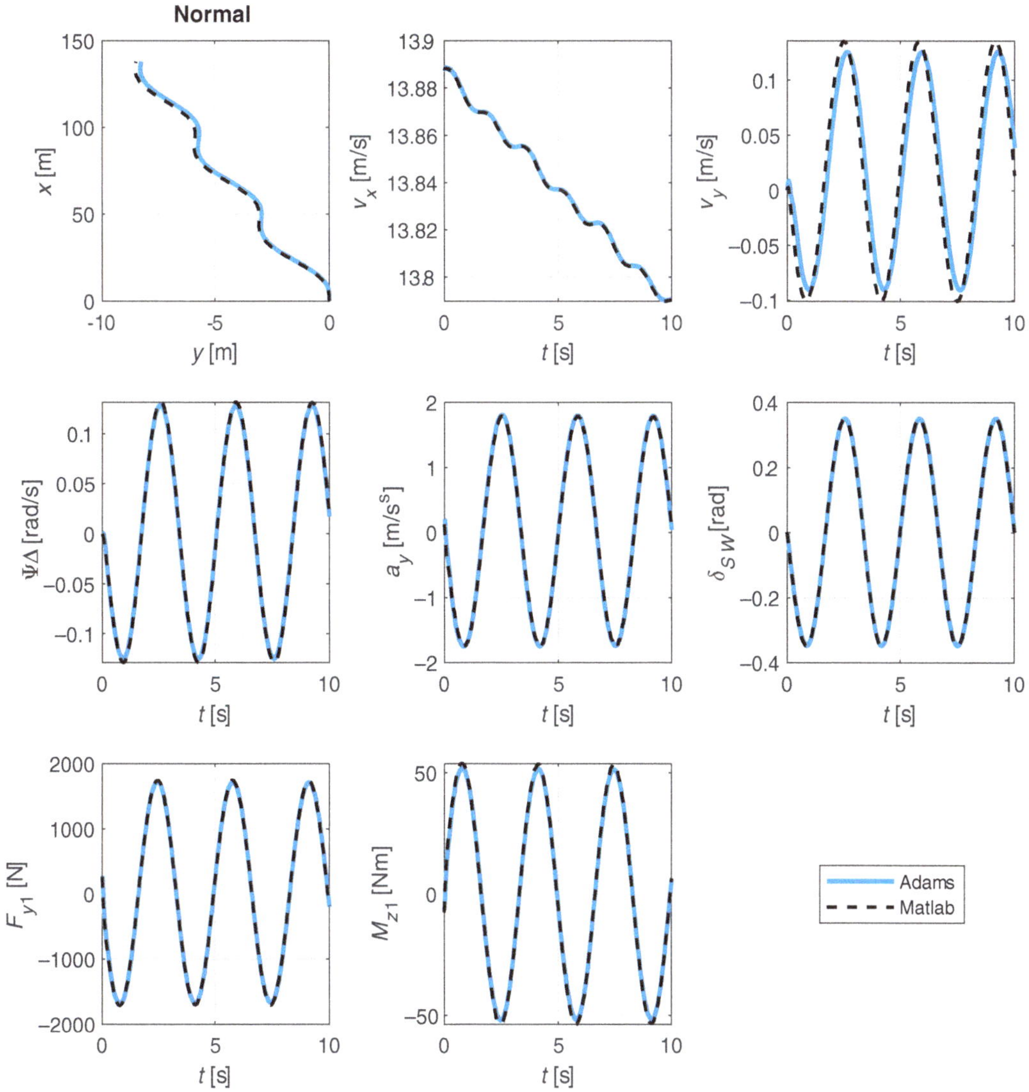

The steering wheel angle δ_{SW} is represented with respect to vehicle response (in terms of lateral acceleration a_y) in **Figure 12.4**, in the range $\delta_{SW} = \pm 5$ deg. The effect on the *weave test* results of the straight-driving steering angle δ_{st} (computed according to Equation 6.11) is represented in these plots: it causes an offset in the diagram $\delta_{SW} - a_y$, that is however imperceptible. Thus, during a *weave test*, the driver is more likely to provide subjective judgment on the steering torque (described with the objective metrics M_S^* represented in Figure 12.3) rather than on the steering wheel angle offset δ_{SW} (0).

The straight-driving steering angle δ_{st} and steering torque $M_{S,st}^*$ effects on the results of the *weave test* are represented together in **Figure 12.5**: they produce an offset with respect to the origin of the $\delta_{SW} - M_S^*$ diagram, that has impact on the steering wheel centering felt by the driver.

FIGURE 12.3 Weave test simulation, steering torque vs lateral acceleration. Road slope $\varphi_r = 0$. The cross mark "*X*" represents the steering torque offset $M^*_{S,st}$ (analytically computed according to Equation 6.15), and shows correspondence with the center of the loop. The closer is the cross with respect to the center of the axes, the better is the driver feeling. Data in Tables 10.1, A.1-A.4.

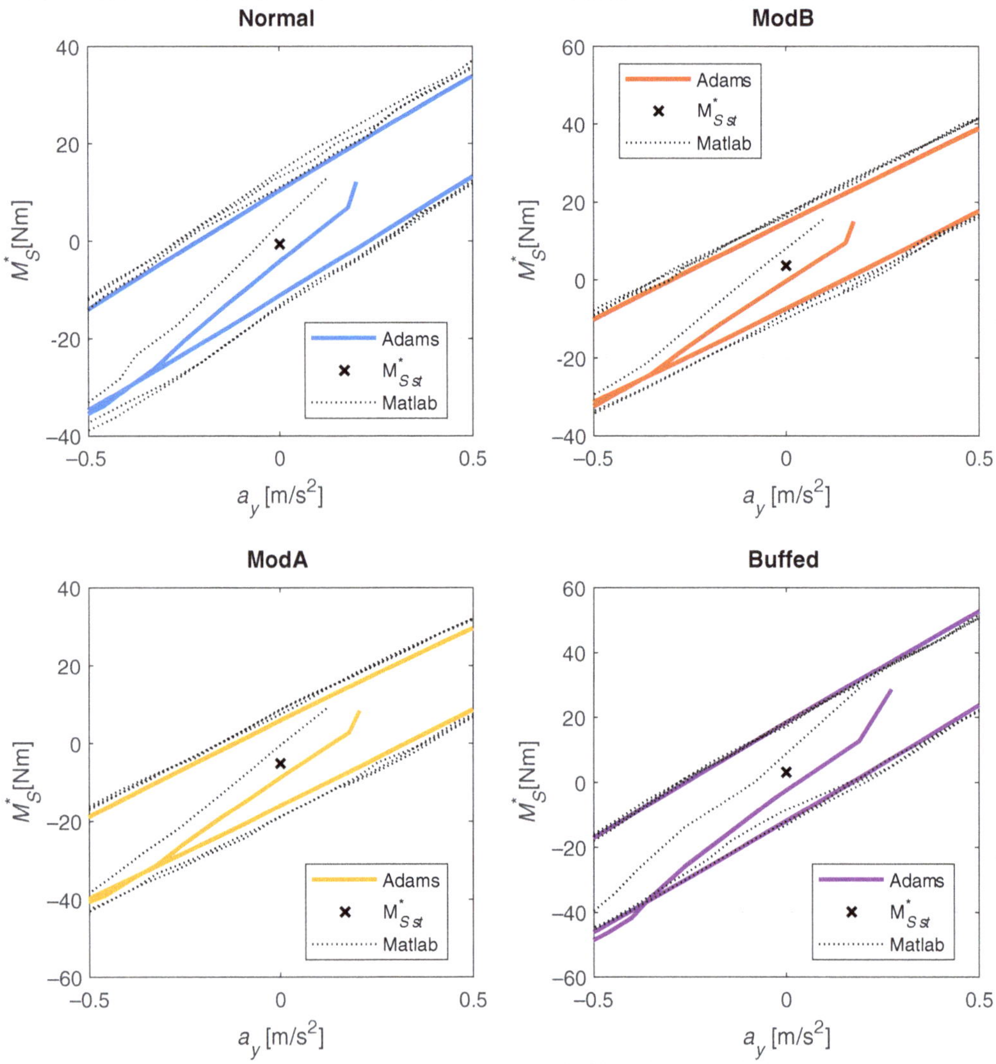

TABLE 12.1 Values of steering torque M^*_S at null lateral acceleration for left and right turn, from *MSC Adams*™ weave test simulation. Comparison with $M^*_{S,st}$ offset (analytically computed according to Equation 6.15). Road cross slope $\varphi_r = 0$. Data in Tables 10.1, A.1-A.4.

Tire set	$M^*_{S,st}$ [Nm]	*MSC Adams*™ $M^*_S(0)$ right [Nm]	*MSC Adams*™ $M^*_S(0)$ left [Nm]
Normal	-0.605	10.2	-11.3
Mod-A	-5.135	5.8	-16.1
Mod-B	3.606	14.7	-7.5
Buffed	3.138	18.3	-12.0

FIGURE 12.4 Weave test simulation, steering wheel angle vs lateral acceleration. Road slope $\varphi_r = 0$. The cross mark "X" represents the steering wheel angle offset $\delta_{SW,st} = \delta_{st} \cdot n_{st}$ (analytically computed according to Equation 6.11), and shows correspondence with the center of the loop. Data in Tables 10.1, A.1-A.4.

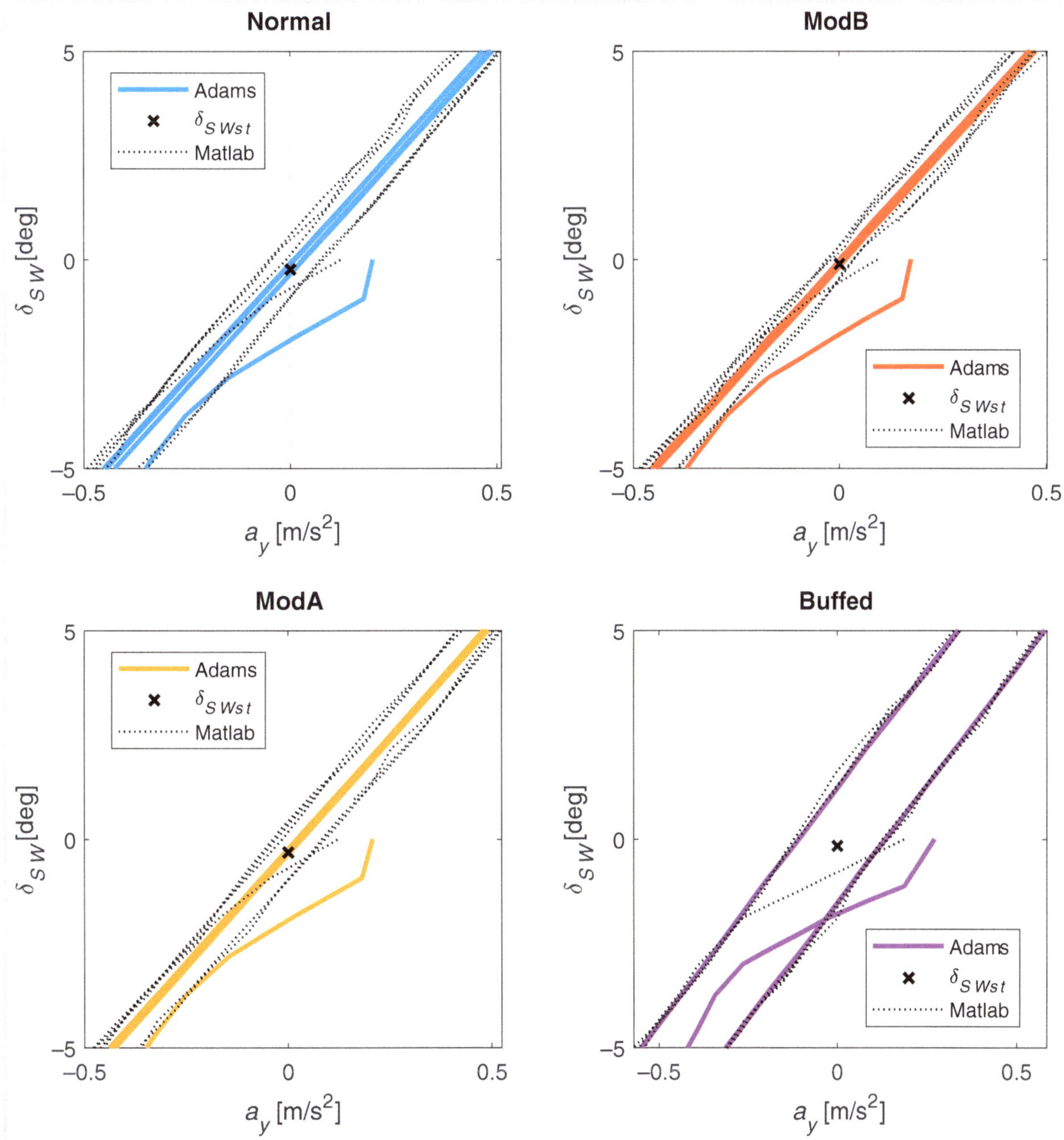

FIGURE 12.5 Weave test simulation, steering wheel angle vs steering torque. Road slope $\varphi_r = 0$. The cross mark "*X*" represents the steering torque offset $M^*_{S,st}$ (analytically computed according to Equation 6.15) and the steering wheel angle offset $\delta_{SW,st} = \delta_{st} \cdot n_{st}$ (analytically computed according to Equation 6.11), and shows correspondence with the center of the loop. Data in Tables 10.1, A.1–A.4.

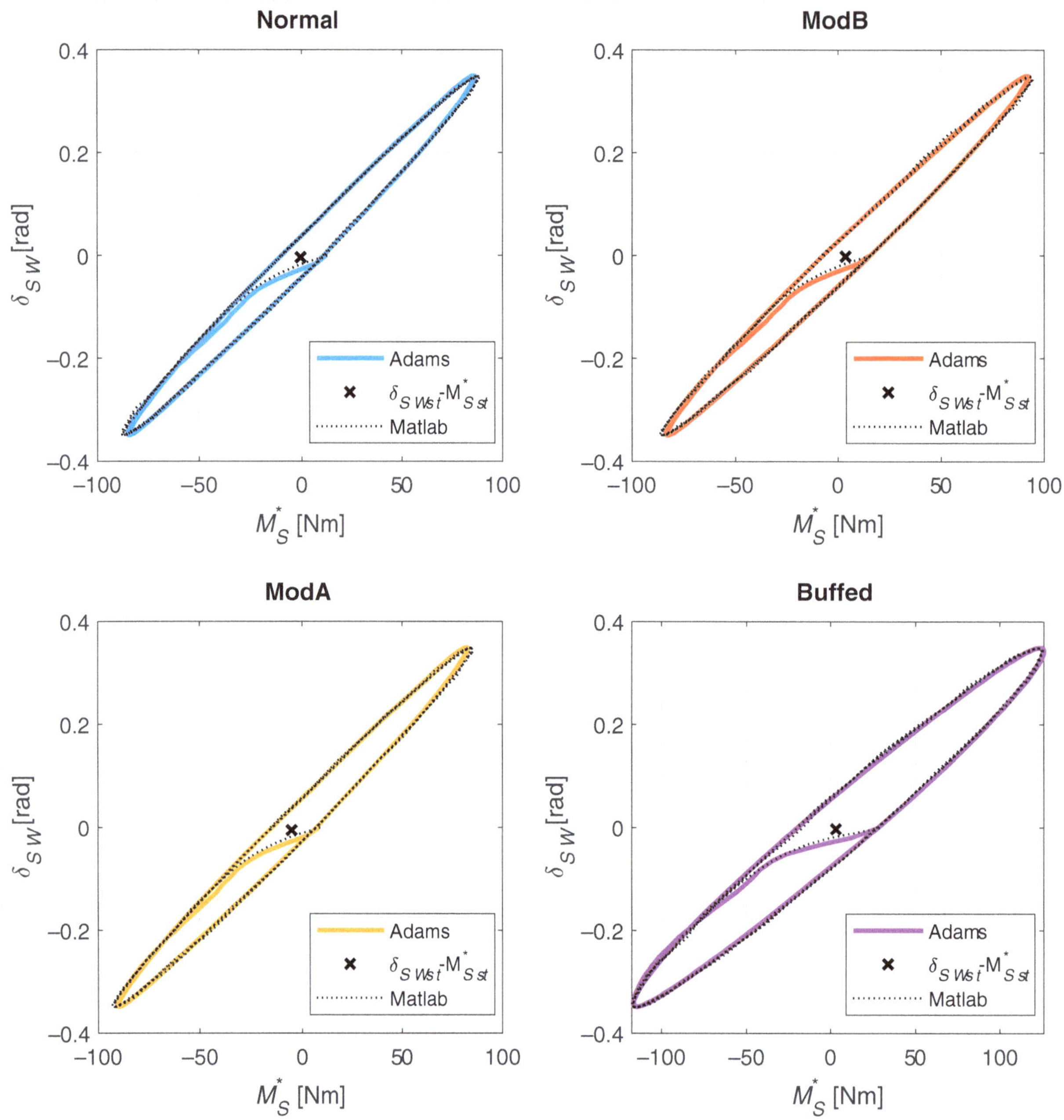

12.2 With Road Cross Slope (φ_r = 0.02 rad)

A second set of simulations is performed with road cross slope (φ_r = 0.02 rad), both with singletrack model in *MATLAB*® and full-vehicle model in *MSC Adams*™. A selection of results is reported in **Figures 12.6–12.9** for the different tire sets, in order to show the agreement between full-vehicle model and single-track model results.

FIGURE 12.6 Weave test simulation, set of results with *Normal* tires. Road slope φ_r = 0.02 rad.

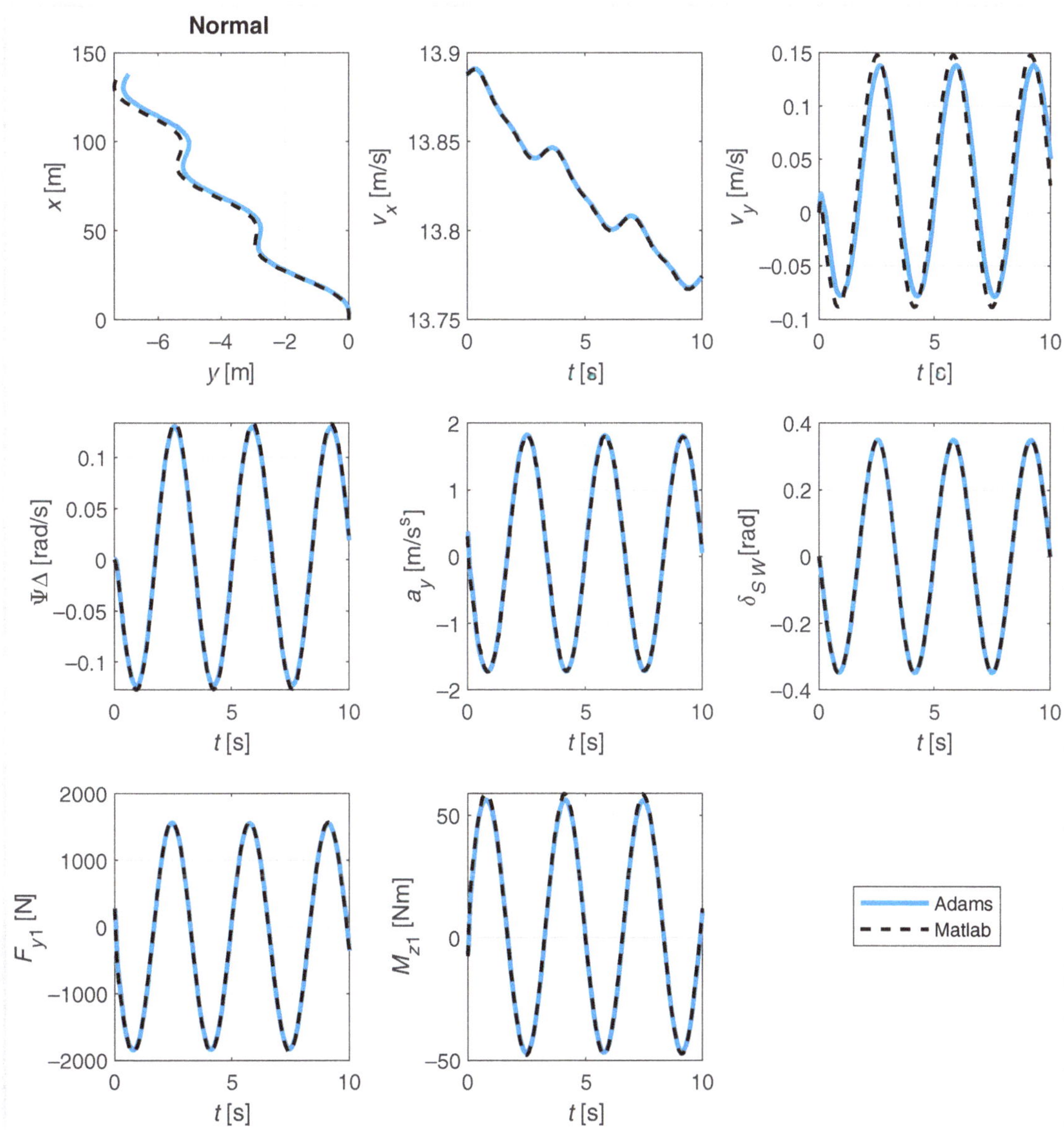

FIGURE 12.7 Weave test simulation, set of results with *Mod-A* tires. Road slope φ_r = 0.02 rad.

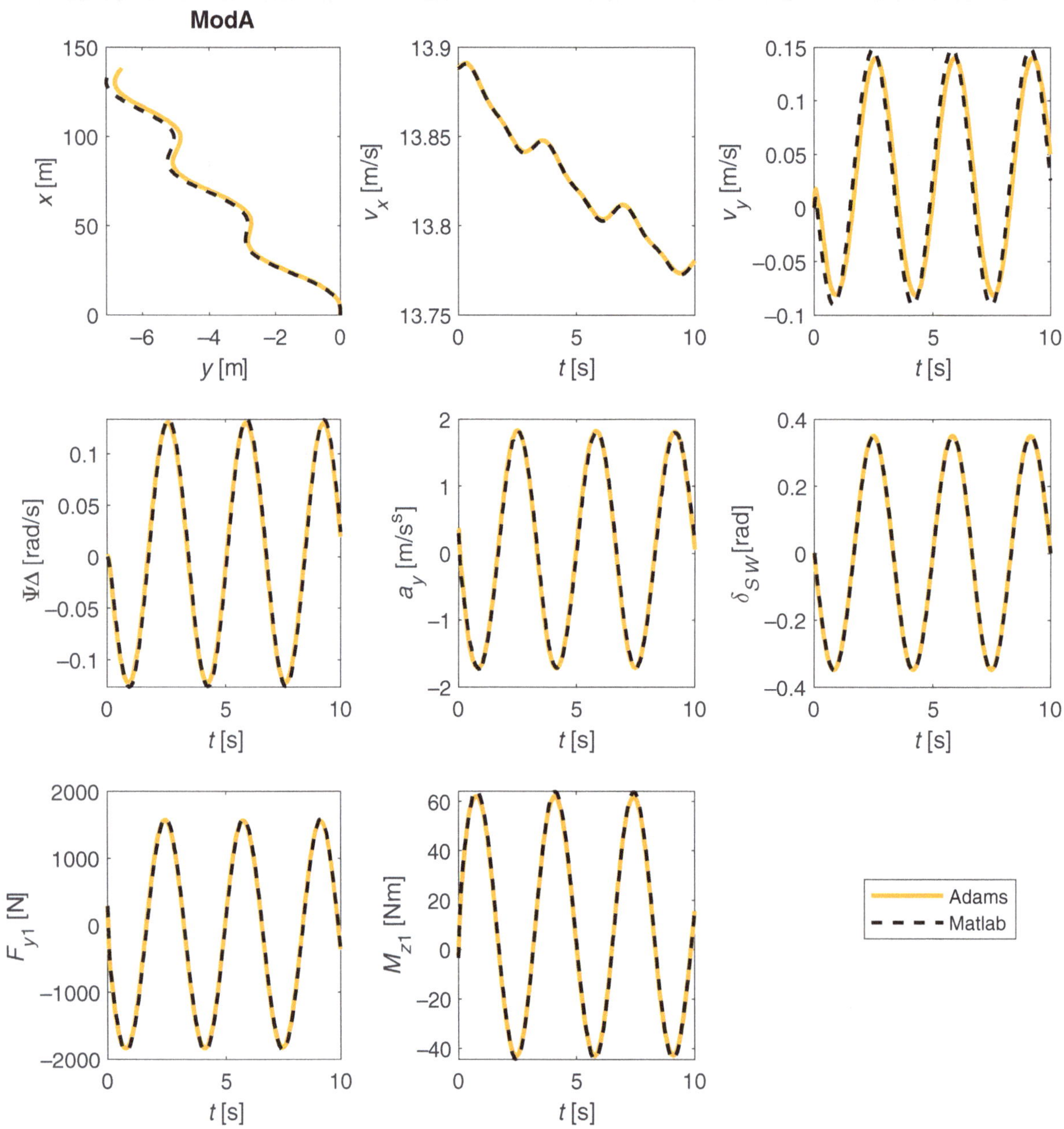

FIGURE 12.8 Weave test simulation, set of results with *Mod-B* tires. Road slope φ_r = 0.02 rad.

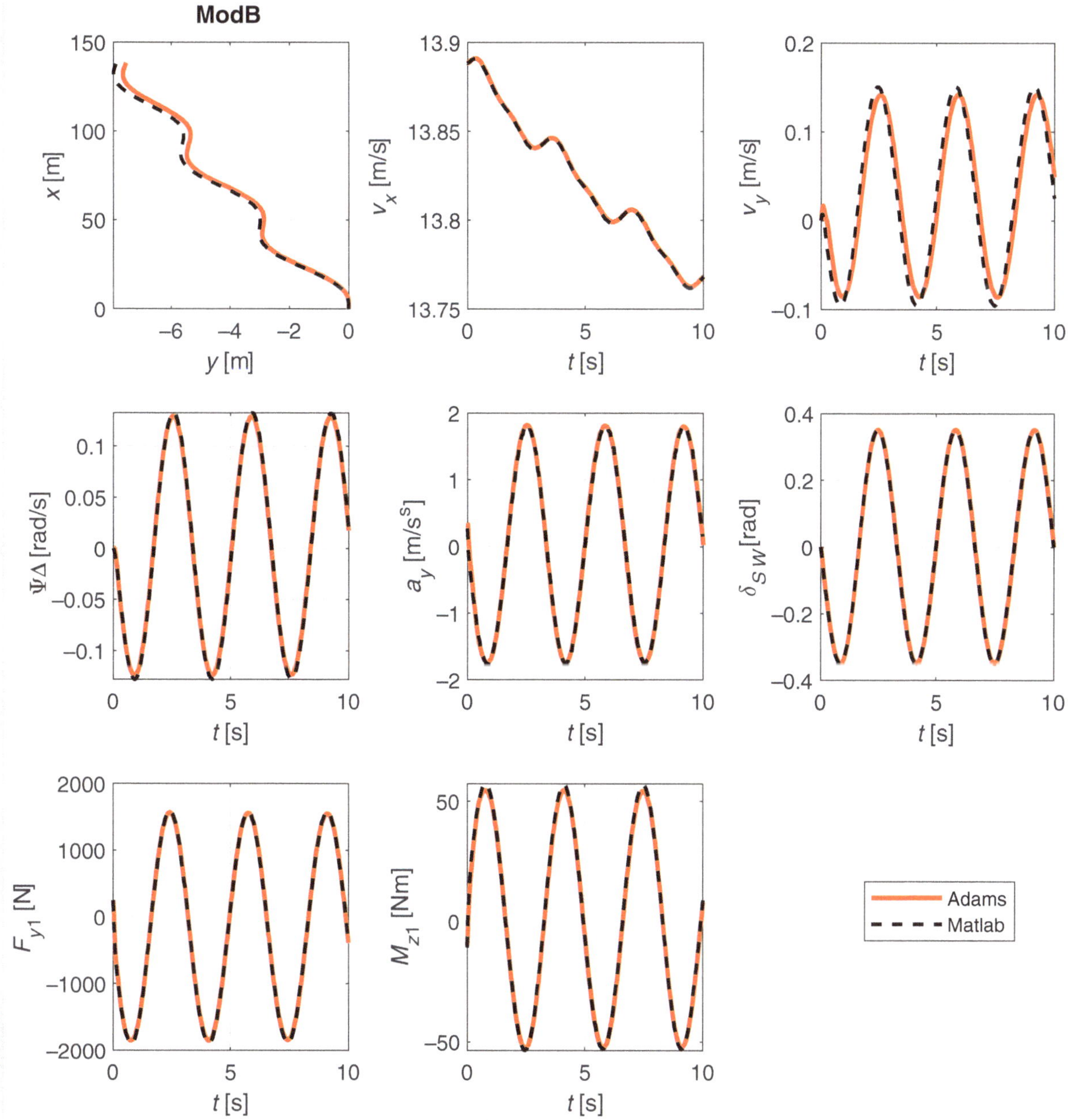

FIGURE 12.9 Weave test simulation, set of results with *Buffed* tires. Road slope φ_r = 0.02 rad.

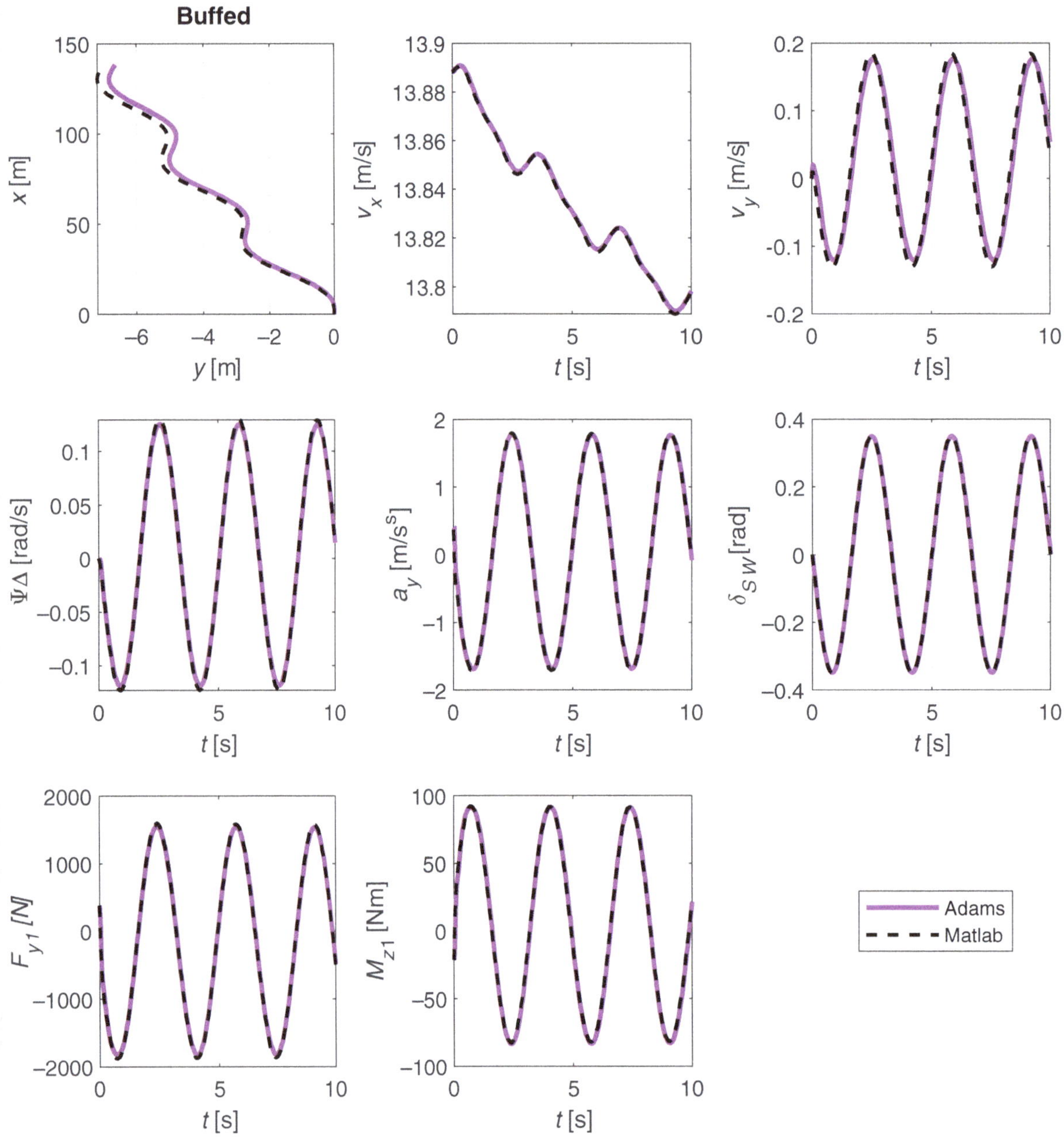

The positive road cross slope $\varphi_r = 0.02$ rad has the effect of moving downwards the steering torque-lateral acceleration $\left(M_S^* - a_y\right)$ curves represented in Figure 12.3, as it can be observed in the resulting curves of Figure 12.10. The diagram has an offset that matches with the value of $M_{S,st}^*$ (computed according to Equation 6.15), as reported in Table 12.2.

As it can be observed, the worst performing tire set, namely *Mod-A*, have steering torque with counterclockwise direction both for turning left and right, that is perceived as a substantially different behavior (or feedback) of the steering system of the vehicle when turning towards the two directions.

The steering wheel angle-lateral acceleration (δ_{SW}-a_y) plots of Figure 12.11 show a downwards shift of the curves with respect to Figure 12.4. Nevertheless, the steering

FIGURE 12.10 Weave test simulation, steering torque vs lateral acceleration. Road slope $\varphi_r = 0.02$ rad. The cross mark "X" represents the steering torque offset $M_{S,st}^*$ (analytically computed according to Equation 6.15), and shows correspondence with the center of the loop. The closer is the cross with respect to the center of the axes, the better is the driver feeling. Data in Tables 10.1, A.1–A.4.

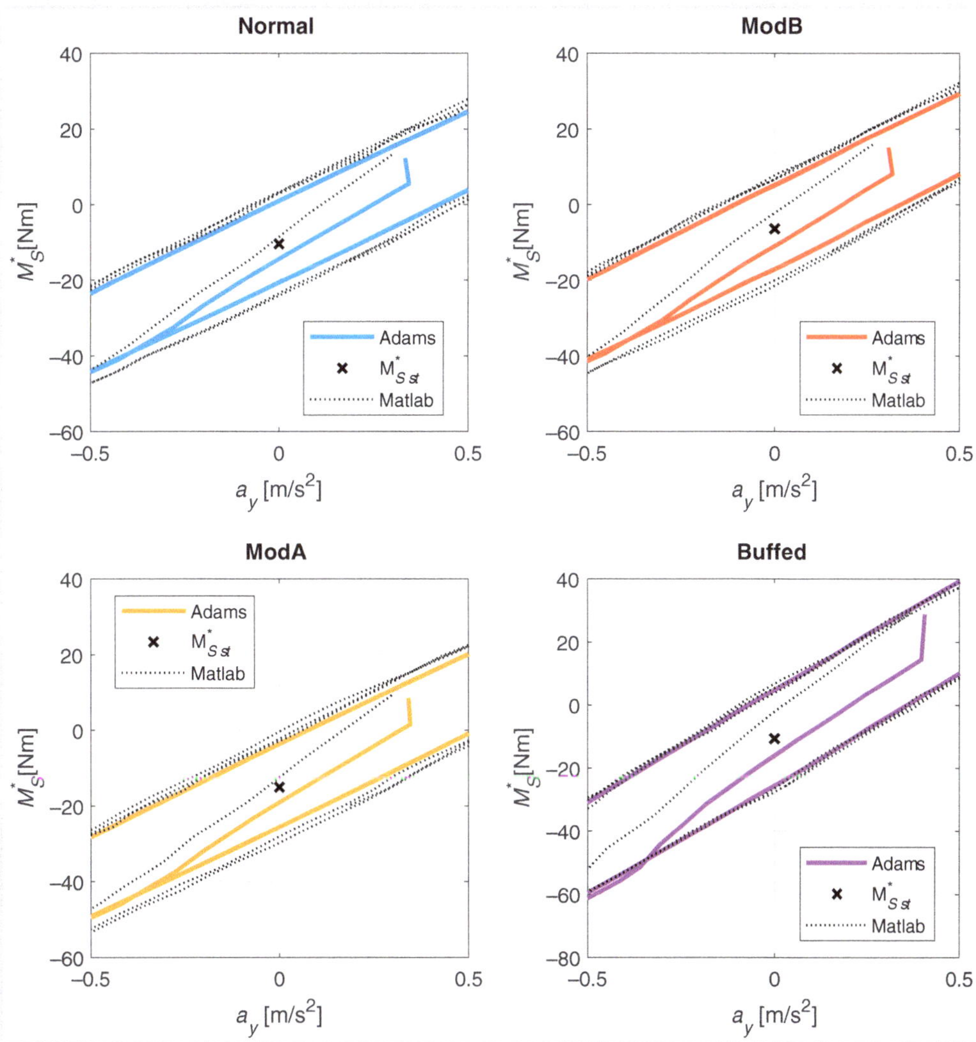

TABLE 12.2 Values of steering torque M^*_S at null lateral acceleration for left and right turn, from *MSC Adams*™ weave test simulation. Comparison with $M^*_{S,st}$ offset (analytically computed according to Equation 6.15). Road cross slope φ_r = 0.02 rad. Data in Tables 10.1, A.1-A.4.

Tire set	$M^*_{S,st}$ [Nm]	*MSC Adams*™ $M^*_S(0)$ right [Nm]	*MSC Adams*™ $M^*_S(0)$ left [Nm]
Normal	-9.958	1.0	-20.6
Mod-A	-14.695	-3.56	-25.5
Mod-B	-6.069	5.2	-17.0
Buffed	-10.763	4.85	-25.5

FIGURE 12.11 Weave test simulation, steering wheel angle vs lateral acceleration. Road slope φ_r = 0.02 rad. The cross mark "X" represents the steering wheel angle offset $\delta_{SW,st} = \delta_{st} \cdot n_{st}$ (analytically computed according to Equation 6.11), and shows correspondence with the center of the loop. Data in Tables 10.1, A.1-A.4.

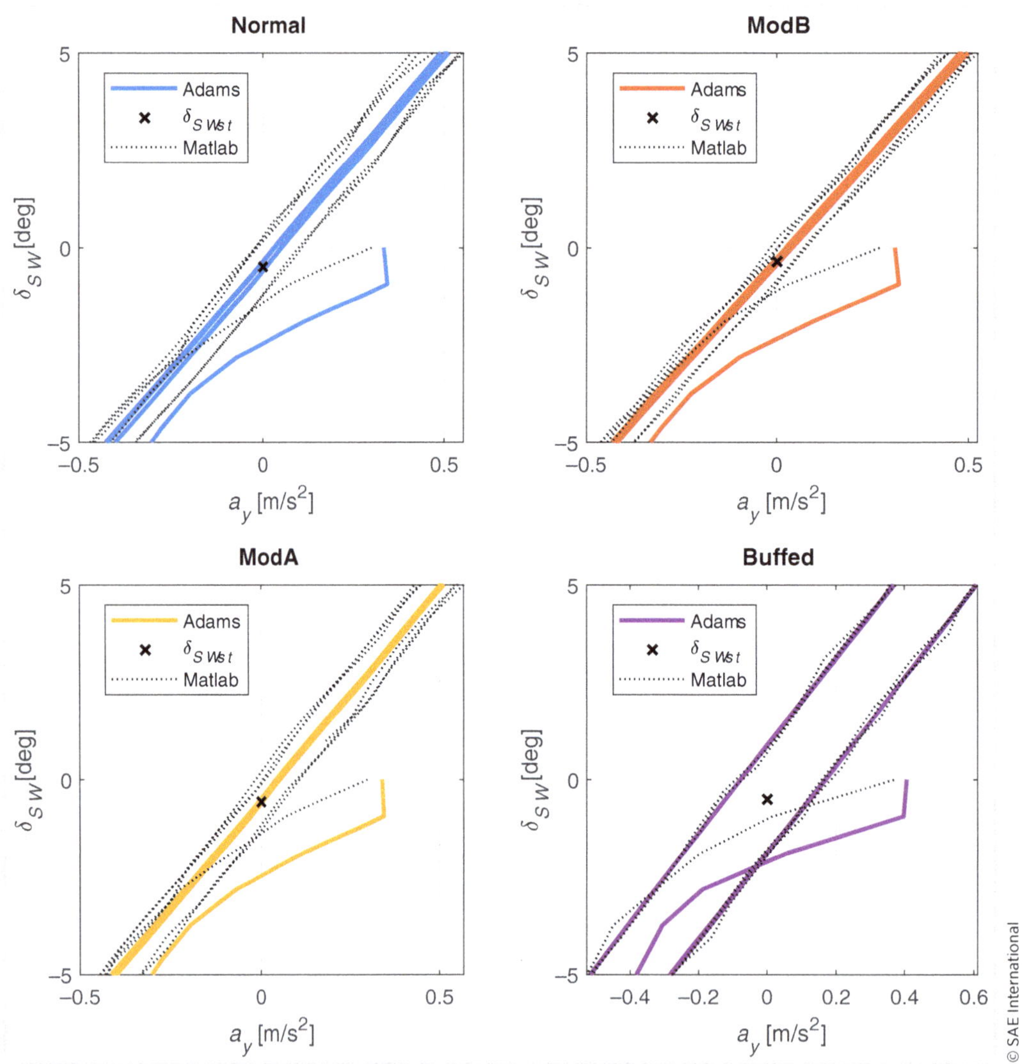

FIGURE 12.12 Weave test simulation, steering wheel angle vs steering torque. Road slope φ_r = 0.02 rad. The cross mark "X" represents the steering torque offset $M^*_{S,st}$ (analytically computed according to Equation 6.15) and the steering wheel angle offset $\delta_{SW,st} = \delta_{st} \cdot n_{st}$ (analytically computed according to Equation 6.11), and shows correspondence with the center of the loop. Data in Tables 10.1, A.1–A.4.

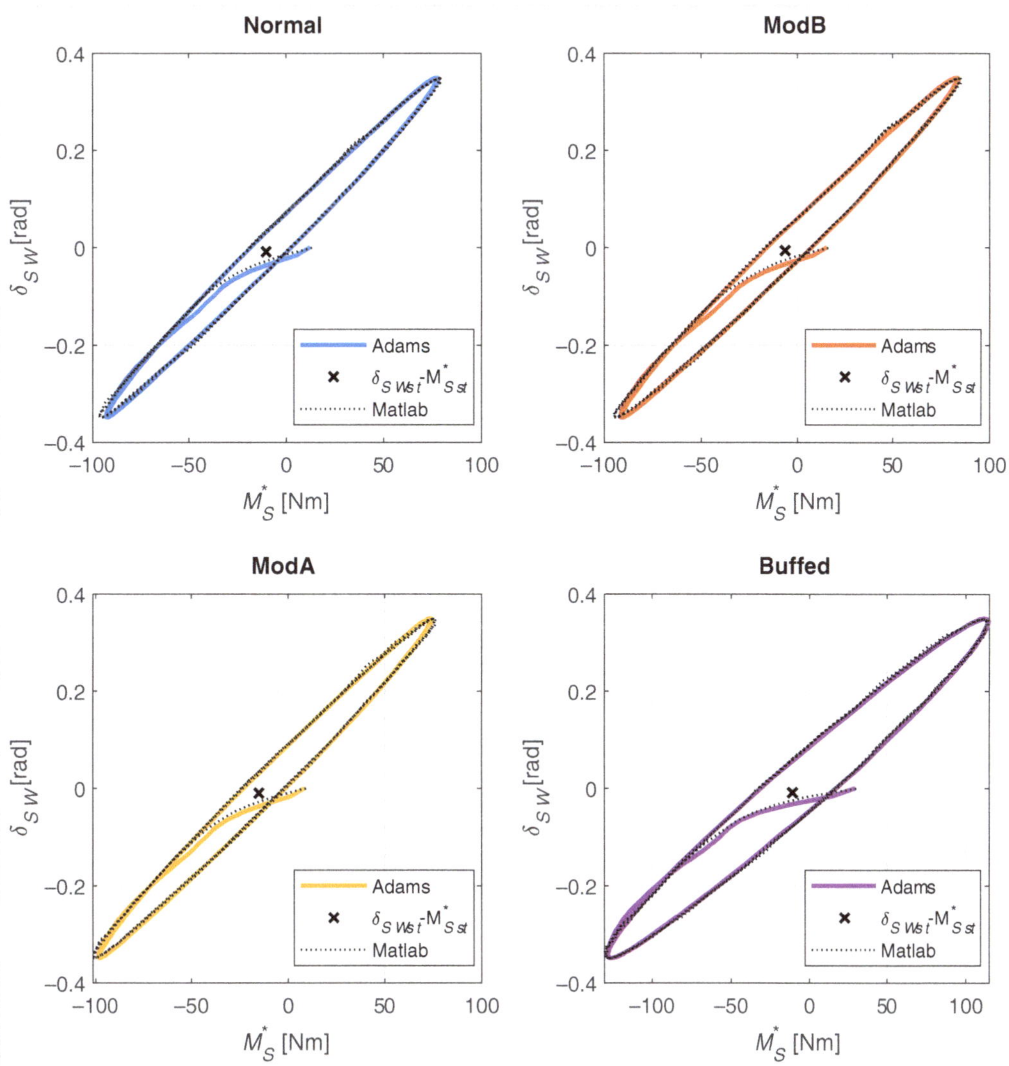

wheel offset at null lateral acceleration is still smaller than 1 deg with all the tire sets, thus it is difficult to be perceived by the driver.

The effect of the straight-driving steering angle δ_{st} and steering torque $M^*_{S,st}$ on the results of the *weave test* with non-null road cross slope is represented in Figure 12.12: it produces an offset with respect to the origin of the $\delta_{SW} - M^*_S$ diagram, of larger value with respect to the one of Figure 12.5.

12.3 Discussion

The results presented in this chapter are meaningful, since they give evidence to the fact that tire characteristics' offsets (such as the ones described by ply-steer and conicity) do have an influence on vehicle's on-center handling.

This effect can be observed by examining the steering torque as function of the vehicle response. In this book, it has been decided to represent the plots of torque on the steering linkage side of the steering gear $\left(M_S^*\right)$ as function of the lateral acceleration (a_y). Other possibilities that can be found in literature [37], aimed at analyzing the on-center behavior, are representing the steering wheel torque $M_S = M_S^* \cdot n_{st} \cdot V_S$ (see Equation 4.3) as function of either lateral acceleration a_y or yaw rate r.

The mentioned plots, with upper and lower curves of the loop representing respectively right and left turn, show an offset with respect to the origin, as reported in Figure 12.3 and Figure 12.10. As a consequence, the steering torque does not behave symmetrically, but assumes different magnitudes for left and right turn when crossing the null lateral acceleration state $a_y = 0$. This is experienced by the driver as a different feedback torque on the steering wheel for left and right turn when performing the *weave test*. In particular, the objective results for *Mod-A* tire set may provide an explanation to the subjective feedback of the driver, that was perceiving different steering "feels" for the two steering directions when using this tire set. The situation is worsened by the introduction of road cross slope, that increases the difference between left and right turn steering torque values.

Furthermore, the analytical results - in terms of straight-driving steering torque offset $M_{S,st}^*$ - turn out to be in agreement with the simulation results: indeed, the offset value is almost centered with respect to the left and right turn values of steering torque M_S^* at null lateral acceleration $a_y = 0$.

Finally, the results obtained with *MATLAB*® single-track model are accurate with respect to the *MSC Adams*™ full-vehicle model simulations.

CHAPTER 13

Wheel Alignment Setting Numerical Analysis

This chapter is the main one in the book. In this chapter, the equations and solutions introduced in Chapter 7 are evaluated and discussed numerically. In particular, the wheel alignment angles and tire parameters required to obtain vanishing straight motion offsets (i.e. steering angle, steering torque, and vehicle attitude) are found numerically. The usefulness of the analytical formulae derived in the preceding chapters in this book is assessed. Moreover, the compliance of the analytical expressions and of the *MATLAB*® single-track vehicle model is assessed by means of comparative simulations with the *MSC Adams*™ full-vehicle model, pertaining to the null steer test and the weave test.

The straight motion offsets (Equations 7.8–7.10) are rewritten here for the reader's convenience

Straight motion offsets

- Steering angle

$$\begin{aligned}\delta_{st} = &-\Psi_{10} \cdot A_1 - \gamma_{10} \cdot B_1 - C_1 + \\ &+\Psi_{20} \cdot A_2 + \gamma_{20} \cdot B_2 + C_2 + \\ &+(-D_1 + D_2)\varphi_r\end{aligned} \tag{13.1}$$

- **Steering torque**

$$M^*_{S,st} = -2C_{M\alpha,10} \cdot \left[\Psi_{10} \cdot (A_1 + A_{1M}) + \gamma_{10} \cdot (B_1 + B_{1M}) + C_1 + C_{1M} + E_{1M} \cdot \varphi_r\right] \tag{13.2}$$

- **Vehicle attitude**

$$\beta_{st} = \Psi_{20} \cdot A_2 + \gamma_{20} \cdot B_2 + C_2 + D_2\varphi_r \tag{13.3}$$

The coefficients introduced in Equations 7.11–7.14 are reported below

Coefficients

A: coefficients multiplying toe (front Ψ_{10}, rear Ψ_{20})

$$A_1 = \left(\frac{C_{F\alpha,10R} - C_{F\alpha,10L}}{C_{F\alpha,10R} + C_{F\alpha,10L}}\right) + \frac{F_{z1}}{2C_{F\alpha,10}} \cdot C_{el,\Psi,1} \cdot \varphi_r$$

$$A_2 = \left(\frac{C_{F\alpha,20R} - C_{F\alpha,20L}}{C_{F\alpha,20R} + C_{F\alpha,20L}}\right) + \frac{F_{z2}}{2C_{F\alpha,20}} \cdot C_{el,\Psi,2} \cdot \varphi_r$$

$$A_{1M} = -\left(\frac{C_{M\alpha,10R} - C_{M\alpha,10L}}{C_{M\alpha,10R} + C_{M\alpha,10L}}\right) - \frac{F_{z1}}{C_{M\alpha,10R} + C_{M\alpha,10L}} \cdot C_{M\alpha,el,\Psi,1} \cdot \varphi_r$$

B: coefficients multiplying camber (front γ_{10}, rear γ_{20})

$$B_1 = \left(\frac{C_{F\gamma,10R} - C_{F\gamma,10L}}{C_{F\alpha,10R} + C_{F\alpha,10L}}\right) + \frac{F_{z1}}{2C_{F\alpha,10}} \cdot C_{el,\gamma,1} \cdot \varphi_r$$

$$B_2 = \left(\frac{C_{F\gamma,20R} - C_{F\gamma,20L}}{C_{F\alpha,20R} + C_{F\alpha,20L}}\right) + \frac{F_{z2}}{2C_{F\alpha,20}} \cdot C_{el,\gamma,2} \cdot \varphi_r$$

$$B_{1M} = \left(\frac{C_{M\gamma,10R} - C_{M\gamma,10L}}{C_{M\alpha,10R} + C_{M\alpha,10L}}\right) - \frac{F_{z1}}{C_{M\alpha,10R} + C_{M\alpha,10L}} \cdot C_{M\alpha,el,\gamma,1} \cdot \varphi_r$$

C: coefficients related to tire characteristics' offsets ($F_{y0,i}$ and $M_{z0,i}$), yielded by ply-steer and conicity

$$C_1 = \left(\frac{F_{y0,1R} + F_{y0,1L}}{C_{F\alpha,10R} + C_{F\alpha,10L}}\right)$$

$$C_2 = \left(\frac{F_{y0,2R} + F_{y0,2L}}{C_{F\alpha,20R} + C_{F\alpha,20L}}\right)$$

$$C_{1M} = \left(\frac{M_{z0,1R} + M_{z0,1L}}{C_{M\alpha,10R} + C_{M\alpha,10L}}\right)$$

D, E: coefficients multiplying road cross slope φ_r

$$D_1 = \frac{F_{z1}}{C^*_{eff,1,AL}} = F_{z1} \cdot \frac{C_{el,AL,1}}{2C_{F\alpha,10}}$$

$$D_2 = \frac{F_{z2}}{C^*_{eff,2,AL}} = F_{z2} \cdot \frac{C_{el,AL,2}}{2C_{F\alpha,20}}$$

$$E_{1M} = \frac{F_{z1}}{2C_{M\alpha,10}} \cdot \left(n - C_{M\alpha,roll,1} + C_{M\alpha,susp,1} - C_{M\alpha,steer,1} - C_{M\alpha,el,AL,1} + \frac{2C_{M\alpha,10}}{C^*_{eff,1,AL}}\right)$$

The values of the aforementioned coefficients, pertaining to the different sets of tires, are reported in **Table 13.1** for null road cross slope $\varphi_r = 0$, and in **Table 13.2** for nonnull road cross slope $\varphi_r = 0.02$ rad.

As it can be observed, the coefficients multiplying static camber γ_{i0}, namely B_1 and B_{1M} for front and B_2 for rear, are null in case the road cross slope is null: indeed, the same load is applied to left and right wheel of a same axle, provided that the vehicle mass and the mass of its occupants are symmetrically distributed with respect to the longitudinal axis. Moreover, even in case a load transfer between left and right side of the vehicle is produced (e.g. by a nonnull road cross slope φ_r), the coefficients B_1, B_{1M}, and B_2 are significantly smaller than the coefficients multiplying static toe angles Ψ_{i0} (respectively A_1, A_{1M}, and A_2). Therefore, the influence of static camber γ_{i0} on straight-driving offsets is neglected henceforward. Only the influence of front and rear static toe angles (respectively, Ψ_{10} and Ψ_{20}) is considered henceforth.

As far as the analytical solutions presented in Chapter 7 are concerned, let us start by discussing about the wheel alignment settings required to cancel out the straight motion steering torque $M^*_{S,st}$ and vehicle attitude β_{st}. Indeed, Equations (7.9–7.10) for these two offsets have independent unknowns: the steering torque offset depends on

TABLE 13.1 Values of wheel alignment coefficients. Road cross slope $\varphi_r = 0$. Coefficients defined in Equations 7.11–7.14. Data in Tables 10.1, A.1–A.4.

Tire ID	A_1	A_{1M}	B_1	B_{1M}	C_1	C_{1M}	$D_1 \cdot \varphi_r$	$E_{1M} \cdot \varphi_r$
Normal	2.64e-3	4.14e-3	0	0	1.47e-3	−1.36e-3	0	0
Mod-A	12.47e-3	−15.17e-3	0	0	1.51e-3	−0.58e-3	0	0
Mod-B	−1.86e-3	1.97e-3	0	0	1.25e-3	−1.87e-3	0	0
Buffed	−4.36e-3	3.99e-3	0	0	1.53e-3	−1.78e-3	0	0

Tire ID	A_2	B_2	C_2	$D_2 \cdot \varphi_r$
Normal	0.98e-3	0	1.16e-3	0
Mod-A	13.82e-3	0	1.12e-3	0
Mod-B	11.25e-3	0	1.09e-3	0
Buffed	−4.31e-3	0	1.30e-3	0

TABLE 13.2 Values of wheel alignment coefficients. Road cross slope $\varphi_r = 0.02$ rad. Coefficients defined in Equations 7.11–7.14. Data in Tables 10.1, A.1– A.4.

Tire ID	A_1	A_{1M}	B_1	B_{1M}	C_1	C_{1M}	$D_1 \cdot \varphi_r$	$E_{1M} \cdot \varphi_r$
Normal	14.20e-3	−30.12e-3	1.16e-3	0.28e-3	1.48e-3	−1.38e-3	1.02e-3	1.77e-3
Mod-A	22.68e-3	−37.15e-3	1.15e-3	−0.59e-3	1.51e-3	−0.59e-3	1.01e-3	1.71e-3
Mod-B	9.33e-3	−23.22e-3	1.11e-3	0.35e-3	1.25e-3	−1.87e-3	0.97e-3	1.65e-3
Buffed	−2.58e-3	−15.45e-3	0.88e-3	−0.42e-3	1.52e-3	−1.77e-3	0.74e-3	1.04e-3

Tire ID	A_2	B_2	C_2	$D_2 \cdot \varphi_r$
Normal	22.22e-3	1.54e-3	1.16e-3	0.74e-3
Mod-A	33.35e-3	1.53e-3	1.12e-3	0.73e-3
Mod-B	33.20e-3	1.48e-3	1.09e-3	0.70e-3
Buffed	10.00e-3	0.91e-3	1.29e-3	0.35e-3

front static toe angle Ψ_{10}, while vehicle attitude depends on rear static toe angle Ψ_{20}. On the other hand, the straight motion steering angle offset δ_{st} (Equation 7.8) depends on both front and rear static toe angles.

The front toe angle Ψ_{10} required to have null steering torque offset $M^*_{S,st} = 0$ is computed according to Equation 7.9 (neglecting front static camber γ_{10} contribution)

$$M^*_{S,st} = -2C_{M\alpha,10} \cdot \left[\Psi_{10} \cdot (A_1 + A_{1M}) + C_1 + C_{1M} + E_{1M} \cdot \varphi_r\right] = 0$$
$$\Rightarrow \Psi_{10} = -\frac{C_1 + C_{1M} + E_{1M} \cdot \varphi_r}{A_1 + A_{1M}}$$

The rear toe angle Ψ_{20} required to have null straight-driving vehicle attitude $\beta_{st} = 0$ is computed according to Equation 7.10 (neglecting rear static camber γ_{20} contribution)

$$\beta_{st} = \Psi_{20} \cdot A_2 + C_2 + D_2 \cdot \varphi_r = 0$$
$$\Rightarrow \Psi_{20} = -\frac{C_2 + D_2 \cdot \varphi_r}{A_2}$$

The numerical solutions, pertaining to the different sets of tires, are reported in **Table 13.3** for null road cross slope $\varphi_r = 0$, and in **Table 13.4** for nonnull road cross slope $\varphi_r = 0.02$ rad. Most of these analytically computed toe values – required in order to obtain vanishing straight motion offsets of steering torque and vehicle attitude – are not physically meaningful: actually, too large toe values - besides being unattainable - both increase the motion resistance and modify profoundly the effective axle cornering stiffness $C^*_{eff,i}$ and aligning stiffness $C_{M\alpha,i}$, due to the effect of the elastokinematic coefficients $C^*_{\Delta Fz,i}$ and $C_{M\alpha,\Delta Fz,i}$. The trend of these parameters as function of static toe angle is reported for the different sets of tires in **Figures 13.1–13.4**.

TABLE 13.3 Static toe angle solutions for null steering torque offset $\left(M^*_{S,st} = 0\right)$ and null straight-driving vehicle attitude ($\beta_{st} = 0$). Road cross slope $\varphi_r = 0$. Data in Tables 10.1, A.1–A.4.

Tire ID	Ψ_{10} [deg] $\left(M^*_{S,st} = 0\right)$	Ψ_{20} [deg] ($\beta_{st} = 0$)
Normal	-0.957	-67.833
Mod-A	19.654	-4.645
Mod-B	348.923	-5.539
Buffed	-38.435	17.292

TABLE 13.4 Static toe angle solutions for null steering torque offset $\left(M^*_{S,st} = 0\right)$ and null straight-driving vehicle attitude ($\beta_{st} = 0$). Road cross slope $\varphi_r = 0.02$ rad. Data in Tables 10.1, A.1–A.4.

Tire ID	Ψ_{10} [deg] $\left(M^*_{S,st} = 0\right)$	Ψ_{20} [deg] ($\beta_{st} = 0$)
Normal	6.727	-4.908
Mod-A	10.432	-3.171
Mod-B	4.247	-3.087
Buffed	2.527	-9.435

FIGURE 13.1 Effective axle cornering stiffness $C^*_{eff,i}$, aligning stiffness $C_{M\alpha,i}$, and elastokinematic coefficient $C^*_{\Delta Fz,i}$ dependence on static toe angle Ψ_{i0}. For *Normal* set of tires. Road cross slope $\varphi_r = 0$. Data in Tables 10.1, A.1–A.4.

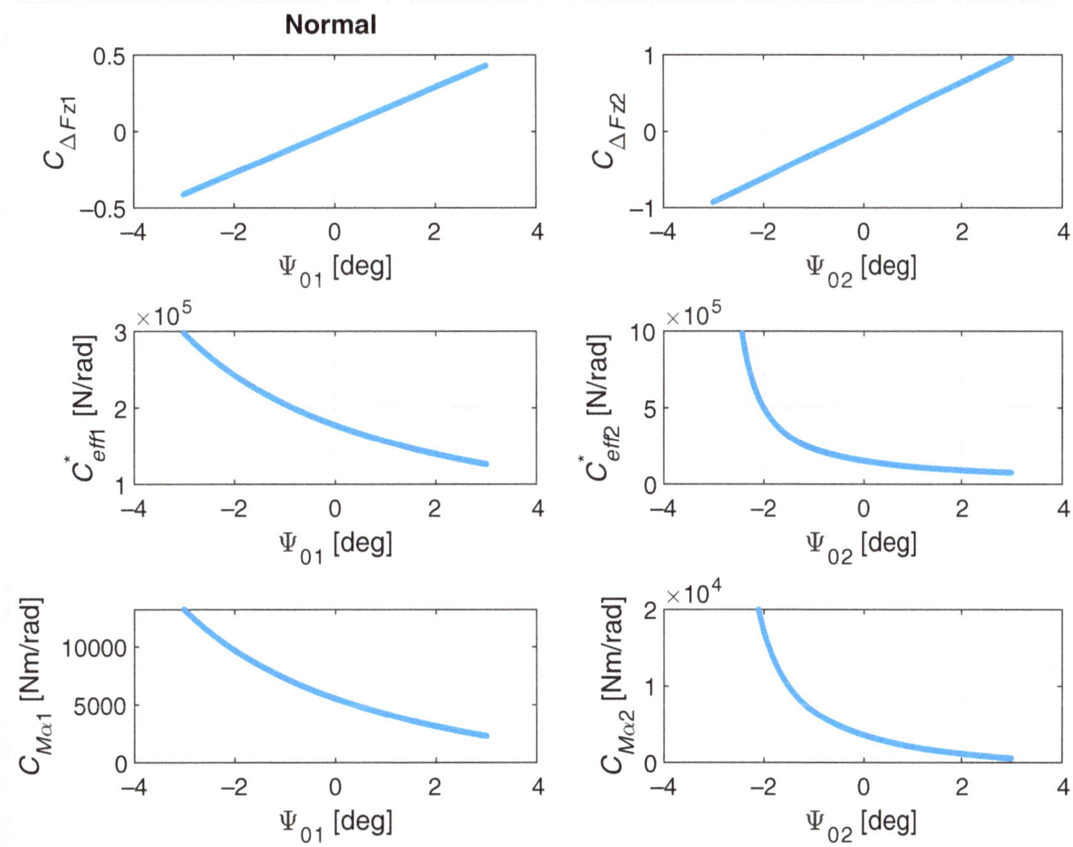

Therefore, it can be concluded that, with the data of the tire sets considered, it *is not possible* to correct the steering torque offset and straight-driving vehicle attitude by acting solely on the wheel alignment.

Suitable adjustments in tire parameters, on the other hand, may allow a proper setting of wheel alignment parameters and a proper solution to the steering pull issue!

To demonstrate this statement and to show the applicability of the analytical expressions presented, an example of tire parameter modification is reported in Table 13.5, aimed at a vanishing $M^*_{S,st}$. We have seen that the larger the steering torque offset expressed in Equation 7.9, the worse the steering feedback perceived by the driver due to a torque difference when turning left and right (as discussed in Chapters 5 and 12). With the modified tire characteristics shifts, the required corrective front toe angle Ψ_{10} is reduced from 4.25 to 0.13 deg.

$$M^*_{S,st} = -2C_{M\alpha,10} \cdot \left[\Psi_{10} \cdot (A_1 + A_{1M}) + C_1 + C_{1M} + E_{1M} \cdot \varphi_r\right] = 0$$

$$\Rightarrow \Psi_{10} = -\frac{C_1 + C_{1M} + E_{1M} \cdot \varphi_r}{A_1 + A_{1M}} = -\frac{0.997 - 2.615 + 1.649}{9.33 - 23.22} = 2.2 \cdot 10^{-3} \text{ rad} = 0.13 \text{ deg}$$

FIGURE 13.2 Effective axle cornering stiffness $C^*_{eff,i}$, aligning stiffness $C_{M\alpha,i}$, and elastokinematic coefficient $C^*_{\Delta Fz,i}$ dependence on static toe angle Ψ_{i0}. For *Mod-A* set of tires. Road cross slope $\varphi_r = 0$. Data in Tables 10.1, A.1–A.4.

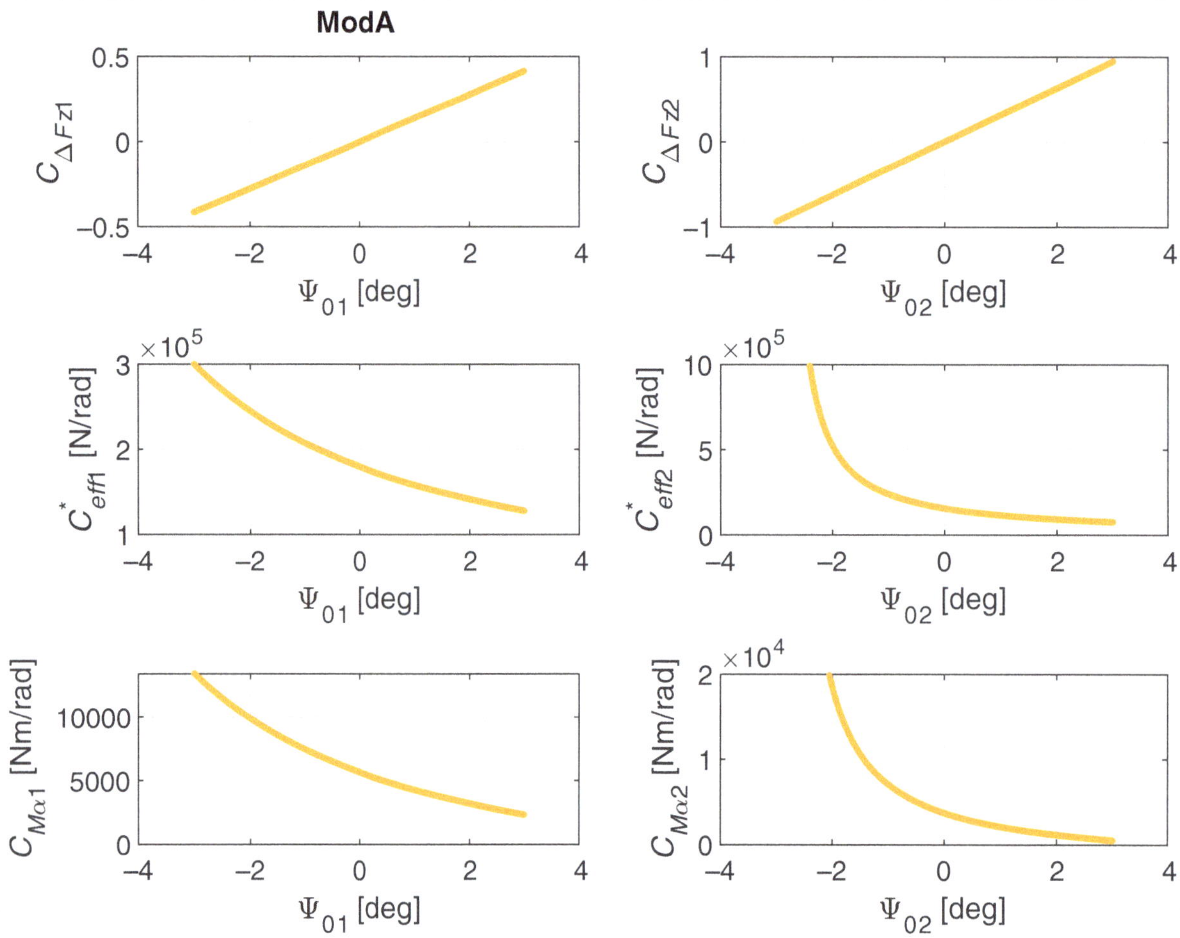

Let us now consider the analysis of the straight motion steering angle offset δ_{st} (Equation 7.8). As already stated, its expression depends both on the front static toe angle Ψ_{10} and on the rear static toe angle Ψ_{20}. Therefore, an attempt is made to find the combination of static toe angles for front (Ψ_{10}) and rear wheels (Ψ_{20}) making the steering angle offset $|\delta_{st}|$ vanish. The *Global Optimization Toolbox* of *MATLAB®* is used for this task. The absolute value of the nonapproximated version of Equation 6.11 is set as the objective function to be minimized by acting on the static toe angle values Ψ_{10} and Ψ_{20}, bounded in the range ±0.02 rad ≈ ±1.15 deg. The results, which referred to the vehicle equipped with *Mod-A* set of tires with null road cross slope $\varphi_r = 0$, are represented in **Figure 13.5**.

FIGURE 13.3 Effective axle cornering stiffness $C^*_{eff,i}$, aligning stiffness $C_{M\alpha,i}$, and elastokinematic coefficient $C^*_{\Delta Fz,i}$ dependence on static toe angle Ψ_{i0}. *For Mod-B* set of tires. Road cross slope $\varphi_r = 0$. Data in Tables 10.1, A.1–A.4.

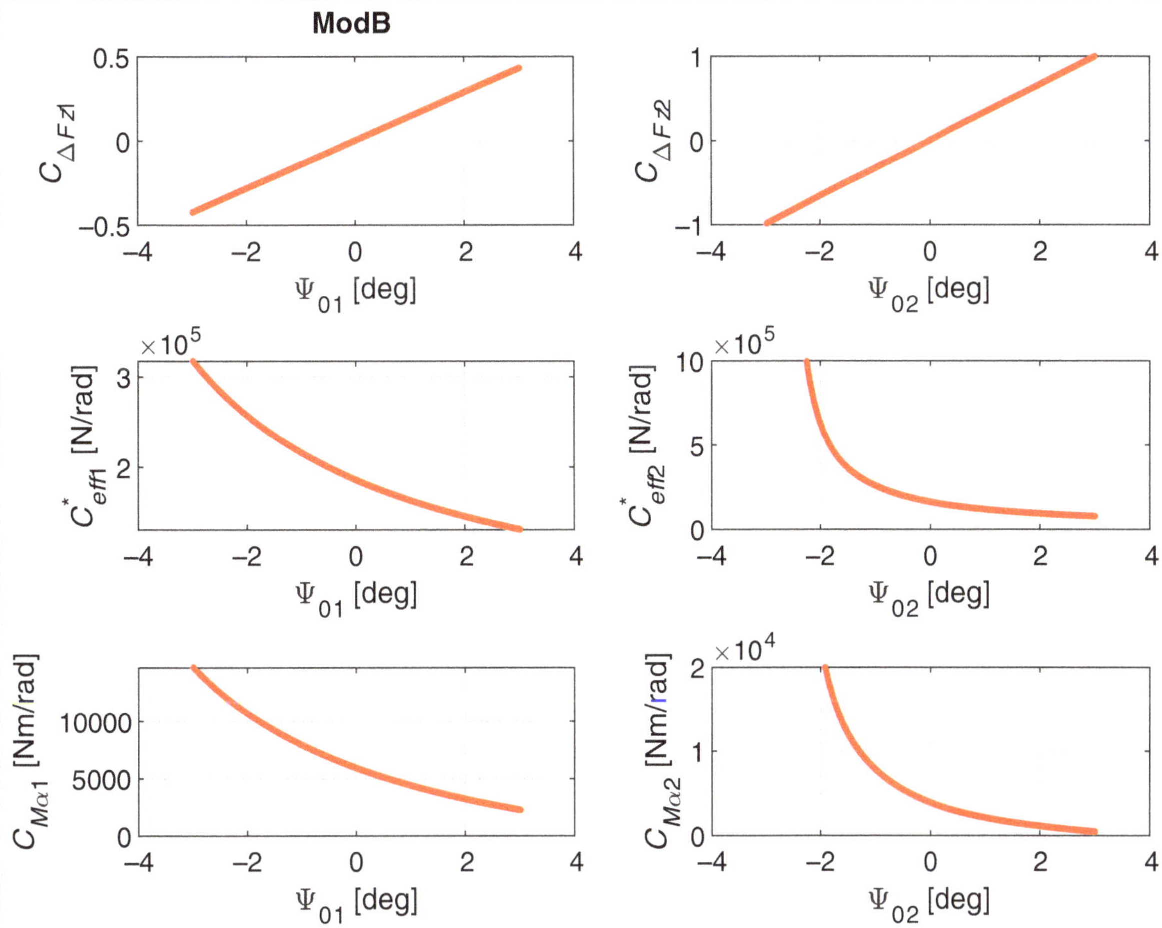

The combination of static toe angle values minimizing the amplitude of the straight-driving steering angle $|\delta_{st}|$ is

$$\Psi_{10} = -0.0160 \text{ rad} \approx -0.9 \text{ deg}, \quad \Psi_{20} = +0.0156 \text{ rad} \approx +0.9 \text{ deg}$$

Once again, these values are not applicable from a practical perspective: for a typical passenger car, static wheel alignment setting for toe angle usually ranges from −0.2 to +0.2 deg.

As already displayed in Table 13.5, suitable adjustments to tire parameters may allow re-ducing the wheel alignment angles required to cancel out the straight motion offsets. Further evaluations can be made by inspecting the values of the coefficients in Tables 13.1–13.2.

FIGURE 13.4 Effective axle cornering stiffness $C^*_{eff,i}$, aligning stiffness $C_{M\alpha,i}$, and elastokinematic coefficient $C^*_{\Delta Fz,i}$ dependence on static toe angle Ψ_{i0}. For *Buffed* set of tires. Road cross slope φ_r = 0. Data in Tables 10.1, A.1–A.4.

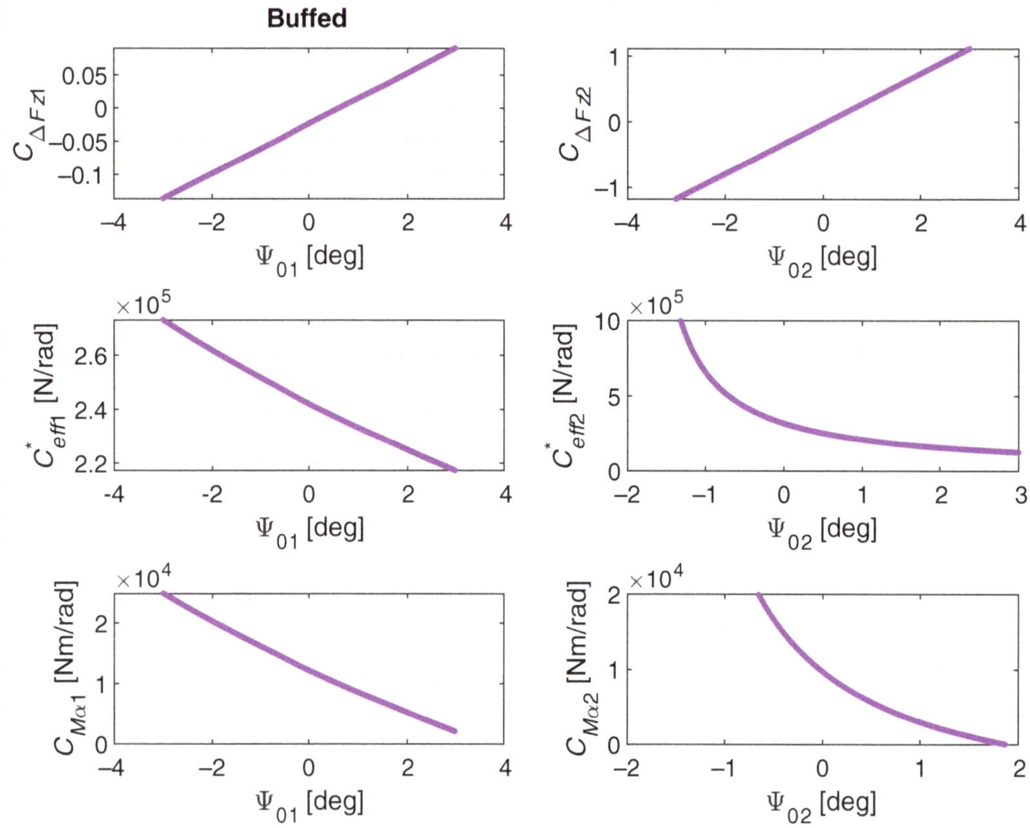

TABLE 13.5 Modified tire characteristics for null steering torque offset $\left(M^*_{S,st} = 0\right)$ at a reduced front static toe angle Ψ_{10} value. Road cross slope φ_r = 0.02 rad. Comparison with the solution and coefficients for *Mod-B* set of tires. Data in Table 10.1. The parameters of the *Modified* set of tires are obtained from the *Mod-B* set of tires with $|F_{y0}|$ reduced by 20% and $|M_{z0}|$ increased by 40%.

	Mod-B			Modified		
Parameter	***L***	***R***	**Axle (*L* + *R*)**	***L***	***R***	**Axle (*L* + *R*)**
$C_{F\alpha}$ [N/rad]	95834.7	96154.9	191989.6	95834.7	96154.9	191989.6
F_{y0} [N]	89.9	149.4	239.3	71.9	119.5	191.4
$C_{M\alpha}$ [Nm/rad]	2806.8	2844.4	5651.2	2806.8	2844.4	5651.2
M_{z0} [Nm]	-4.97	-5.59	-10.56	-6.96	-7.82	-14.78
Ψ_{10} [deg]	4.25			0.13		

Tire ID	A_1	A_{1M}	C_1	C_{1M}	$E_{1M} \cdot \varphi_r$
Mod-B	9.33e-3	-23.22e-3	1.25e-3	-1.87e-3	1.65e-3
Modified	9.33e-3	-23.22e-3	0.99e-3	-2.61e-3	1.65e-3

FIGURE 13.5 *MATLAB® Global Optimization Toolbox* results for the combination of Ψ_{10} and Ψ_{20} static toe angle values that make the steering angle offset amplitude $|\delta_{st}|$ vanish. For *Mod-A* set of tires, road cross slope φ_r = 0 rad.

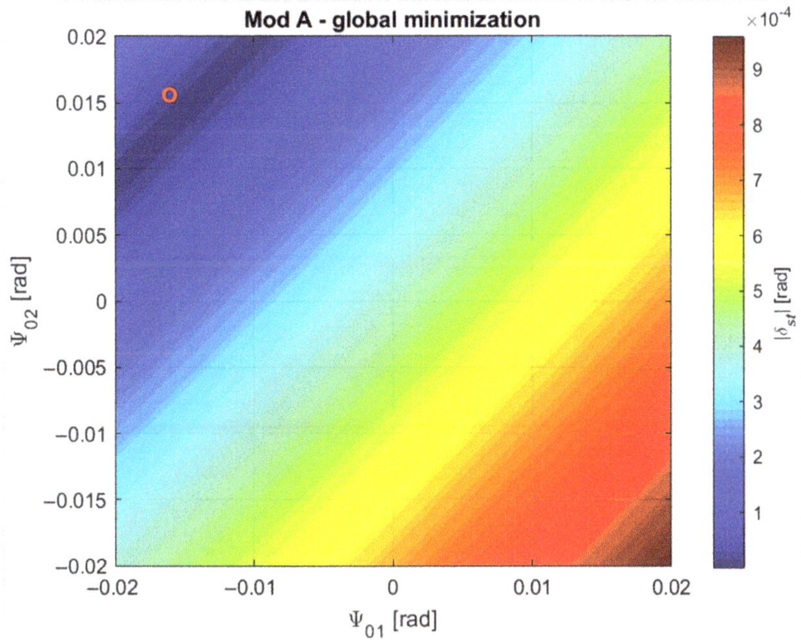

13.1 Static Toe Angle Variation: Null Steer Test

Let us consider the *MSC Adams*™ full-vehicle model equipped with the *Mod-A* set of tires (the worst performing one), together with the *MATLAB®* single-track model. Let us consider the null steer test (discussed in Chapter 10) and let us modify the wheel alignment. In order to compare the results of the two simulations, both the models are provided with the same small static toe angle of 0.005 rad ≈ 0.3 deg. Two different configurations are tested, with opposite static toe angles for front and rear wheels, alternatively convergent (toe-in, $\Psi_{i0} < 0$) and divergent (toe-out, $\Psi_{i0} > 0$)

1. $\Psi_{10} = -0.005$ rad, $\Psi_{20} = +0.005$ rad
2. $\Psi_{10} = +0.005$ rad, $\Psi_{20} = -0.005$ rad

After the null steer test is conducted, the resulting trajectories are compared. The results with null road cross slope $\varphi_r = 0$ are reported in **Figure 13.6**, while results with nonnull road cross slope $\varphi_r = 0.02$ rad are reported in **Figure 13.7**.

It is important to underline that trajectory is affected by wheel alignment also in the null road cross slope case, since slightly different left and right tire characteristics are used in *MSC Adams*™, as reported in Chapter 10. This gives evidence to the fact that – in the presence of manufacturing tolerances, resulting in different values for left

FIGURE 13.6 Trajectories of a null steer test with full-vehicle model and single-track model with *Mod-A* tires. Front and rear static toe angles $\Psi_{10} = \pm 0.005$ rad, $\Psi_{20} = \mp 0.005$ rad. Road cross slope $\varphi_r = 0$. Data in Tables 10.1, A.1–A.4.

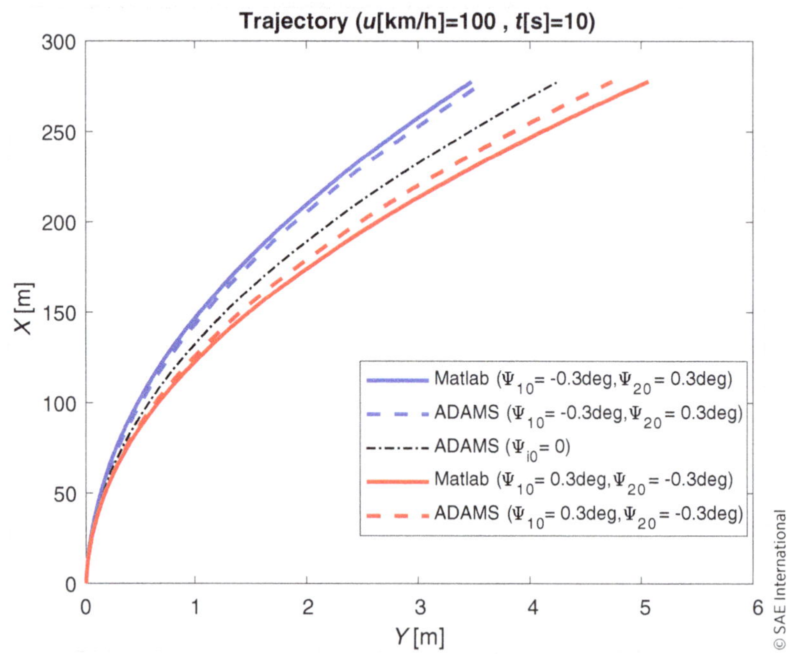

FIGURE 13.7 Trajectories of a null steer test with full-vehicle model and single-track model with *Mod-A* tires. Front and rear static toe angles $\Psi_{10} = \pm 0.005$ rad, $\Psi_{20} = \mp 0.005$ rad. Road cross slope $\varphi_r = 0.02$ rad. Data in Tables 10.1, A.1–A.4.

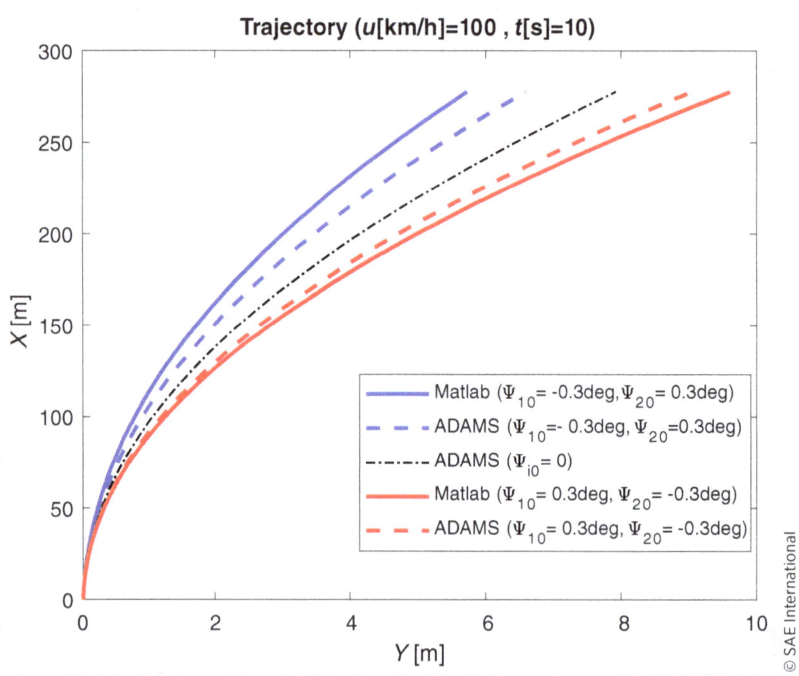

TABLE 13.6 Results of the trajectory simulated with single-track model with and without toe variation. *Mod-A* set of tires. δ_{st} is the steering angle that would be necessary to run straight ahead, computed according to Equation 6.11. Data in Tables 10.1, A.1-A.4.

Ψ_{10} [rad]	Ψ_{20} [rad]	φ_r [rad]	Δy drift at x = 100m [m]	δ_{st} [10^{-4} rad]
0	0	0	0.585	-4.259
-0.005	+0.005	0	0.505	-2.935
+0.005	-0.005	0	0.641	-5.518
0	0	0.02	1.042	-7.653
-0.005	+0.005	0.02	0.911	-4.800
+0.005	-0.005	0.02	1.202	-10.504

and right tires of a same axle – wheel alignment does have an influence on the trajectory and steering pull issue.

The trend of the trajectory variation is correctly predicted by the single-track model, as it can be observed in Figures 13.6 and 13.7. Moreover, the link already evidenced in Chapter 10 between the analytically computed steering angle offset δ_{st} and the lateral drift Δy can be observed, as it can be deduced from values in **Table 13.6**: the road cross slope being equal, the larger the steering angle offset δ_{st}, the larger the lateral drift Δy. Indeed, the straight-driving steering angle δ_{st} (computed according to Equation 6.11) is the steering angle that would be needed to run straight ahead.

13.2 Static Toe Angle Variation: Weave Test

The same adjusted *weave test* adopted in Chapter 12 is considered to test the compliance of both the analytical result of the steering torque offset $M^*_{S,st}$ and the *MATLAB*® single-track model in case a wheel alignment change regarding static toe angles ($\Psi_{i0} \neq 0$) is applied to the vehicle. The validation is carried out by comparison with the results of the simulations performed with the full-vehicle model in *MSC Adams*™, equipped with *Mod-A* set of tires, identified as the worst performing one in the previous chapters.

The same convergent static toe angles $\Psi_{10} = \Psi_{20} = -1$ deg is adopted in this case for both front and rear wheels.

13.2.1 Results with Effective Axle Aligning Moment

The elastokinematic and wheel alignment influence on the aligning moment is considered in this section. Therefore, the effective axle aligning moment $M_{z,i}$ is computed according to Equation 2.52, and the steering torque offset $M^*_{S,st}$ is computed according to Equation 7.9, namely

$$M^*_{S,st} = -2C_{M\alpha,10} \cdot \left[\Psi_{10} \cdot (A_1 + A_{1M}) + \gamma_{10} \cdot (B_1 + B_{1M}) + C_1 + C_{1M} + E_{1M} \cdot \varphi_r\right]$$

The results of the *weave test* simulations are reported in **Figures 13.8** and **13.9**, respectively, for the case with null road cross slope $\varphi_r = 0$ and nonnull road cross slope

FIGURE 13.8 Weave test simulation, *Mod-A* set of tires, steering torque vs lateral acceleration. Road cross slope φ_r = 0 rad, Ψ_{i0} = −1 deg, with elastokinematic correction on aligning moment.

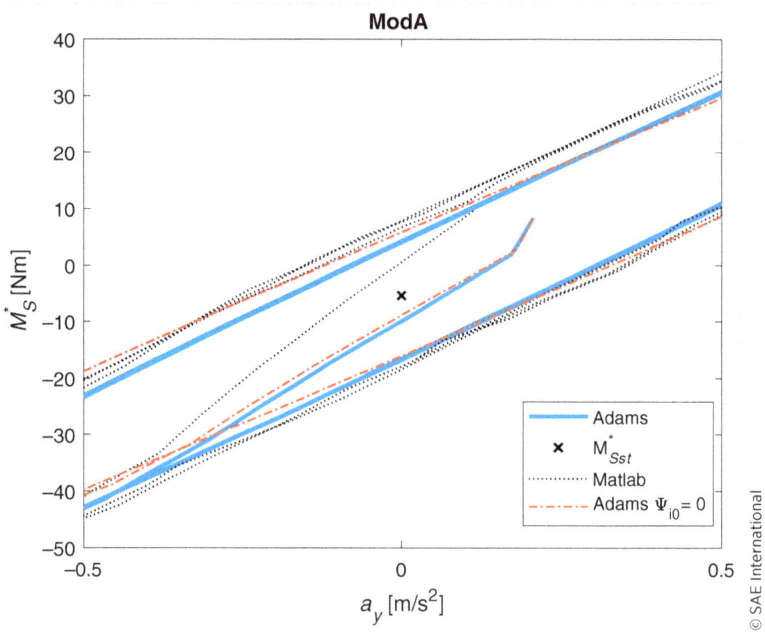

FIGURE 13.9 Weave test simulation, *Mod-A* set of tires, steering torque vs lateral acceleration. Road cross slope φ_r = 0.02 rad, Ψ_{i0} = −1 deg, with elastokinematic correction on aligning moment.

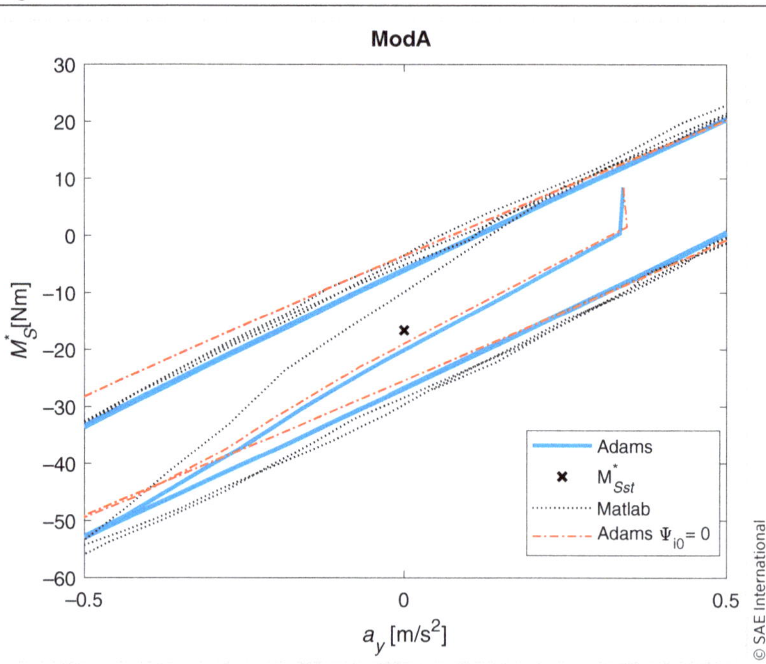

TABLE 13.7 Values of steering torque M_S^* at null lateral acceleration for left and right turn, from MSC Adams™ weave test simulation. Comparison with the analytically computed $M_{S,st}^*$ offset (according to Equation 7.9). Data in Tables 10.1, A.1–A.4.

Tire set	Ψ_{10} [deg]	φ_r [rad]	$M_{S,st}^*$ [Nm]	*MSC Adams™* $M_S^*(0)$ right [Nm]	*MSC Adams™* $M_S^*(0)$ left [Nm]
Mod-A	0	0	-5.139	5.8	-16.1
Mod-A	-1	0	-5.401	4.4	-16.8
Mod-A	0	0.02	-14.957	-3.56	-25.5
Mod-A	-1	0.02	-16.503	-6.1	-26.9

φ_r = 0.02 rad. Sufficient agreement is noticeable between *MSC Adams*™ full-vehicle simulation results and *MATLAB*® single-track model and analytical expression $M_{S,st}^*$, in both cases with null and nonnull road cross slope φ_r, as it can be observed from the values reported in **Table 13.7**.

A selection of results is reported in **Figures 13.10** and **13.11** for null and nonnull road cross slope simulations, in order to show the agreement between full-vehicle model and single-track model results.

As far as the *Mod-A* set of tire is concerned, the dependence of the steering torque offset $M_{S,st}^*$ (computed according to Equation 7.9) on the front static toe angle Ψ_{10} is represented in **Figures 13.12** and **13.13**, respectively, for the case with null road cross slope $\varphi_r = 0$ rad and with nonnull road cross slope $\varphi_r = 0.02$ rad. As it can be observed, it is not possible to reduce significantly the steering torque offset with this set of tires since the analytically computed corrective toe angles are too large (and unattainable).

13.2.2 Results Without Effective Axle Aligning Moment

An additional simulation is presented in this section to show the effects on the results, when the aligning moment is computed without the elastokinematic and wheel alignment influence discussed in Chapter 7. In other words, in this section the axle aligning moment $M_{z,i}$ is computed according to Equation 3.6, and the steering torque offset $M_{S,st}^*$ is computed according to Equations 6.15 and 6.16, namely

$$M_{z,i} = -C_{M\alpha,i} \cdot \alpha_i + M_{z0,iR} + M_{z0,iL}$$

where

$$-C_{M\alpha,i} = -\left(C_{M\alpha,iL} + C_{M\alpha,iR}\right)$$

Thus

$$M_{S,st}^* = -2C_{M\alpha,10} \cdot \left[\Psi_{10} \cdot A_1 + \gamma_{10} \cdot B_1 + C_1 + C_{1M} + \right. \\ \left. + \frac{F_{z1}}{2C_{M\alpha,10}} \cdot \left(n + \frac{2C_{M\alpha,10}}{C_{eff,1,AL}^*}\right) \cdot \varphi_r\right]$$

The comparison between the *MSC Adams*™ full-vehicle model results and the outcomes of both the *MATLAB*® single-track model and the $M_{S,st}^*$ analytical expression

FIGURE 13.10 Weave test simulation, set of results with *Mod-A* tires. Road slope $\varphi_r = 0$ rad. Static toe angles $\Psi_{10} = \Psi_{20} = -1$ deg.

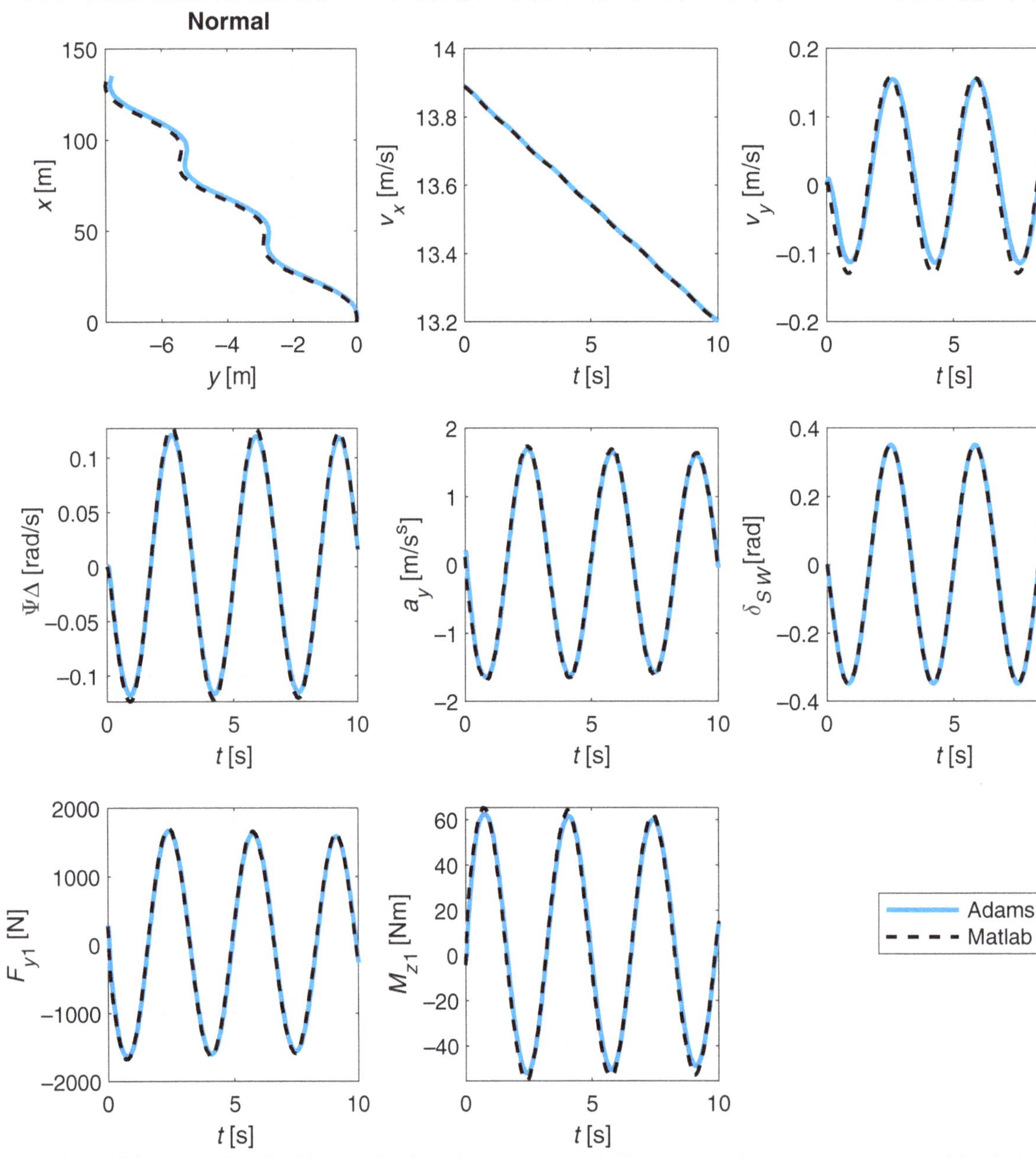

FIGURE 13.11 Weave test simulation, set of results with *Mod-A* tires. Road slope φ_r = 0.02 rad. Static toe angles $\Psi_{10} = \Psi_{20} = -1$ deg.

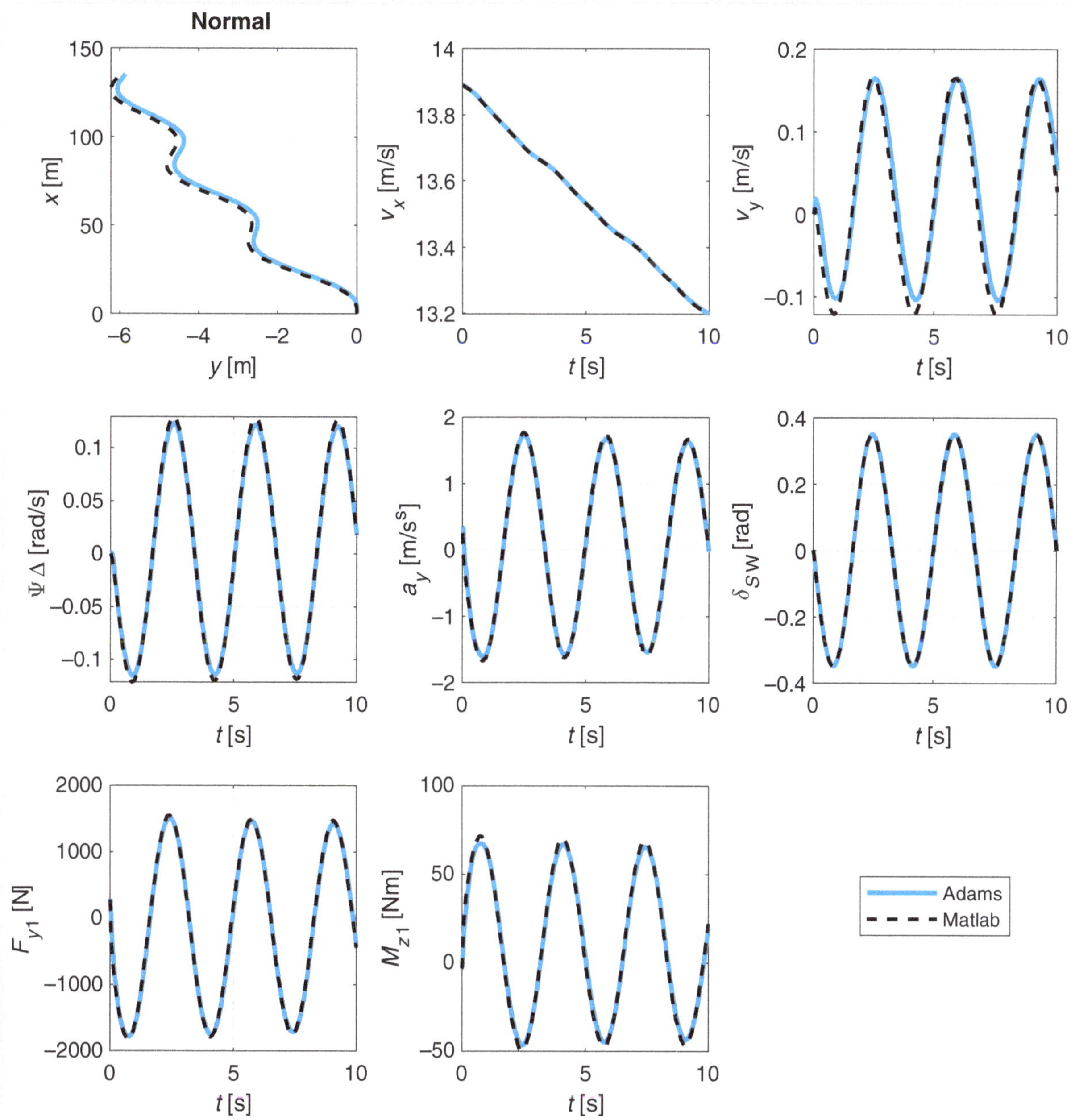

FIGURE 13.12 Steering torque offset $M^*_{S,st}$ dependence on front static toe angle Ψ_{10}. Road cross slope φ_r = 0 rad, *Mod-A* set of tires.

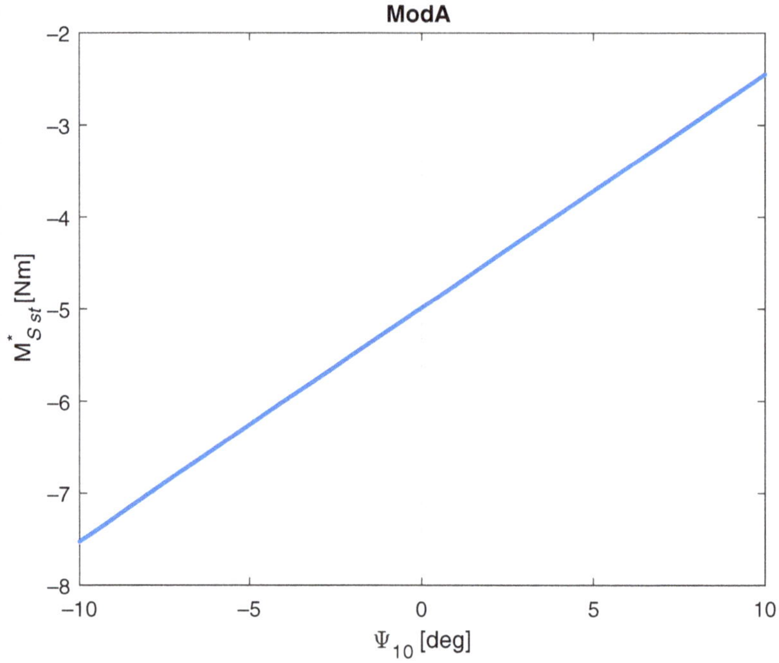

FIGURE 13.13 Steering torque offset $M^*_{S,st}$ dependence on front static toe angle Ψ_{10}. Road cross slope φ_r = 0.02 rad, *Mod-A* set of tires.

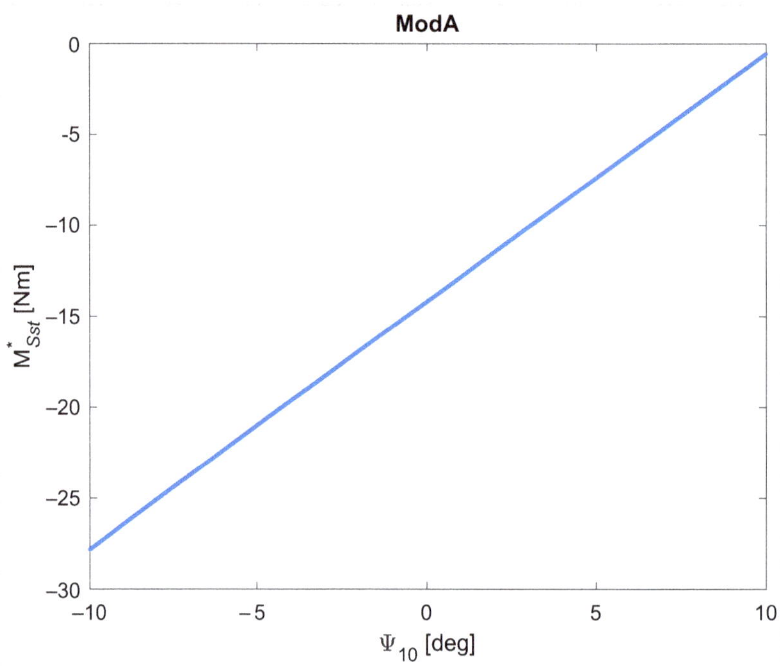

FIGURE 13.14 Weave test simulation, *Mod-A* set of tires, steering torque vs lateral acceleration. Road cross slope φ_r = 0.02 rad, Ψ_{i0} = −1 deg, without elastokinematics considered by means of the effective axle aligning moment: scarce correspondence can be noticed between *MSC Adams*™ full-vehicle model and *MATLAB*® single-track model results. Data in Tables 10.1, A.1–A.4.

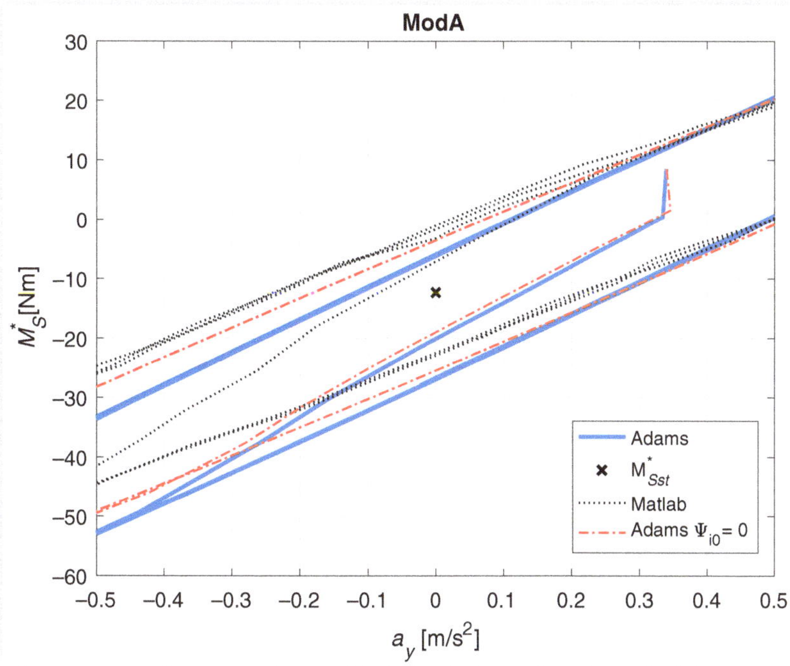

is shown in Figure 13.14. It can be inferred that without the effective axle characteristics considerations on aligning moment, the analytical expressions and the single-track model provide an improper description of the results obtained with the *MSC Adams*™ full-vehicle model.

13.3 Discussion

First of all, the inspection of the values of the different coefficients (reported in Tables 13.1 and 13.2) highlights that static camber γ_{i0} plays a negligible role in the steering pull issue with respect to static toe angles Ψ_{i0}. Static camber is multiplied by coefficients B_1, B_{1M}, and B_2, which are significantly smaller than the coefficients A_1, A_{1M}, and A_2 multiplying static toe angles. Therefore, the effects of reasonably small static camber values can be disregarded.

Furthermore, the trajectories reported in Figure 13.6 prove that static toe angles Ψ_{i0} does have an influence on the steering pull issue, particularly evident on the trajectory of a null steer test, also in case the road surface is flat (φ_r = 0). This happens since left and right tire cornering stiffnesses are different, which may be due to tire manufacturing tolerances. When the road cross slope φ_r assumes a nonnull value, a further difference between left and right tire characteristics is provoked by the generated load transfer, as it can be observed by comparing the A_1 and A_2 coefficients in Tables 13.1 and 13.2.

Then, the static toe angle values required to have null steering torque offset $M^*_{S,st}$ and straight-driving vehicle attitude β_{st} are computed, according to the analytical expressions, and reported in Tables 13.3 and 13.4. They turn out to be too large to be physically meaningful in most of the cases evaluated. Therefore, with the data of the tire sets considered, it is not possible to have null $M^*_{S,st}$ or β_{st} by acting solely on wheel alignment. To correct the steering pull related issues, the tire characteristics and shifts are to be modified, to the permitted extent, in order to limit the offsets without significant alterations of static toe angle values, as exemplified in the case presented in Table 13.5. This conclusion is in line with the viewpoint of Oh et al. [16]: vehicle and tire manufacturers should cooperate from initial stages of product development, to enhance vehicle pull performances.

The *MATLAB*® single-track model and the analytical expressions are validated for the cases including toe angle modifications by means of comparative simulations of null steer test and weave test performed with the full-vehicle model in *MSC Adams*™. The trend of the trajectories obtained with the *MSC Adams*™ full-vehicle model is suitably predicted by the single-track model, as shown in Figures 13.6 and 13.7. Moreover, the weave test results presented in Figures 13.8-13.11 suggest that the single-track model and analytical expressions results are compliant with the *MSC Adams*™ full-vehicle model simulation results.

Finally, the need to consider the effective axle aligning moment characteristics is pointed out by comparing the results of Figures 13.9 and 13.14: the analytical steering torque offset $M^*_{S,st}$ and the single-track results are closer to the *MSC Adams*™ full-vehicle simulation outputs when the elastokinematics and wheel alignment effects are considered by means of the effective axle aligning moment, as in Figures 13.8 and 13.9. A substantial difference is instead observable in Figure 13.14, where these effects are not included.

bibliography

1. Gim, G., "ch. Subjective and Objective Evaluations of Car Handling and Ride." In *Road and Off-Road Vehicle System Dynamics* (Boca Raton, FL: CRC Press/Taylor & Francis Group, 2014), 1395–1476.
2. Badiru, I. A., "Assessment of the Capability of EPS to Reduce Steering Wheel Pull and Vehicle Misalignment," *SAE Int. J. Passeng. Cars - Mech. Syst.* 8, no. 2 2015: 624–629, https://doi.org/10.4271/2015-01-1505.
3. Pottinger, M. G., "Pull: The Science of a Nuisance," *Tire Science and Technology* 41, no. 1 (Jan–Mar 2013): 40–59.
4. Pottinger, M. G., "Tire/Vehicle Pull: An Introduction Emphasizing Plysteer Effects," *Tire Science and Technology* 18, no. 3 (Jul–Sep 1990): 170–190.
5. Matyja, F. E., "Steering Pull and Residual Aligning Torque," *Tire Science and Technology* 15, no. 3 (Jul–Sep 1987): 207–240.
6. Bert, C. W., "Simplified Prediction of Ply Steer in Radial Tires," *Tire Science and Technology* 8, no. 1–2 (Jan–Jun 1980): 3–9.
7. Pacejka, H. B., *Tire and Vehicle Dynamics*, 3rd ed. (Amsterdam, the Netherlands: Elsevier, 2012).
8. SAE International, "J670 Vehicle Dynamics Terminology," SAE International Standard, Jan 2008.
9. Mundl, R., Fischer, M., Wajroch, M., and Lee, S. W., "Simulation and Validation of the Ply Steer Residual Aligning Torque Induced by the Tyre Tread Pattern," *Vehicle System Dynamics* 43, no. Supplement (2005): 434–443.
10. Ohishi, K., Suita, H., and Ishihara, K., "The Finite Element Approach to Predict the Plysteer Residual Cornering Force of Tires," *Tire Science and Technology* 30, no. 2 (Apr–Jun 2002): 122–133.
11. Murakoshi, H., Ide, H., and Nishihata, S., "An Approach to Vehicle Pull Using a Tire Finite Element Model," *Tire Science and Technology* 20, no. 4 (Oct–Dec 1992): 212–229.
12. Lindenmuth, B. E., "Tire Conicity and Ply Steer Effects on Vehicle Performance," SAE Technical Paper 740074, 1974, https://doi.org/10.4271/740074.
13. Cho, Y. G., "Steering Pull and Drift Considering Road Wheel Alignment Tolerance during High Speed Driving," *International Journal of Vehicle Design* 54, no. 1 (2010): 73–91.
14. Topping, R. W., "Tire Induced Steering Pull," SAE Technical Paper 750406, 1975, https://doi.org/10.4271/750406.
15. Murari, T. B., Lima, D. M., Zebende, G. F., and Moret, M. A., "Vehicle Steering Pull: From Product Development to Manufacturing," *Product: Management & Development* 14, no. 1 (Jun 2016): 22–31.
16. Oh, S. H., Cho, Y. H., and Gim, G., "Identification of a Vehicle Pull Mechanism," in *Seoul 2000 FISITA World Automotive Congress*, Seoul, South Korea, Jun 2000.
17. Lee, J. H., "Analysis of Tire Effect on the Simulation of Vehicle Straight Line Motion," *Vehicle System Dynamics*, 33, no. 6 (2000): 373–390.
18. Yamazaki, S., Fujikawa, T., Suzuki, T., and Yamaguchi, I., "Influence of Wheel Alignment and Tire Characteristics on Vehicle Drift," *Tire Science and Technology* 26, no. 3 (Jul–Sep 1998): 186–205.

19. Yu, H. J., "Tire/Suspension Aligning Moment and Vehicle Pull," *Tire Science and Technology* 28, no. 3, (Jul–Sep 2000): 157–177.
20. Park, K., Heo, S. J., Kang, D. O., Jeong, J. I, Yi, J. H., Lee, J. H., and Kim, K. W., "Robust Design Optimization of Suspension System Considering Steering Pull Reduction," *International Journal of Automotive Technology* 14, no. 6 (Dec 2013): 927–933.
21. Kubota, M., Yoshizawa, M., and Mouri, H., "An Investigation of a Steering-Pull Reduction Method Using the Electric Power Steering System," in *14th Asia Pacific Automotive Engineering Conference*, Hollywood, United States, August 5–8, 2007.
22. Lattuada, A., "Straight Ahead Running of Road Vehicles," Master's thesis, Politecnico di Milano, 2018, supervisor: G. R. M. Mastinu.
23. Lattuada, A., "Tire-Induced Vehicle Pull," Master's thesis, KTH Royal Institute of Technology, 2018, examiner: J. Jerrelind.
24. Mastinu, G. R. M., ch. Conceptual Design of Road Vehicles Related to Dynamics. In *Road and Off-Road Vehicle System Dynamics* (Boca Raton, FL: CRC Press/Taylor & Francis Group, 2014), pp. 329–394.
25. Boiocchi, M. and Matrascia, G., ch. Pneumatic Tire, Construction and Testing. In *Road and Off-Road Vehicle System Dynamics* (Boca Raton, FL: CRC Press/Taylor & Francis Group, 2014), pp. 657–696.
26. Clark, S. K., "Mechanics of Pneumatic Tires," U.S. Department of Transportation, National Highway Traffic Safety Administration, Nov 1971.
27. Sala, G., Di Landro, L., Airoldi, A., and Bettini, P., *Tecnologie e Materiali Aerospaziali* (Politecnico di Milano: Dipartimento di Ingegneria Aerospaziale, 2009).
28. Pottinger, M. G., "Tire with No Ply Steer Belt," US Patent 3,945,422, Mar 1976.
29. Michelin, "XAS Michelin Classics," 1965, http://classic.michelin.com/en/Classic-Tyre-range/XAS.
30. MTS Systems, "Flat-Trac Tire Test Systems," Nov 2017, https://www.mts.com/cs/groups/public/documents/library/dev_002227.pdf.
31. Mitschke, M., *Dynamik der Kraftfahrzeuge - Band C: Fahrverhalten* (Wiesbaden, Germany: Springer Vieweg, 1990).
32. Matschinsky, W., ch. Suspension Systems. In *Road and Off-Road Vehicle System Dynamics* (Boca Raton, FL: CRC Press/Taylor & Francis Group, 2014), 727–768.
33. Kageyama, I., ch. Steering System. In *Road and Off-Road Vehicle System Dynamics* (Boca Raton, FL: CRC Press/Taylor & Francis Group, 2014), 919–942.
34. Pfeffer, P. E. and Harrer, M., "On-Centre Steering Wheel Torque Characteristics during Steady State Cornering," SAE Technical Paper 2008-01-0502, 2008, https://doi.org/10.4271/2008-01-0502.
35. Koegeler, M., Schick, B., and Ecker, J., "Continuous Development of Steering Feel Using Objective and Model Based Methods," in *7th International Munich Chassis Symposium*, Germany, Springer Fachmedien Wiesbaden, 2016.
36. Farrer, D. G., "An Objective Measurement Technique for the Quantification of On-Centre Handling Quality," SAE Technical Paper 930827, 1993, https://doi.org/10.4271/930827.
37. Salaani, M. K., Heydinger, G. J., and Grygier, P. A., "Experimental Steering Feel Performance Measures," SAE Technical Paper 2004-01-1074, 2004, https://doi.org/10.4271/2004-01-1074.
38. Kim, J., "Objectification of On-Center Handling Characteristics Based on Spring-Mass-Damper System," *International Journal of Automotive Technology* 12, no. 6 (Dec 2011): 857–864.

39. Nybacka, M., He, X., Su, Z., Drugge, L., and Bakker, E., "Links between Subjective Assessments and Objective Metrics for Steering, and Evaluation of Driver Ratings," *Vehicle System Dynamics* 52, no. sup 1 (2014): 31–50.

40. "Road Vehicles - Test Method for the Quantification of On-Centre Handling - Part 1: Weave Test," SS-ISO13674-1, 2010.

41. Ljungberg, M., Nybacka, M., Gómez, G. G., and Katzourakis, D., "Electric Power Assist Steering System Parametrization and Optimization Employing Computer-Aided Engineering," SAE Technical Paper 2015-01-1500, Apr 2015, https://doi.org/10.4271/2015-01-1500.

42. Yum, K., "Analysis of Influence of Tire F & M on Improvement of Vehicle On-Center Steering," *SAE Int. J. Passeng. Cars – Mech. Syst.* 9, no. 2 (Jun 2016): 546-550. https://doi.org/10.4271/2016-01-1569.

43. Papagiannakis, A. T. and Masad, E. A., *Pavement Design and Materials* (Hoboken, NJ: John Wiley & Sons, 2008).

44. Pennsylvania State University, "Crown and Cross-Slope," Jan 2005, https://www.dirtandgravel.psu.edu/sites/default/files/General%20Resources/Technical%20Bulletins/IB_Crown_and_Cross_Slope.pdfhttps://www.dirtandgravel.psu.edu/sites/default/files/General%20Resources/Technical%20Bulletins/IB_Crown_and_Cross_Slope.pdf.

45. Ministero delle Infrastrutture e dei Trasporti, Ispettorato Generale per la Circolazione e la Sicurezza Stradale, "Norme funzionali e geometriche per la costruzione delle strade," Decreto Ministeriale 5, Nov 2001.

46. State of California, Caltrans, "Highway Design Manual," Dec 2016, http://www.dot.ca.gov/design/manuals/hdm/chp0300.pdf.

47. Commonwealth of Massachusetts, "521 CMR: Architectural Access Board - Accessible Route," Jan 2006, https://www.mass.gov/files/documents/2016/09/uz/521020.pdf.

48. Schuring, D. J., "Energy Loss of Pneumatic Tires under Freely Rolling, Braking and Driving Conditions," *Tire Science and Technology* 4, no. 1 (Feb 1976): 3–15.

49. Guzzella, L. and Sciarretta, A., *Vehicle Propulsion Systems*, 3rd ed. (Berlin Heidelberg: Springer-Verlag, 2013).

50. Schuring, D. J., "Effect of Tire Rolling Loss on Vehicle Fuel Consumption," *Tire Science and Technology* 22, no. 3 (Jul–Sep 1994): 148–161.

51. Mitschke, M., *Dynamik der Kraftfahrzeuge - Band B: Schwingungen* (Wiesbaden, Germany: Springer Vieweg, 1990).

52. "Mechanical Vibration - Road Surface Profiles - Reporting of Measured Data," ISO 8608, Nov 2016.

53. Wennerström, E., S. Nordmark, B. Thorvald, *Stochastic Analysis and Comfort* (Stockholm, Sweden: KTH Royal Institute of Technology, 2011). (SD2225, Ground Vehicle Dynamics).

54. Hrovat, D., ch. Active and Semi-Active Suspension Control. In *Road and Off-Road Vehicle System Dynamics* (Boca Raton, FL: CRC Press/Taylor & Francis Group, 2014), 1179–1226.

55. Popp, K., ch. Ride Comfort and Road Holding. In *Road and Off-Road Vehicle System Dynamics* (Boca Raton, FL: CRC Press/Taylor & Francis Group, 2014), 1059–1092.

56. Bogsjö, K., "Road Profile Statistics Relevant for Vehicle Fatigue," Ph.D. dissertation, Lund University, 2007, Mathematical Statistics.

57. Johannesson, P. and Rychlik, I., "Laplace Processes for Describing Road Profiles," *Procedia Engineering* 66(Supplement C):464–473, 2013, Fatigue Design 2013, International Conference Proceedings.

58. Ammon, D., ch. Detailed Modeling, Simulation, and Analysis of Vehicle Dynamics. In *Road and Off-Road Vehicle System Dynamics* (Boca Raton, FL: CRC Press/Taylor & Francis Group, 2014), 1058–1023.

59. Gent, A.N. and Walter, J.D., *The Pneumatic Tire* (Washington, D.C.: National Highway Traffic Safety Administration (NHTSA) - US Department of Transportation, August 2005), available online at: www.nhtsa.gov/staticfiles/safercar/pdf/PneumaticTire_HS-810-561.pdf.

60. Rhyne, T. B., "Development of a Vertical Stiffness Relationship for Belted Radial Tires," *Tire Science and Technology* 33, no. 3 (Jul–Sep 2005): 136-155.

61. Previati, G., Gobbi, M., and Mastinu, G., "Mass Properties of Cars - An Investigation," in M. Spiryagin, T. Gordon, C. Cole, and T. McSweeney, (Eds.), *Dynamics of Vehicles on Roads and Tracks: Proceedings of the 25th International Symposium on Dynamics of Vehicles on Roads and Tracks (IAVSD 2017), 14-18 August 2017, Rockhampton, Queensland, Australia*, 1st ed. (Boca Raton, FL: CRC Press LLC, Taylor & Francis Group, 2017), https://www.crcpress.com/Dynamics-of-Vehicles-on-Roads-and-Tracks-Proceedings-of-the-25th-International/Spiryagin-Gordon-Cole-McSweeney/p/book/9781138035713.

62. Ammon, D., "Problems in Road Surface Modelling," *Vehicle System Dynamics: International Journal of Vehicle Mechanics and Mobility*, 20, no. sup 1 (1992): 28–41.

63. Shinozuka, M., "Simulation of Multivariate and Multidimensional Random Processes," *Journal of the Acoustical Society of America* 49, no. 1 (1971): 357–368.

64. Shinozuka, M., and Kobori, T., "Fatigue Analysis of Highway Bridges," *Japan Society of Civil Engineers* 208 (Dec 1972): 137–148.

65. Abe, M., *Vehicle Handling Dynamics*, 2nd ed. (Amsterdam, the Netherlands: Elsevier, 2015).

66. *Adams/Tire*, MSC Software, 2014, https://simcompanion.mscsoftware.com/infocenter/index?page=content&cat=1VMO50&channel=DOCUMENTATION.

67. *Adams/Car*, MSC Software, 2014, https://simcompanion.mscsoftware.com/infocenter/index?page=content&cat=1VMO50&channel=DOCUMENTATION.

68. Wennerström, E., Nordmark, S., and Thorvald, B., *General Rigid Body Motion* (Stockholm, Sweden: KTH Royal Institute of Technology, 2011), (SD2225, Ground Vehicle Dynamics).

69. Cheli, F. and Diana, G., *Advanced Dynamics of Mechanical Systems* (Cham, Switzerland: Springer, 2015).

70. Jazar, R. N., *Vehicle Dynamics* (New York: Springer Science+Business Media, LLC, 2008).

71. Besselink, I., ch. Appendix 3 - MF-Tire/MF-Swift Parameters and Estimation Methods. In *Tire and Vehicle Dynamics*, 3rd ed. (Amsterdam, the Netherlands: Elsevier, 2012), 613–625.

72. Wennerström, E., Nordmark, S., and Thorvald, B., *Vehicle Stability* (Stockholm, Sweden: KTH Royal Institute of Technology, 2011), (SD2225, Ground Vehicle Dynamics).

73. Guiggiani, M., *The Science of Vehicle Dynamics*, 2nd ed. (Cham, Switzerland: Springer International Publishing, 2014).

appendix A

A man should keep his little brain attic stocked with all the furniture that he is likely to use, and the rest he can put away in the lumber-room of his library, where he can get it if he wants it.

—Sir Arthur Conan Doyle,
The Five Orange Pips (1891)

A.1 Derivation of 3DoFs Model Equations of Motion

The coordinates chosen to describe the vehicle motions are expressed as functions of the inertial frame coordinates

$$\begin{gathered} u = \dot{X}cos\Psi + \dot{Y}sin\Psi \\ v = -\dot{X}sin\Psi + \dot{Y}cos\Psi \\ r = \dot{\Psi} \quad (\text{yawrate}) \\ \varphi \quad (\text{roll angle}) \end{gathered} \tag{A.1}$$

The Lagrange's equation [69] is written as in Equation A.2, where $\underline{q}$ is the vector of variables chosen to describe the motion, E_k, D, and V are, respectively, the kinetic energy, the dissipation function, and the potential energy. Finally, $\underline{Q}$ is the vector of the nonconservative forces acting on the system, obtained from the virtual work δ^*L with virtual displacements $\underline{q}$.

$$\begin{aligned} &\left\{\frac{d}{dt}\left(\frac{\partial E_k}{\partial \dot{\underline{q}}}\right)\right\}^T - \left\{\frac{\partial E_k}{\partial \underline{q}}\right\}^T + \left\{\frac{\partial D}{\partial \dot{\underline{q}}}\right\}^T + \left\{\frac{\partial V}{\partial \underline{q}}\right\}^T = \underline{Q} \\ &\delta^* L = \underline{Q}^T \delta^* \underline{q} \end{aligned} \tag{A.2}$$

Modified Lagrange's equation pertaining to the specific coordinate expressed in Equation A.1 can be written as [7]

$$\begin{gathered} \frac{\partial E_k}{\partial \dot{X}} = \frac{\partial E_k}{\partial u}\frac{\partial u}{\partial \dot{X}} + \frac{\partial E_k}{\partial v}\frac{\partial v}{\partial \dot{X}} = \frac{\partial E_k}{\partial u}cos\Psi - \frac{\partial E_k}{\partial v}sin\Psi \\ \frac{\partial E_k}{\partial \dot{Y}} = \frac{\partial E_k}{\partial u}\frac{\partial u}{\partial \dot{Y}} + \frac{\partial E_k}{\partial v}\frac{\partial v}{\partial \dot{Y}} = \frac{\partial E_k}{\partial u}sin\Psi + \frac{\partial E_k}{\partial v}cos\Psi \\ \frac{\partial E_k}{\partial \dot{\Psi}} = \frac{\partial E_k}{\partial r} \\ \frac{\partial E_k}{\partial \Psi} = \frac{\partial E_k}{\partial u}v - \frac{\partial E_k}{\partial v}u \end{gathered} \tag{A.3}$$

Therefore, the set of modified Lagrangian equations is reported for the chosen coordinates

$$\begin{aligned}
&\frac{d}{dt}\frac{\partial E_k}{\partial u} - r\frac{\partial E_k}{\partial v} = Q_u \\
&\frac{d}{dt}\frac{\partial E_k}{\partial v} + r\frac{\partial E_k}{\partial u} = Q_v \\
&\frac{d}{dt}\frac{\partial E_k}{\partial r} - v\frac{\partial E_k}{\partial u} + u\frac{\partial E_k}{\partial v} = Q_r \\
&\frac{d}{dt}\frac{\partial E_k}{\partial \dot{\varphi}} - \frac{\partial E_k}{\partial \varphi} + \frac{\partial D}{\partial \dot{\varphi}} + \frac{\partial V}{\partial \varphi} = 0
\end{aligned} \tag{A.4}$$

The equations of motion reported in Equation 2.57 are obtained by substituting the following terms into Equation A.4.

A.1.1 Kinetic Energy

The generalized expression of the kinetic energy of a rotating rigid body can be written as [70]

$$\begin{aligned}
E_k = &\frac{1}{2}\left[m\left(v_x^2 + v_y^2 + v_z^2\right) + J_{11}\omega_x^2 + J_{22}\omega_y^2 + J_{33}\omega_z^2\right] - \\
&-\left(J_{12}\omega_x\omega_y + J_{23}\omega_y\omega_z + J_{13}\omega_x\omega_z\right)
\end{aligned} \tag{A.5}$$

With the coordinate axes as indicated in Figure 2.6, the Euler's angles (see **Figure A.1**) are used in order to transform the coordinates to the vehicle-fixed reference system [68–70].

- **Velocity.** The velocity of the center of gravity of vehicle – in the body-fixed coordinate system – is given by the components u and v, and the contributions due to the angular velocities and position of the center of gravity with respect to the rotation axes intersection. The pitch angle can be considered as the constant inclination of the roll axis

$$\theta = \frac{h_2 - h_1}{l} \tag{A.6}$$

FIGURE A.1 Euler's angle rotations (from [68]).

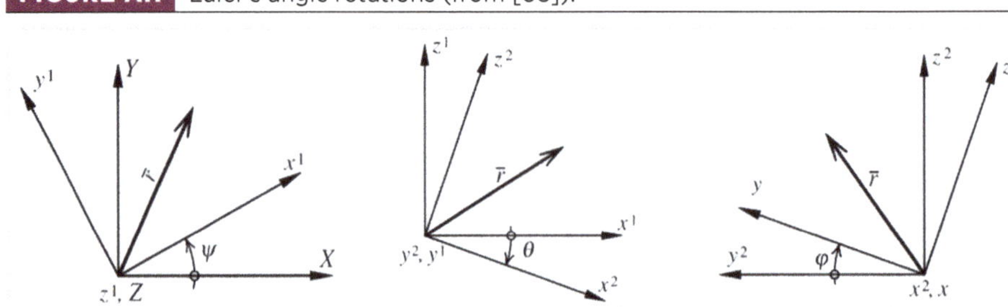

Thus

$$\underline{r_G} = \begin{Bmatrix} -h'\theta \\ h'\varphi \\ -h' \end{Bmatrix} \tag{A.7}$$

Therefore, the velocity of the center of gravity is [69]

$$\underline{v_G} = \begin{Bmatrix} u \\ v \\ 0 \end{Bmatrix} + \begin{Bmatrix} \dot{\varphi} \\ 0 \\ r \end{Bmatrix} \wedge \underline{r_G} = \begin{Bmatrix} u - h'\varphi r \\ v + h'\dot{\varphi} \\ h'\varphi\dot{\varphi} \end{Bmatrix} \tag{A.8}$$

- **Angular velocity.** The angular velocity of the system, in the vehicle reference system, is obtained with the Euler's angles, representing the following rotations, in order (as in Figure A.1)
  - yaw Ψ around $\hat{Z}$
  - pitch θ around $\widehat{y_1}$
  - roll φ around $\widehat{x_2}$

Using the proper transformation matrices that can be found in literature [68, 69], the angular velocity is expressed in the body-fixed coordinate system (for small angles Ψ, θ, and φ, and $\dot{\theta} = 0$)

$$\underline{\omega_b} = \begin{Bmatrix} \dot{\varphi} - rsin\theta \\ rsin\varphi cos\theta \\ rcos\varphi cos\theta \end{Bmatrix} \approx \begin{Bmatrix} \dot{\varphi} - r\theta \\ r\varphi \\ r \end{Bmatrix} \tag{A.9}$$

The kinetic energy of the rigid body can finally be computed as reported by Pacejka [7]

$$E_k = \frac{1}{2} m\left[\left(u - h'\varphi r\right)^2 + \left(v + h'\dot{\varphi}\right)^2 \right] + \frac{1}{2}\left[I_x\dot{\varphi}^2 + I_y\left(r\varphi\right)^2 + I_z r^2 \right] - \\ - I_{xz} r\dot{\varphi} + \frac{1}{2} I_z\left(-\varphi^2 r^2 + 2\theta r\dot{\varphi}\right) \tag{A.10}$$

A.1.2 Dissipative Function

The dissipative function accounts for the roll damping given by suspensions (with roll damping coefficients indicated as $c_{\varphi 1}$ and $c_{\varphi 2}$, respectively, for front and rear suspensions)

$$D = \frac{1}{2}\left(c_{\varphi 1} + c_{\varphi 2}\right)\dot{\varphi}^2 \tag{A.11}$$

A.1.3 Potential Energy

The potential energy accounts for both the contribution from roll stiffness and the contribution from center of gravity height. The latter is expanded in a Taylor's series in order to obtain a quadratic form of potential energy, suitable for the derivation in Lagrange's equation

$$\begin{aligned} h_G &= h'cos\varphi = h' + \left(\left.\frac{\partial h_G}{\partial \varphi}\right|_0\right)\varphi + \frac{1}{2}\left(\left.\frac{\partial^2 h_G}{\partial \varphi^2}\right|_0\right)\varphi^2 = h' - \frac{1}{2}h'\varphi^2 \\ V &= \frac{1}{2}\left(k_{\varphi 1} + k_{\varphi 2}\right)\dot{\varphi}^2 + mgh'\left(1 - \frac{1}{2}\varphi^2\right) \end{aligned} \tag{A.12}$$

A.1.4 Nonconservative Forces

The contributions of nonconservative forces and moments (given by road–tires contact) are reported in the different directions

$$\begin{aligned} Q_u &= F_{x,1}cos\delta - F_{y,1}sin\delta + F_{x,2} \\ Q_v &= F_{x,1}sin\delta + F_{y,1}cos\delta + F_{y,2} \\ Q_r &= a_1F_{x,1}sin\delta + a_1F_{y,1}cos\delta + M_{z,1} + a_2F_{y,2} + M_{z,2} \end{aligned} \tag{A.13}$$

A.2 Sign Conventions for Force, Moment, and Wheel Slip

A.2.1 Modified SAE (Pacejka)

The sign convention chosen for this book is consistent with the *Pacejka's modified SAE* [7], represented in **Figure A.2**.

With this convention, when no offsets are considered, side force is positive when slip angle is positive, while the opposite happens with aligning moment.

FIGURE A.2 *Pacejka's modified SAE* sign convention for force, moment, and wheel slip angle [7].

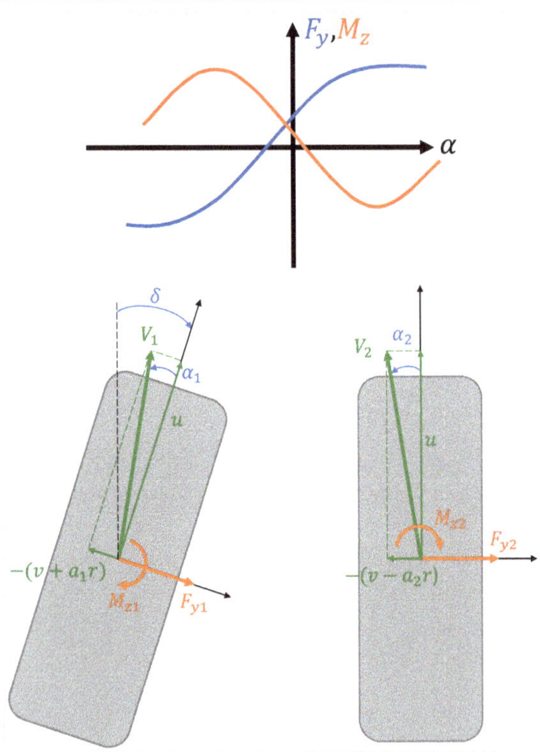

A.2.2 SAE

Another sign convention, here reported but not adopted for this book, is the *SAE* [7], represented in **Figure A.3**.

With this convention, when no offsets are considered, side force is positive when slip angle is negative, while the opposite happens with aligning moment.

A.2.3 ISO

Another sign convention, reported here but not adopted for this book, is the *ISO* [7], represented in **Figure A.4**.

This convention has the *z-axis* upward and the *y-axis* leftward.

FIGURE A.3 *SAE* sign convention for force, moment, and wheel slip angle [7].

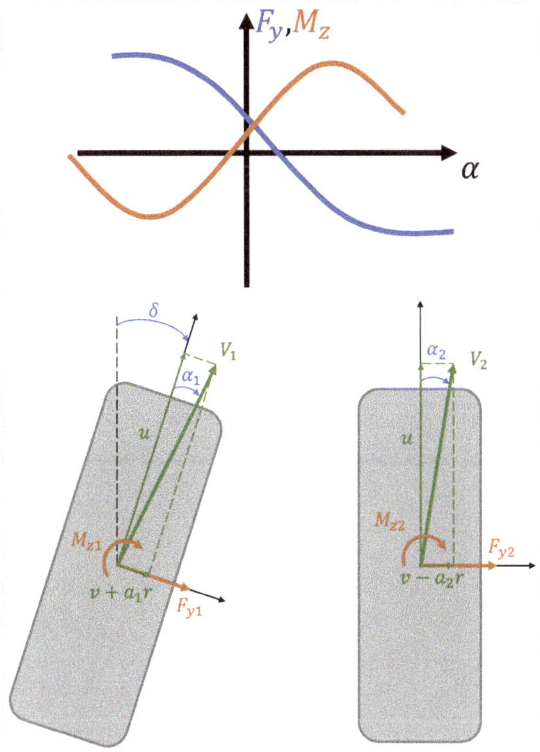

FIGURE A.4 *ISO* sign convention for force, moment, and wheel slip angle [7].

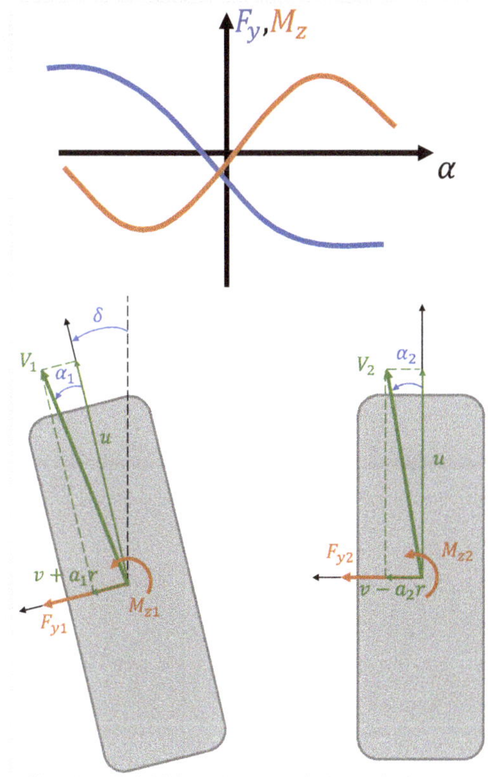

A.3 Approximation of Tire Side Force

A comparison is made between a linearization of tire side force $F_y = F_y(\alpha, \gamma, F_z)$ and the approximation adopted by Pacejka [7].

- **Linearization**

$$F_y = \left(F_y \Big|_{\substack{\alpha=0 \\ \gamma=0 \\ F_z=F_{z0}}} \right) + \left(\frac{\partial F_y}{\partial \alpha} \Big|_{\substack{\alpha=0 \\ \gamma=0 \\ F_z=F_{z0}}} \right) \cdot \alpha + \left(\frac{\partial F_y}{\partial \gamma} \Big|_{\substack{\alpha=0 \\ \gamma=0 \\ F_z=F_{z0}}} \right) \cdot \gamma + \left(\frac{\partial F_y}{\partial F_z} \Big|_{\substack{\alpha=0 \\ \gamma=0 \\ F_z=F_{z0}}} \right) \cdot \left(F_z - F_{z0} \right) =$$
$$= F_{y0} + C_\alpha \cdot \alpha + C_\gamma \cdot \gamma + C_{Fz} \cdot \Delta F_z \qquad \text{(A.14)}$$

- **Approximation from Pacejka [7]**

$$F_y = F_{y0} + \left(C_\alpha + \xi_\alpha \cdot \Delta F_z \right) \cdot \alpha + \left(C_\gamma + \xi_\gamma \cdot \Delta F_z \right) \cdot \gamma \qquad \text{(A.15)}$$

FIGURE A.5 Error of the linearization (*Lin*, solid lines) and Pacejka's approximation (*Pac*, dashed lines) of F_y with respect to Magic Formula curves, for different normal loads, camber angles and slip angles.

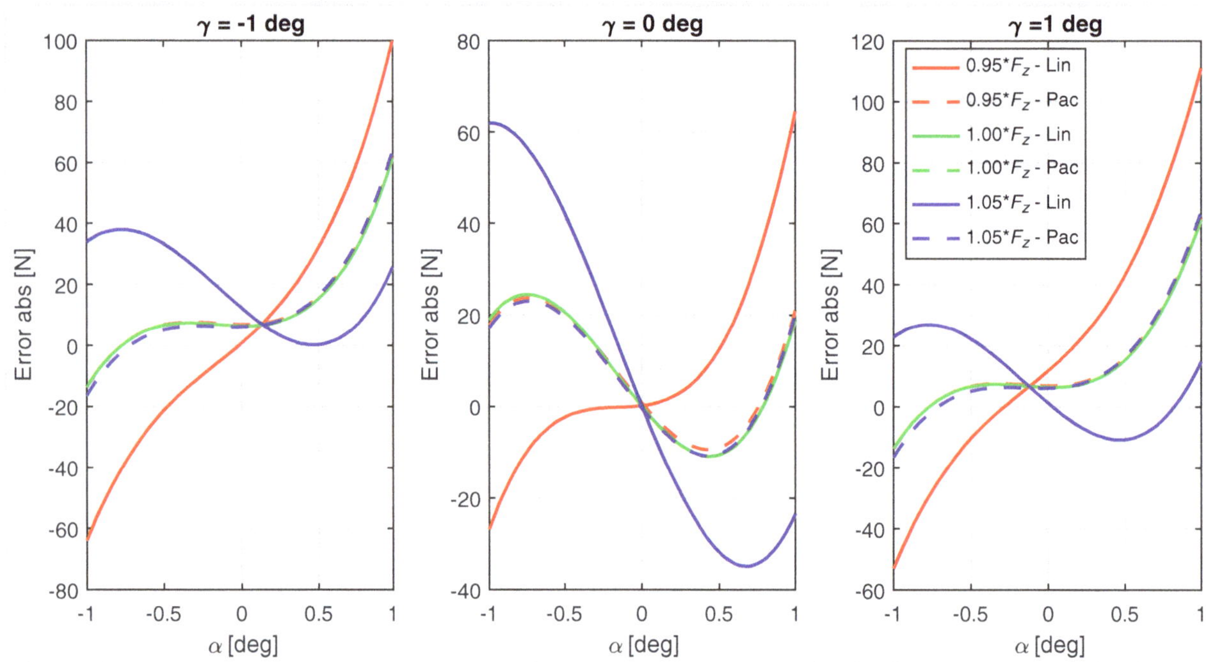

On comparing the approximations with a Magic Formula characteristic curve (for different slip angles $\alpha \in [-1 \text{ deg}, 1 \text{ deg}]$, three camber angles $\gamma = [-1 \text{ deg}, 0 \text{ deg}, 1 \text{ deg}]$, and three loads $F_z = [0.95, 1, 1.05] \cdot F_{z0}$), the approximation of Pacejka results in a smaller error with respect to the linearization, as it can be observed in **Figure A.5**. The error is defined using the difference with respect to the Magic Formula curve

$$err_{abs} = \frac{F_{y,MF} - F_{y,approx}}{F_{y,MF}}$$

A.4 Effective Axle Side Force

In this section of the Appendix, the full derivation of the effective axle side force (Equations 2.15–2.36) is reported.

The linear equation of tire side force, for the axle i (1 = front, 2 = rear), and left (L) or right (R) tire, reads

$$F_{y,iL/R} = C_{F\alpha,iL/R}\left(F_z\right) \cdot \alpha_{iL/R} + C_{F\gamma,iL/R}\left(F_z\right) \cdot \gamma_{iL/R} \tag{A.16}$$

The ply-steer equivalent side slip angle and conicity equivalent camber angle are introduced in Equation 1.2. They are reported here, for the axle i (1 = front, 2 = rear), and left (L) or right (R) tire:

$$\gamma_{con,iL/R} = \frac{\alpha_{FM0,iL/R}}{\frac{C_{M\gamma,i0L/R}}{C_{M\alpha,i0L/R}} + \frac{C_{F\gamma,i0L/R}}{C_{F\alpha,i0L/R}}} = \frac{\frac{F_{y0,iL/R}}{C_{F\alpha,i0L/R}} + \frac{M_{z0,iL/R}}{C_{M\alpha,i0L/R}}}{\frac{C_{M\gamma,i0L/R}}{C_{M\alpha,i0L/R}} + \frac{C_{F\gamma,i0L/R}}{C_{F\alpha,i0L/R}}}$$

$$\alpha_{ply,iL/R} = \Delta\alpha_{0,iL/R} - \frac{C_{F\gamma,i0L/R}}{C_{F\alpha,i0L/R}} \cdot \gamma_{con,iL/R} = \frac{F_{y0,iL/R}}{C_{F\alpha,i0L/R}} - \frac{C_{F\gamma,i0L/R}}{C_{F\alpha,i0L/R}} \cdot \gamma_{con,iL/R} \tag{A.17}$$

Hypotheses. The linearization coefficients $\xi_{\alpha,i}$ and $\xi_{\gamma,i}$, describing the dependence of the cornering stiffness $C_{F\alpha,i}$ and camber coefficient $C_{F\gamma,i}$ on vertical load variation $\Delta F_{z,i}$, are assumed to be equal for left and right wheels

$$\xi_{\alpha,iL} = \xi_{\alpha,iR} = \xi_{\alpha,i}$$
$$\xi_{\gamma,iL} = \xi_{\gamma,iR} = \xi_{\gamma,i}$$

On the other hand, the cornering stiffness and camber coefficient are considered to be different for left and right tires, in order to accommodate the description of asymmetric tire characteristics. With the Equation 2.2, they can be written as

$$C_{F\alpha,iL/R} = C_{F\alpha,i0L/R} \pm \left(\left. \frac{\partial C_{F\alpha}}{\partial F_z} \right|_{F_{z0,iL/R}} \right) \cdot \Delta F_{z,i} = C_{F\alpha,i0L/R} \pm \xi_{\alpha,i} \cdot \Delta F_{z,i}$$

$$C_{F\gamma,iL/R} = C_{F\gamma,i0L/R} \pm \left(\left. \frac{\partial C_{F\gamma}}{\partial F_z} \right|_{F_{z0,iL/R}} \right) \cdot \Delta F_{z,i} = C_{F\gamma,i0L/R} \pm \xi_{\gamma,i} \cdot \Delta F_{z,i} \tag{A.18}$$

Derivation. Introducing into Equation A.16 the ply-steer and conicity equivalent angles, the cornering stiffness and camber coefficient and the static toe and camber angles, the side forces for axle i (for left L and right R side) can be written as

$$\begin{aligned} F_{y,iL} &= \left(C_{F\alpha,i0L} + \xi_{\alpha,i}\Delta F_{z,i}\right)\left(\alpha_i - \Psi_{i0} + \alpha_{ply,iL}\right) + \\ &+ \left(C_{F\gamma,i0L} + \xi_{\gamma,i}\Delta F_{z,i}\right)\left(\gamma_i - \gamma_{i0} + \gamma_{con,iL}\right) \\ F_{y,iR} &= \left(C_{F\alpha,i0R} - \xi_{\alpha,i}\Delta F_{z,i}\right)\left(\alpha_i + \Psi_{i0} + \alpha_{ply,iR}\right) + \\ &+ \left(C_{F\gamma,i0R} - \xi_{\gamma,i}\Delta F_{z,i}\right)\left(\gamma_i + \gamma_{i0} + \gamma_{con,iR}\right) \end{aligned} \tag{A.19}$$

It is possible to introduce the average of left and right cornering stiffness and camber coefficient as

$$C_{F\alpha,i0} = \frac{C_{F\alpha,i0L} + C_{F\alpha,i0R}}{2}$$
$$C_{F\gamma,i0} = \frac{C_{F\gamma,i0L} + C_{F\gamma,i0R}}{2} \tag{A.20}$$

The side force for the axle i is given by the sum of left and right forces, namely

$$
\begin{aligned}
F_{y,i} &= F_{y,iL} + F_{y,iR} = \\
&= 2C_{F\alpha,i0}\cdot\alpha_i + 2C_{F\gamma,i0}\cdot\gamma_i - 2\xi_{\alpha,i}\Delta F_{z,i}\cdot\Psi_{i0} - 2\xi_{\gamma,i}\Delta F_{z,i}\cdot\gamma_{i0} + \\
&+ \xi_{\alpha,i}\Delta F_{z,i}\cdot(\alpha_{ply,iL} - \alpha_{ply,iR}) + \xi_{\gamma,i}\Delta F_{z,i}\cdot(\gamma_{con,iL} - \gamma_{con,iR}) + \\
&+ [C_{F\alpha,i0L}\cdot(\alpha_{ply,iL} - \Psi_{i0}) + C_{F\alpha,i0R}\cdot(\alpha_{ply,iR} + \Psi_{i0})] + \\
&+ [C_{F\gamma,i0L}\cdot(\gamma_{con,iL} - \gamma_{i0}) + C_{F\gamma,i0R}\cdot(\gamma_{con,iR} + \gamma_{i0})]
\end{aligned}
\tag{A.21}
$$

where the last two terms can be written in a more compact fashion

$$
\begin{aligned}
F_{y,i} &= 2C_{F\alpha,i0}\cdot\alpha_i + 2C_{F\gamma,i0}\cdot\gamma_i - 2\xi_{\alpha,i}\Delta F_{z,i}\cdot\Psi_{i0} - 2\xi_{\gamma,i}\Delta F_{z,i}\cdot\gamma_{i0} + \\
&+ \xi_{\alpha,i}\Delta F_{z,i}\cdot\left(\alpha_{ply,iL} - \alpha_{ply,iR}\right) + \xi_{\gamma,i}\Delta F_{z,i}\cdot\left(\gamma_{con,iL} - \gamma_{con,iR}\right) + \\
&+ F_{y0,i,ply} + \\
&+ F_{y0,i,con}
\end{aligned}
\tag{A.22}
$$

The following terms are now to be substituted in the previous equation

$$\gamma_i = \gamma_{r,i} = \tau_i\cdot\varphi \tag{A.23}$$

$$\alpha_i = \alpha_{ai} + \Psi_{r,i} + \Psi_{sf,i} + \Psi_{c,i} \tag{A.24}$$

$$\Psi_{r,i} = \varepsilon_i\cdot\varphi \tag{A.25}$$

$$\Psi_{sf,i} = c_{sf,i}\cdot F_{y,i} \tag{A.26}$$

$$\Psi_{c,i} = -\frac{F_{y,i}\left(n_i - t_{0,i}\right)}{c_{\Psi,i}} \tag{A.27}$$

where the terms of roll and load transfer follow from the steady-state cornering assumption, and are defined as function of the lateral force (or lateral acceleration)

$$\varphi = \frac{-h'}{k_{\varphi1} + k_{\varphi2} - mgh'}\cdot\frac{l}{l - a_i}F_{y,i} \tag{A.28}$$

$$
\begin{aligned}
\Delta F_{zi} &= F_{y,i}\frac{l}{l - a_i}\sigma_i \\
\sigma_i &= \frac{1}{2s_i}\left(\frac{k_{\varphi i}}{k_{\varphi1} + k_{\varphi2} - mgh'}h' + \frac{l - a_i}{l}h_i\right)
\end{aligned}
\tag{A.29}
$$

By substituting all the previously listed terms into the axle's side force equation, it is possible to write

$$F_{y,i} = 2C_{F\alpha,i0} \cdot \alpha_{ai} + F_{y,i} \cdot (2C_{F\alpha,i0}) \cdot \left[\varepsilon_i \cdot \frac{-h'}{k_{\varphi 1} + k_{\varphi 2} - mgh'} \cdot \frac{l}{l - a_i} c_{sf,i} - \frac{(n_i - t_{0,i})}{c_{\Psi,i}} \right] +$$
$$+F_{y,i} \cdot (2C_{F\gamma,i0}) \cdot \left[\tau_i \cdot \frac{-h'}{k_{\varphi 1} + k_{\varphi 2} - mgh'} \cdot \frac{l}{l - a_i} \right] +$$
$$+F_{y,i} \cdot \sigma_i \cdot \frac{l}{l - a_i} \cdot [\xi_{\alpha,i} \cdot (-2\Psi_{i0} + \alpha_{ply,iL} - \alpha_{ply,iR}) + \xi_{\gamma,i} \cdot (-2\gamma_{i0} + \gamma_{con,iL} - \gamma_{con,iR})] +$$
$$+F_{y0,i,ply} + F_{y0,i,con}$$

which, by grouping all the terms multiplying lateral force $F_{y,i}$ and dividing by $2C_{F\alpha,i0}$, becomes

$$\alpha_{ai} + \frac{F_{y0,i,ply} + F_{y0,i,con}}{2C_{F\alpha,i0}} = \frac{F_{y,i}}{2C_{F\alpha,i0}} \cdot \left\{ 1 - \left[\frac{-h'}{k_{\varphi 1} + k_{\varphi 2} - mgh'} \cdot \frac{l}{l - a_i} \cdot (\varepsilon_i \cdot 2C_{F\alpha,i0} + \tau_i \cdot 2C_{F\gamma,i0}) \right] - \right.$$
$$-[c_{sf,i} \cdot 2C_{F\alpha,i0}] + \left[\frac{(n_i - t_{0,i})}{c_{\Psi,i}} \cdot 2C_{F\alpha,i0} \right] -$$
$$\left. -\left[\sigma_i \cdot \frac{l}{l - a_i} \cdot [\xi_{\alpha,i} \cdot (-2\Psi_{i0} + \alpha_{ply,iL} - \alpha_{ply,iR}) + \xi_{\gamma,i} \cdot (-2\gamma_{i0} + \gamma_{con,iL} - \gamma_{con,iR})] \right] \right\}$$

Defining the effective axle cornering stiffness as the slope

$$C^*_{eff,i} = \frac{dF_{y,i}}{d\alpha_{ai}} \tag{A.30}$$

its reciprocal is directly found from the coefficient multiplying the side force $F_{y,i}$ in the previous equation

$$\frac{1}{C^*_{eff,i}} = \frac{1}{2C_{F\alpha,i0}} \cdot \left\{ 1 + \overbrace{\left[\frac{+h'}{k_{\varphi 1} + k_{\varphi 2} - mgh'} \cdot \frac{l}{l - a_i} \cdot (\varepsilon_i \cdot 2C_{F\alpha,i0} + \tau_i \cdot 2C_{F\gamma,i0}) \right]}^{C_{roll,i}} - \right.$$
$$-\overbrace{[c_{sf,i} \cdot 2C_{F\alpha,i0}]}^{C_{susp,i}} + \overbrace{\left[\frac{(n_i - t_{0,i})}{c_{\Psi,i}} \cdot 2C_{F\alpha,i0} \right]}^{C_{steer,i}} +$$
$$\left. + \overbrace{\left[\sigma_i \cdot \frac{l}{l - a_i} \cdot [\xi_{a,i} \cdot (+2\Psi_{i0} - \alpha_{ply,iL} + \alpha_{ply,iR}) + \xi_{\gamma,i} \cdot (+2\gamma_{i0} - \gamma_{con,iL} + \gamma_{con,iR})] \right]}^{C^*_{\Delta F_z,i}} \right\}$$

$$\frac{1}{C^*_{eff,i}} = \frac{1}{2C_{F\alpha,i0}} \cdot \{1 + C_{roll,i} - C_{susp,i} + C_{steer,i} + C^*_{\Delta F_z,i}\} \tag{A.31}$$

$$C^*_{eff,i} = \frac{2C_{F\alpha,i0}}{1 + C_{roll,i} - C_{susp,i} + C_{steer,i} + C^*_{\Delta F_z,i}} \tag{A.32}$$

with

$$C_{roll,i} \frac{2l}{l-a_i} \cdot \frac{h'}{k_{\varphi 1} + k_{\varphi 2} - mgh'} \left(\varepsilon_i C_{F\alpha,i0} + \tau_i C_{F\gamma,i0} \right) \tag{A.33}$$

$$C_{susp,i} = 2C_{F\alpha,i0} \cdot c_{sf,i} \tag{A.34}$$

$$C_{steer,i} = 2C_{F\alpha,i0} \cdot \frac{n_i - t_{0,i}}{c_{\Psi,i}} \tag{A.35}$$

$$\begin{aligned} C^*_{\Delta F_{z,i}} = \sigma_i \frac{l}{l-a_i} [& \xi_{\alpha,i} \left(2\Psi_{i0} - \alpha_{ply,iL} + \alpha_{ply,iR} \right) + \\ & + \xi_{\gamma,i} \left(2\gamma_{i0} - \gamma_{con,iL} + \gamma_{con,iR} \right)] \end{aligned} \tag{A.36}$$

The axle side force becomes

$$F_{y,i} = C^*_{eff,i} \cdot \left(\alpha_{ai} + \frac{F_{y0,i,ply} + F_{y0,i,con}}{2C_{F\alpha,i0}} \right) = C^*_{eff,i} \cdot \alpha_{ai} + F_{y0,i} \tag{A.37}$$

where the shift is given by

$$\begin{aligned} F_{y0,i} &= C^*_{eff,i} \cdot \frac{F_{y0,i,ply} + F_{y0,i,con}}{2C_{F\alpha,i0}} = C^*_{eff,i} \cdot \frac{F_{y0,i,ply} + F_{y0,i,con}}{C_{F\alpha,i0R} + C_{F\alpha,i0L}} = \\ &= C^*_{eff,i} \cdot \left\{ \frac{[C_{F\alpha,i0L} \cdot (\alpha_{ply,iL} - \Psi_{i0}) + C_{F\alpha,i0R} \cdot (\alpha_{ply,iR} + \Psi_{i0})]}{C_{F\alpha,i0R} + C_{F\alpha,i0L}} + \right. \\ &\quad \left. + \frac{[C_{F\gamma,i0L} \cdot (\gamma_{con,iL} - \gamma_{i0}) + C_{F\gamma,i0R} \cdot (\gamma_{con,iR} + \gamma_{i0})]}{C_{F\alpha,i0R} + C_{F\alpha,i0L}} \right\} = \\ &= C^*_{eff,i} \cdot \left\{ \Psi_{i0} \cdot \left(\frac{C_{F\alpha,i0R} - C_{F\alpha,i0L}}{C_{F\alpha,i0R} + C_{F\alpha,i0L}} \right) + \gamma_{i0} \cdot \left(\frac{C_{F\gamma,i0R} - C_{F\gamma,i0L}}{C_{F\alpha,i0R} + C_{F\alpha,i0L}} \right) + \right. \\ &\quad \left. + \frac{C_{F\alpha,i0L} \cdot \alpha_{ply,iL} + C_{F\alpha,i0R} \cdot \alpha_{ply,iR} + C_{F\gamma,i0L} \cdot \gamma_{con,iL} + C_{F\gamma,i0R} \cdot \gamma_{con,iR}}{C_{F\alpha,i0R} + C_{F\alpha,i0L}} \right\} = \\ &= C^*_{eff,i} \cdot \left\{ \Psi_{i0} \cdot \left(\frac{C_{F\alpha,i0R} - C_{F\alpha,i0L}}{C_{F\alpha,i0R} + C_{F\alpha,i0L}} \right) + \gamma_{i0} \cdot \left(\frac{C_{F\gamma,i0R} - C_{F\gamma,i0L}}{C_{F\alpha,i0R} + C_{F\alpha,i0L}} \right) + \right. \\ &\quad \left. + \frac{F_{y0,iL} - C_{F\gamma,i0L} \cdot \gamma_{con,iL} + F_{y0,iR} - C_{F\gamma,i0R} \cdot \gamma_{con,iR} + C_{F\gamma,i0L} \cdot \gamma_{con,iL} + C_{F\gamma,i0R} \cdot \gamma_{con,iR}}{C_{F\alpha,i0R} + C_{F\alpha,i0L}} \right\} = \\ &= C^*_{eff,i} \cdot \left\{ \Psi_{i0} \cdot \left(\frac{C_{F\alpha,i0R} - C_{F\alpha,i0L}}{C_{F\alpha,i0R} + C_{F\alpha,i0L}} \right) + \gamma_{i0} \cdot \left(\frac{C_{F\gamma,i0R} - C_{F\gamma,i0L}}{C_{F\alpha,i0R} + C_{F\alpha,i0L}} \right) + \left(\frac{F_{y0,iR} + F_{y0,iL}}{C_{F\alpha,i0R} + C_{F\alpha,i0L}} \right) \right\} \end{aligned} \tag{A.38}$$

A.4.1 Effective Axle Side Force with Cross-toe and Cross-camber

It is possible to include the description of cross-toe and cross-camber in the effective axle side force formulation. This means introducing different static toe and static camber angles for left and right wheel, i.e., $\Psi_{i0,L} \neq \Psi_{i0,R}$ and $\gamma_{i0,L} \neq \gamma_{i0,R}$.

Therefore, Equation A.19 becomes

$$\begin{aligned} F_{y,iL} &= \left(C_{F\alpha,i0L} + \xi_{\alpha,i}\Delta F_{z,i}\right)\left(\alpha_i - \Psi_{i0,L} + \alpha_{ply,iL}\right) + \\ &\quad + \left(C_{F\gamma,i0L} + \xi_{\gamma,i}\Delta F_{z,i}\right)\left(\gamma_i - \gamma_{i0,L} + \gamma_{con,iL}\right) \\ F_{y,iR} &= \left(C_{F\alpha,i0R} - \xi_{\alpha,i}\Delta F_{z,i}\right)\left(\alpha_i + \Psi_{i0,R} + \alpha_{ply,iR}\right) + \\ &\quad + \left(C_{F\gamma,i0R} - \xi_{\gamma,i}\Delta F_{z,i}\right)\left(\gamma_i + \gamma_{i0,R} + \gamma_{con,iR}\right) \end{aligned} \tag{A.39}$$

It is possible to introduce the average of left and right static toe and camber angles

$$\begin{aligned} \Psi_{i0} &= \frac{\Psi_{i0,L} + \Psi_{i0,R}}{2} \\ \gamma_{i0} &= \frac{\gamma_{i0,L} + \gamma_{i0,R}}{2} \end{aligned} \tag{A.40}$$

The side force for the axle i is given by the sum of left and right forces, namely

$$\begin{aligned} F_{y,i} &= F_{y,iL} + F_{y,iR} = \\ &= 2C_{F\alpha,i0}\cdot\alpha_i + 2C_{F\gamma,i0}\cdot\gamma_i - \xi_{\alpha,i}\Delta F_{z,i}\cdot(\Psi_{i0,L} + \Psi_{i0,R}) - \xi_{\gamma,i}\Delta F_{z,i}\cdot(\gamma_{i0,L} + \gamma_{i0,R}) + \\ &\quad + \xi_{\alpha,i}\Delta F_{z,i}\cdot(\alpha_{ply,iL} - \alpha_{ply,iR}) + \xi_{\gamma,i}\Delta F_{z,i}\cdot(\gamma_{con,iL} - \gamma_{con,iR}) + \\ &\quad + [C_{F\alpha,i0L}\cdot(\alpha_{ply,iL} - \Psi_{i0,L}) + C_{F\alpha,i0R}\cdot(\alpha_{ply,iR} + \Psi_{i0,R})] + \\ &\quad + [C_{F\gamma,i0L}\cdot(\gamma_{con,iL} - \gamma_{i0,L}) + C_{F\gamma,i0R}\cdot(\gamma_{con,iR} + \gamma_{i0,R})] \\ &\Rightarrow F_{y,i} = 2C_{F\alpha,i0}\cdot\alpha_i + 2C_{F\gamma,i0}\cdot\gamma_i - 2\xi_{\alpha,i}\Delta F_{z,i}\cdot\Psi_{i0} - 2\xi_{\gamma,i}\Delta F_{z,i}\cdot\gamma_{i0} + \\ &\quad + \xi_{\alpha,i}\Delta F_{z,i}\cdot(\alpha_{ply,iL} - \alpha_{ply,iR}) + \xi_{\gamma,i}\Delta F_{z,i}\cdot(\gamma_{con,iL} - \gamma_{con,iR}) + \\ &\quad + F_{y0,i,ply} + \\ &\quad + F_{y0,i,con} \end{aligned} \tag{A.41}$$

As it can be observed, Equation A.41 has the same formulation as Equation A.22. This is a direct consequence of the definitions introduced in Equation A.40. Thus, following the same approach adopted previously, it is possible to demonstrate that the effective axle cornering stiffness $C^*_{eff,i}$ maintains the same formulation as expressed by Equation A.32 even when cross-toe and cross-camber are introduced.

On the other hand, the expression of the axle side force shift $F_{y0,i}$ is affected by the cross-toe and cross-camber. Its expression, with respect to Equation A.38, becomes

$$
\begin{aligned}
F_{y0,i} &= C^*_{eff,i}\cdot\frac{F_{y0,i,ply}+F_{y0,i,con}}{2C_{F\alpha,i0}} = C^*_{eff,i}\cdot\frac{F_{y0,i,ply}+F_{y0,i,con}}{C_{F\alpha,i0R}+C_{F\alpha,i0L}} = \\
&= C^*_{eff,i}\cdot\left\{\frac{\left[C_{F\alpha,i0L}\cdot(\alpha_{ply,iL}-\Psi_{i0,L})+C_{F\alpha,i0R}\cdot(\alpha_{ply,iR}+\Psi_{i0,R})\right]}{C_{F\alpha,i0R}+C_{F\alpha,i0L}}+\right.\\
&\quad\left.+\frac{\left[C_{F\gamma,i0L}\cdot(\gamma_{con,iL}-\gamma_{i0,L})+C_{F\gamma,i0R}\cdot(\gamma_{con,iR}+\gamma_{i0,R})\right]}{C_{F\alpha,i0R}+C_{F\alpha,i0L}}\right\} = \\
&= C^*_{eff,i}\cdot\left\{\left(\frac{\Psi_{i0,R}\cdot C_{F\alpha,i0R}-\Psi_{i0,L}\cdot C_{F\alpha,i0L}}{C_{F\alpha,i0R}+C_{F\alpha,i0L}}\right)+\left(\frac{\gamma_{i0,R}\cdot C_{F\gamma,i0R}-\gamma_{i0,L}\cdot C_{F\gamma,i0L}}{C_{F\alpha,i0R}+C_{F\alpha,i0L}}\right)+\right.\\
&\quad\left.+\frac{C_{F\alpha,i0L}\cdot\alpha_{ply,iL}+C_{F\alpha,i0R}\cdot\alpha_{ply,iR}+C_{F\gamma,i0L}\cdot\gamma_{con,iL}+C_{F\gamma,i0R}\cdot\gamma_{con,iR}}{C_{F\alpha,i0R}+C_{F\alpha,i0L}}\right\} = \\
&= C^*_{eff,i}\cdot\left\{\left(\frac{\Psi_{i0,R}\cdot C_{F\alpha,i0R}-\Psi_{i0,L}\cdot C_{F\alpha,i0L}}{C_{F\alpha,i0R}+C_{F\alpha,i0L}}\right)+\left(\frac{\gamma_{i0,R}\cdot C_{F\gamma,i0R}-\gamma_{i0,L}\cdot C_{F\gamma,i0L}}{C_{F\alpha,i0R}+C_{F\alpha,i0L}}\right)+\right.\\
&\quad\left.+\frac{F_{y0,iL}-C_{F\gamma,i0L}\cdot\gamma_{con,iL}+F_{y0,iR}-C_{F\gamma,i0R}\cdot\gamma_{con,iR}+C_{F\gamma,i0L}\cdot\gamma_{con,iL}+C_{F\gamma,i0R}\cdot\gamma_{con,iR}}{C_{F\alpha,i0R}+C_{F\alpha,i0L}}\right\} = \\
&= C^*_{eff,i}\cdot\left\{\left(\frac{\Psi_{i0,R}\cdot C_{F\alpha,i0R}-\Psi_{i0,L}\cdot C_{F\alpha,i0L}}{C_{F\alpha,i0R}+C_{F\alpha,i0L}}\right)+\left(\frac{\gamma_{i0,R}\cdot C_{F\gamma,i0R}-\gamma_{i0,L}\cdot C_{F\gamma,i0L}}{C_{F\alpha,i0R}+C_{F\alpha,i0L}}\right)+\right.\\
&\quad\left.+\left(\frac{F_{y0,iR}+F_{y0,iL}}{C_{F\alpha,i0R}+C_{F\alpha,i0L}}\right)\right\} \qquad \text{(A.42)}
\end{aligned}
$$

A.5 Effective Axle Aligning Moment

In this section of the Appendix, the full derivation of the effective axle aligning moment (Equations 2.44–2.53) is reported.

The linear equation of tire aligning moment, for the axle i (1 = front, 2 = rear), and left (L) or right (R) tire, reads

$$M_{z,iL/R} = -C_{M\alpha,iL/R}(F_z)\cdot\alpha_{iL/R}+C_{M\gamma,iL/R}(F_z)\cdot\gamma_{iL/R} \qquad \text{(A.43)}$$

The ply-steer equivalent side slip angle and conicity equivalent camber angle are introduced in Equation 1.2. They are reported here for the axle i (1 = front, 2 = rear), and left (L) or right (R) tire

$$
\begin{aligned}
\gamma_{con,iL/R} &= \frac{\alpha_{FM0,iL/R}}{\dfrac{C_{M\gamma,i0L/R}}{C_{M\alpha,i0L/R}}+\dfrac{C_{F\gamma,i0L/R}}{C_{F\alpha,i0L/R}}} = \frac{\dfrac{F_{y0,iL/R}}{C_{F\alpha,i0L/R}}+\dfrac{M_{z0,iL/R}}{C_{M\alpha,i0L/R}}}{\dfrac{C_{M\gamma,i0L/R}}{C_{M\alpha,i0L/R}}+\dfrac{C_{F\gamma,i0L/R}}{C_{F\alpha,i0L/R}}} \\
\alpha_{ply,iL/R} &= \Delta\alpha_{0,iL/R}-\frac{C_{F\gamma,i0L/R}}{C_{F\alpha,i0L/R}}\cdot\gamma_{con,iL/R} = \frac{F_{y0,iL/R}}{C_{F\alpha,i0L/R}}-\frac{C_{F\gamma,i0L/R}}{C_{F\alpha,i0L/R}}\cdot\gamma_{con,iL/R}
\end{aligned}
\qquad \text{(A.44)}
$$

Hypotheses. The linearization coefficients $\xi_{M\alpha,i}$ and $\xi_{M\gamma,i}$, describing the dependence of the aligning stiffness $C_{M\alpha,i}$ and aligning camber coefficient $C_{M\gamma,i}$ on vertical load variation $\Delta F_{z,i}$, are assumed to be equal for left and right wheels

$$\xi_{M\alpha,iL} = \xi_{M\alpha,iR} = \xi_{M\alpha,i}$$
$$\xi_{M\gamma,iL} = \xi_{M\gamma,iR} = \xi_{M\gamma,i}$$

On the other hand, the aligning stiffness and aligning camber coefficient are considered to be different for left and right tires, in order to accommodate the description of asymmetric tire characteristics. As in Equation 2.4, they can be written as

$$\begin{aligned} C_{M\alpha,iL/R} &= C_{M\alpha,i0L/R} \pm \left(\left.\frac{\partial C_{M\alpha}}{\partial F_z}\right|_{F_{z0,iL/R}} \right) \cdot \Delta F_{z,i} = C_{M\alpha,i0L/R} \pm \xi_{M\alpha,i} \cdot \Delta F_{z,i} \\ C_{M\gamma,iL/R} &= C_{M\gamma,i0L/R} \pm \left(\left.\frac{\partial C_{M\gamma}}{\partial F_z}\right|_{F_{z0,iL/R}} \right) \cdot \Delta F_{z,i} = C_{M\gamma,i0L/R} \pm \xi_{M\gamma,i} \cdot \Delta F_{z,i} \end{aligned} \tag{A.45}$$

Derivation. Introducing into Equation A.43, the ply-steer and conicity equivalent angles, the aligning stiffness and aligning camber coefficient and the static toe and camber angles, the aligning moments for axle i (for left L and right R side) can be written as

$$\begin{aligned} M_{z,iL} &= \left(-C_{M\alpha,i0L} - \xi_{M\alpha,i}\Delta F_{z,i}\right)\left(\alpha_i - \Psi_{i0} + \alpha_{ply,iL}\right) + \\ &\quad + \left(C_{M\gamma,i0L} + \xi_{M\gamma,i}\Delta F_{z,i}\right)\left(\gamma_i - \gamma_{i0} + \gamma_{con,iL}\right) \\ M_{z,iR} &= \left(-C_{M\alpha,i0R} + \xi_{M\alpha,i}\Delta F_{z,i}\right)\left(\alpha_i + \Psi_{i0} + \alpha_{ply,iR}\right) + \\ &\quad + \left(C_{M\gamma,i0R} - \xi_{M\gamma,i}\Delta F_{z,i}\right)\left(\gamma_i + \gamma_{i0} + \gamma_{con,iR}\right) \end{aligned} \tag{A.46}$$

It is possible to express the average of left and right aligning stiffness and aligning camber coefficient as

$$\begin{aligned} C_{M\alpha,i0} &= \frac{C_{M\alpha,i0L} + C_{M\alpha,i0R}}{2} \\ C_{M\gamma,i0} &= \frac{C_{M\gamma,i0L} + C_{M\gamma,i0R}}{2} \end{aligned} \tag{A.47}$$

The aligning moment for the axle i is given by the sum of left and right moments, namely

$$\begin{aligned} M_{z,i} &= M_{z,iL} + M_{z,iR} = \\ &= -2C_{M\alpha,i0} \cdot \alpha_i + 2C_{M\gamma,i0} \cdot \gamma_i + 2\xi_{M\alpha,i}\Delta F_{z,i} \cdot \Psi_{i0} - 2\xi_{M\gamma,i}\Delta F_{z,i} \cdot \gamma_{i0} - \\ &\quad - \xi_{M\alpha,i}\Delta F_{z,i} \cdot (\alpha_{ply,iL} - \alpha_{ply,iR}) + \xi_{M\gamma,i}\Delta F_{z,i} \cdot (\gamma_{con,iL} - \gamma_{con,iR}) + \\ &\quad + [-C_{M\alpha,i0L} \cdot (\alpha_{ply,iL} - \Psi_{i0}) - C_{M\alpha,i0R} \cdot (\alpha_{ply,iR} + \Psi_{i0})] + \\ &\quad + [C_{M\gamma,i0L} \cdot (\gamma_{con,iL} - \gamma_{i0}) + C_{M\gamma,i0R} \cdot (\gamma_{con,iR} + \gamma_{i0})] \end{aligned} \tag{A.48}$$

where the last two terms can be written in a more compact fashion

$$
\begin{aligned}
M_{z,i} = &-2C_{M\alpha,i0} \cdot \alpha_i + 2C_{M\gamma,i0} \cdot \gamma_i + 2\xi_{M\alpha,i} \Delta F_{z,i} \cdot \Psi_{i0} - 2\xi_{M\gamma,i} \Delta F_{z,i} \cdot \gamma_{i0} - \\
&- \xi_{M\alpha,i} \Delta F_{z,i} \cdot \left(\alpha_{ply,iL} - \alpha_{ply,iR}\right) + \xi_{M\gamma,i} \Delta F_{z,i} \cdot \left(\gamma_{con,iL} - \gamma_{con,iR}\right) + \\
&+ M_{z0,i,ply} + \\
&+ M_{z0,i,con}
\end{aligned}
$$

The following terms are now to be substituted in the previous equation

$$\gamma_i = \gamma_{r,i} = \tau_i \cdot \varphi \tag{A.49}$$

$$\alpha_i = \alpha_{ai} + \Psi_{r,i} + \Psi_{sf,i} + \Psi_{c,i} \tag{A.50}$$

$$\Psi_{r,i} = \varepsilon_i \cdot \varphi \tag{A.51}$$

$$\Psi_{sf,i} = c_{sf,i} \cdot F_{y,i} \tag{A.52}$$

$$\Psi_{c,i} = -\frac{F_{y,i}\left(n_i - t_{0,i}\right)}{c_{\Psi,i}} \tag{A.53}$$

where the terms of roll and load transfer follow from the steady-state cornering assumption, and are defined as function of the lateral force (or lateral acceleration)

$$\varphi = \frac{-h'}{k_{\varphi 1} + k_{\varphi 2} - mgh'} \cdot \frac{l}{l - a_i} F_{y,i} \tag{A.54}$$

$$
\begin{gathered}
\Delta F_{zi} = F_{y,i} \frac{l}{l - a_i} \sigma_i \\
\sigma_i = \frac{1}{2s_i}\left(\frac{k_{\varphi i}}{k_{\varphi 1} + k_{\varphi 2} - mgh'} h' + \frac{l - a_i}{l} h_i\right)
\end{gathered} \tag{A.55}
$$

By substituting all the previously listed terms into the axle's aligning moment equation, it is possible to write

$$
\begin{aligned}
M_{z,i} = &-2C_{M\alpha,i0} \cdot \alpha_{ai} + F_{y,i} \cdot (-2C_{M\alpha,i0}) \cdot \left[\varepsilon_i \cdot \frac{-h'}{k_{\varphi 1} + k_{\varphi 2} - mgh'} \cdot \frac{l}{l - a_i} + c_{sf,i} - \frac{(n_i - t_{0,i})}{c_{\Psi,i}}\right] + \\
&+ F_{y,i} \cdot (2C_{M\gamma,i0}) \cdot \left[\tau_i \cdot \frac{-h'}{k_{\varphi 1} + k_{\varphi 2} - mgh'} \cdot \frac{l}{l - a_i}\right] + \\
&+ F_{y,i} \cdot \sigma_i \cdot \frac{l}{l - a_i} \cdot [\xi_{M\alpha,i0} \cdot (2\Psi_{i0} - \alpha_{ply,iL} + \alpha_{ply,iR}) + \xi_{M\gamma,i} \cdot (-2\gamma_{i0} + \gamma_{con,iL} - \gamma_{con,iR})] + \\
&+ M_{z0,i,ply} + M_{z0,i,con}
\end{aligned}
$$

which, by grouping all the terms multiplying lateral force $F_{y,i}$, becomes

$$M_{z,i} = -2C_{M\alpha,i0} \cdot \alpha_{ai} + F_{y,i} \cdot \left\{ \overbrace{\left[\frac{-h'}{k_{\varphi 1} + k_{\varphi 2} - mgh'} \cdot \frac{l}{l - a_i} (-\varepsilon_i \cdot 2C_{M\alpha,i0} + \tau_i \cdot 2C_{M\gamma,i0}) \right]}^{C_{M\alpha,roll,i}} - \overbrace{[c_{sf,i} \cdot 2C_{M\alpha,i0}]}^{C_{M\alpha,susp,i}} + \overbrace{\left[\frac{(n_i - t_{0,i})}{c_{\Psi,i}} \cdot 2C_{M\alpha,i0} \right]}^{C_{M\alpha,steer,i}} + \overbrace{\left[\sigma_i \cdot \frac{l}{l - a_i} \cdot [\xi_{M\alpha,i} \cdot (2\Psi_{i0} - \alpha_{ply,iL} + \alpha_{ply,iR}) + \xi_{M\gamma,i} \cdot (-2\gamma_{i0} + \gamma_{con,iL} - \gamma_{con,iR})] \right]}^{C_{M\alpha,\Delta F_z,i}} \right\} + M_{z0,i,ply} + M_{z0,i,con} \tag{A.56}$$

with

$$C_{M\alpha,roll,i} \frac{2l}{l - a_i} \cdot \frac{h'}{k_{\varphi 1} + k_{\varphi 2} - mgh'} \left(\varepsilon_i C_{M\alpha,i0} - \tau_i C_{M\gamma,i0} \right) \tag{A.57}$$

$$C_{M\alpha,susp,i} = 2C_{M\alpha,i0} \cdot c_{sf,i} \tag{A.58}$$

$$C_{M\alpha,steer,i} = 2C_{M\alpha,i0} \cdot \frac{n_i - t_{0,i}}{c_{\Psi,i}} \tag{A.59}$$

$$C_{M\alpha,\Delta F_z,i} = \sigma_i \frac{l}{l - a_i} [\xi_{M\alpha,i} \left(2\Psi_{i0} - \alpha_{ply,iL} + \alpha_{ply,iR} \right) + \xi_{M\gamma,i} \left(-2\gamma_{i0} + \gamma_{con,iL} - \gamma_{con,iR} \right)] \tag{A.60}$$

The axle side force expression (Equation A.37), reported here for convenience

$$F_{y,i} = C^*_{eff,i} \cdot \alpha_{ai} + F_{y0,i}$$

can be substituted into the axle aligning moment Equation A.56

$$\begin{aligned} M_{z,i} &= -2C_{M\alpha,i0} \cdot \alpha_{ai} + (C^*_{eff,i} \cdot \alpha_{ai} + F_{y0,i}) \cdot (C_{M\alpha,roll,i} - C_{M\alpha,susp,i} + C_{M\alpha,steer,i} + C_{M\alpha,\Delta F_z,i}) + \\ &\quad + M_{z0,i,ply} + M_{z0,i,con} = \\ &= -C_{M\alpha,i} \cdot \alpha_{ai} + M_{z0,i} \end{aligned} \tag{A.61}$$

where the slope (or *effective axle aligning stiffness*) $CM_{\alpha,i}$ is

$$-C_{M\alpha,i} = -2C_{M\alpha,i0} + C^*_{eff,i} \cdot (C_{M\alpha,roll,i} - C_{M\alpha,susp,i} + C_{M\alpha,steer,i} + C_{M\alpha,\Delta F_z,i}) \tag{A.62}$$

and the shift $M_{z0,i}$ is given by

$$\begin{aligned}
M_{z0,i} &= F_{y0,i} \cdot (CM_{\alpha,roll,i} - C_{M\alpha,susp,i} + C_{M\alpha,steer,i} + C_{M\alpha,\Delta F_z,i}) + \\
&+ \Psi_{i0} \cdot (C_{M\alpha,i0L} - C_{M\alpha,i0R}) - \gamma_{i0} \cdot (C_{M\gamma,i0L} - C_{M\gamma,i0R}) - \\
&- C_{M\alpha,i0L} \cdot \alpha_{ply,iL} - C_{M\alpha,i0R} \cdot \alpha_{ply,iR} + \\
&+ C_{M\gamma,i0L} \cdot \gamma_{con,iL} + C_{M\gamma,i0R} \cdot \gamma_{con,iR} = \\
&= F_{y0,i} \cdot (C_{M\alpha,roll,i} - C_{M\alpha,susp,i} + C_{M\alpha,steer,i} + C_{M\alpha,\Delta F_z,i}) + \\
&+ \Psi_{i0} \cdot (C_{M\alpha,i0L} - C_{M\alpha,i0R}) - \gamma_{i0} \cdot (C_{M\gamma,i0L} - C_{M\gamma,i0R}) - \\
&- C_{M\alpha,i0L} \cdot \left(\frac{F_{y0,iL}}{C_{F,i0L}} - \frac{C_{F\gamma,i0L}}{C_{F,i0L}} \cdot \gamma_{con,iL} \right) - C_{M\alpha,i0R} \cdot \left(\frac{F_{y0,iR}}{C_{F,i0R}} - \frac{C_{F\gamma,i0R}}{C_{F\alpha,i0R}} \cdot \gamma_{con,iR} \right) + \\
&+ C_{M\gamma,i0L} \cdot \gamma_{con,iL} + C_{M\gamma,i0R} \cdot \gamma_{con,iR} = \\
&= F_{y0,i} \cdot (C_{M\alpha,roll,i} - C_{M\alpha,susp,i} + C_{M\alpha,steer,i} + C_{M\alpha,\Delta F_z,i}) + \\
&+ \Psi_{i0} \cdot (C_{M\alpha,i0L} - C_{M\alpha,i0R}) - \gamma_{i0} \cdot (C_{M\gamma,i0L} - C_{M\gamma,i0R}) + \\
&+ C_{M\alpha,i0L} \cdot \left[-\frac{F_{y0,iL}}{C_{F\alpha,i0L}} + \left(\frac{C_{F\gamma,i0L}}{C_{F\alpha,i0L}} + \frac{C_{M\gamma,i0L}}{C_{M\alpha,i0L}} \right) \cdot \gamma_{con,iL} \right] + \\
&+ C_{M\alpha,i0R} \cdot \left[-\frac{F_{y0,iR}}{C_{F\alpha,i0R}} + \left(\frac{C_{F\gamma,i0R}}{C_{F\alpha,i0R}} + \frac{C_{M\gamma,i0R}}{C_{M\alpha,i0R}} \right) \cdot \gamma_{con,iR} \right] = \\
&= F_{y0,i} \cdot (C_{M\alpha,roll,i} - C_{M\alpha,susp,i} + C_{M\alpha,steer,i} + C_{M\alpha,\Delta F_z,i}) + \\
&+ \Psi_{i0} \cdot (C_{M\alpha,i0L} - C_{M\alpha,i0R}) - \gamma_{i0} \cdot (C_{M\gamma,i0L} - C_{M\gamma,i0R}) + \\
&+ C_{M\alpha,i0L} \cdot \left[-\frac{F_{y0,iL}}{C_{F\alpha,i0L}} + \left(\frac{C_{F\gamma,i0L}}{C_{F\alpha,i0L}} + \frac{C_{M\gamma,i0L}}{C_{M\alpha,i0L}} \right) \cdot \left(\frac{\frac{F_{y0,iL}}{C_{F\alpha,i0L}} + \frac{M_{z0,iL}}{C_{M\alpha,iL}}}{\frac{C_{M\gamma,iL}}{C_{M\alpha,iL}} + \frac{C_{F\gamma,i0L}}{C_{F\alpha,i0L}}} \right) \right] + \\
&+ C_{M\alpha,i0R} \cdot \left[-\frac{F_{y0,iR}}{C_{F\alpha,i0R}} + \left(\frac{C_{F\gamma,i0R}}{C_{F\alpha,i0R}} + \frac{C_{M\gamma,i0R}}{C_{M\alpha,i0R}} \right) \cdot \left(\frac{\frac{F_{y0,iR}}{C_{F\alpha,i0R}} + \frac{M_{z0,iR}}{C_{M\alpha,iR}}}{\frac{C_{M\gamma,iR}}{C_{M\alpha,iR}} + \frac{C_{F\gamma,i0R}}{C_{F\alpha,i0R}}} \right) \right] = \\
&= F_{y0,i} \cdot (C_{M\alpha,roll,i} - C_{M\alpha,susp,i} + C_{M\alpha,steer,i} + C_{M\alpha,\Delta F_z,i}) + \\
&+ \Psi_{i0} \cdot (C_{M\alpha,i0L} - C_{M\alpha,i0R}) - \gamma_{i0} \cdot (C_{M\gamma,i0L} - C_{M\gamma,i0R}) + \\
&+ C_{M\alpha,i0L} \cdot \left[-\frac{F_{y0,iL}}{C_{F\alpha,i0L}} + \frac{F_{y0,iL}}{C_{F\alpha,i0L}} + \frac{M_{z0,iL}}{C_{M\alpha,i0L}} \right] + \\
&+ C_{M\alpha,i0R} \cdot \left[-\frac{F_{y0,iR}}{C_{F\alpha,i0R}} + \frac{F_{y0,iR}}{C_{F\alpha,i0R}} + \frac{M_{z0,iR}}{C_{M\alpha,i0R}} \right] = \\
&= F_{y0,i} \cdot (C_{M\alpha,roll,i} - C_{M\alpha,susp,i} + C_{M\alpha,steer,i} + C_{M\alpha,\Delta F_z,i}) + \\
&+ \Psi_{i0} \cdot (C_{M\alpha,i0L} - C_{M\alpha,i0R}) - \gamma_{i0} \cdot (C_{M\gamma,i0L} - C_{M\gamma,i0R}) + M_{z0,iL} + M_{z0,iR}
\end{aligned} \tag{A.62}$$

A.6 Handling Diagram Equations Derivation

In this section of the Appendix, the full derivations of Equations 3.7 and 3.8, 3.17 and 3.18, and 6.5 and 6.8 are reported. The computations have been verified also using *Symbolic Math Toolbox*™ of *MATLAB*®.

The following equilibria are written, pertaining to moments around front and rear tire contact points, during steady-state cornering on a cross-sloped road (please refer to Figure 6.4)

- moment equilibrium around *point B* of **Figure A.6a**

$$F_{y,1} \cdot l = ma_y \cdot a_2 - mg\varphi_r \cdot a_2 - \left(M_{z,1} + M_{z,2}\right) \tag{A.64}$$

- moment equilibrium around *point A* of Figure A.6a

$$F_{y,2} \cdot l = ma_y \cdot a_1 - mg\varphi_r \cdot a_1 + \left(M_{z,1} + M_{z,2}\right) \tag{A.65}$$

- moment equilibrium around *point B* of **Figure A.6b**

$$\begin{aligned} &F_{z1} \cdot l + ma_x \cdot h = mg \cdot a_2 \\ &F_{z1} + \frac{ma_x \cdot h}{l} = mg \cdot \frac{a_2}{l} = F_{z1,st} \end{aligned} \tag{A.66}$$

FIGURE A.6 Simple vehicle model schemes.

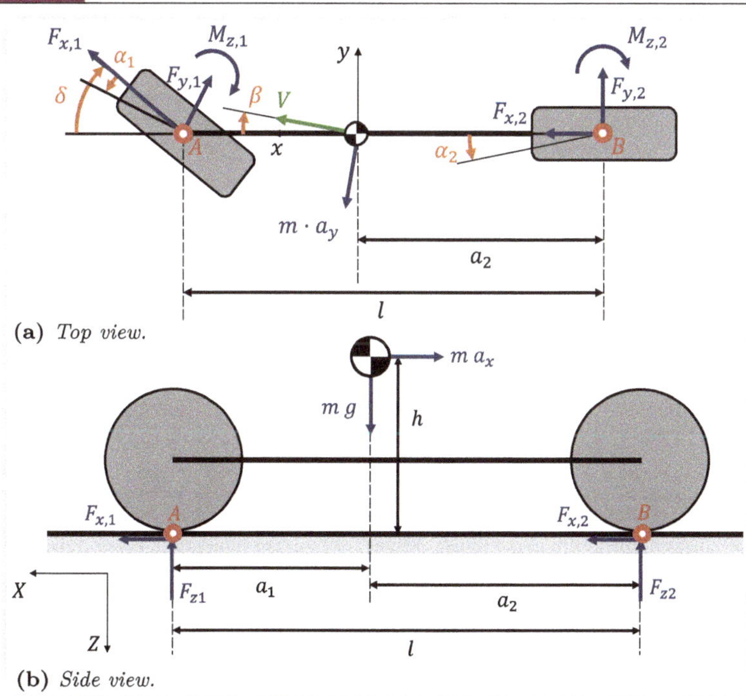

- moment equilibrium around *point A* of Figure A.6b

$$\begin{aligned} &F_{z2}\cdot l - ma_x\cdot h = mg\cdot a_1 \\ &F_{z2} - \frac{ma_x\cdot h}{l} = mg\cdot\frac{a_1}{l} = F_{z2,st} \end{aligned} \tag{A.67}$$

The lateral forces are projected onto the road plane (sloped by angle φ_r, see Figure 6.4). The formulae are obtained by considering small-angle approximation.

Please note that Equations A.66 and A.67 have been rewritten introducing the relation providing the normal load on the axle i

$$F_{zi,st} = mg(l - a_i)/l \tag{A.68}$$

By dividing Equation A.64 by Equation A.66 and Equation A.65 by Equation A.67, Equations A.69 and A.70 are, respectively, obtained

$$\begin{aligned} &\frac{F_{y,1}\cdot l}{F_{z1,st}\cdot l} = \frac{ma_y\cdot a_2}{mg\cdot a_2} - \frac{mg\varphi_r\cdot a_2}{mg\cdot a_2} - \frac{(M_{z,1}+M_{z,2})}{mg\cdot a_2} \\ &\Rightarrow \frac{F_{y,1}}{F_{z1,st}} = \frac{a_y}{g} - \varphi_r - \frac{(M_{z,1}+M_{z,2})/l}{mg\cdot a_2/l} \end{aligned} \tag{A.69}$$

$$\begin{aligned} &\frac{F_{y,2}\cdot l}{F_{z2,st}\cdot l} = \frac{ma_y\cdot a_1}{mg\cdot a_1} - \frac{mg\varphi_r\cdot a_1}{mg\cdot a_1} + \frac{(M_{z,1}+M_{z,2})}{mg\cdot a_1} \\ &\Rightarrow \frac{F_{y,2}}{F_{z2,st}} = \frac{a_y}{g} - \varphi_r + \frac{(M_{z,1}+M_{z,2})/l}{mg\cdot a_1/l} \end{aligned} \tag{A.70}$$

Further, the following expressions are obtained

$$\begin{cases} \dfrac{F_{y,1} + \dfrac{(M_{z,1}+M_{z,2})}{l}}{F_{z1,st}} = \dfrac{a_y}{g} - \varphi_r = \mu - \varphi_r \\ \dfrac{F_{y,2} - \dfrac{(M_{z,1}+M_{z,2})}{l}}{F_{z2,st}} = \dfrac{a_y}{g} - \varphi_r = \mu - \varphi_r \end{cases} \tag{A.71}$$

It is now possible to introduce the linear expressions of tire side forces

$$F_{y,i} = C^*_{eff,i}\cdot\alpha_i + F_{y0,i} \tag{A.72}$$

and aligning moments

$$\begin{aligned} M_{z,i} &= -C_{M\alpha,i}\cdot\alpha_i + M_{z0,i} \\ -C_{M\alpha,i} &= -(C_{M\alpha,iL} + C_{M\alpha,iR}) \\ M_{z0,i} &= M_{z0,iL} + M_{z0,iR} \end{aligned} \tag{A.73}$$

Moreover, in case the longitudinal acceleration is null ($a_x = 0$), the following relation holds: $F_{zi,st} = F_{zi}$. Henceforward, the normal load on the axle i is regardless indicated as F_{zi}, for compactness.

By substituting Equations A.72 and A.73 in Equation A.71, it becomes

$$\begin{cases} \dfrac{C^*_{eff,1}\cdot\alpha_1 + F_{y0,1} + \dfrac{-C_{M\alpha,1}\cdot\alpha_1 - C_{M\alpha,2}\cdot\alpha_2 + \left(M_{z0,1}+M_{z0,2}\right)}{l}}{F_{z1}} = \dfrac{a_y}{g} - \varphi_r = \mu - \varphi_r \\ \dfrac{C^*_{eff,2}\cdot\alpha_2 + F_{y0,2} + \dfrac{+C_{M\alpha,1}\cdot\alpha_1 + C_{M\alpha,2}\cdot\alpha_2 - \left(M_{z0,1}-M_{z0,2}\right)}{l}}{F_{z2}} = \dfrac{a_y}{g} - \varphi_r = \mu - \varphi_r \end{cases} \tag{A.74}$$

$$\Rightarrow \begin{cases} \left(C^*_{eff,1} - \dfrac{C_{M\alpha,1}}{l}\right)\cdot\alpha_1 - \dfrac{C_{M\alpha,2}}{l}\cdot\alpha_2 + F_{y0,1} + \dfrac{\left(M_{z0,1}+M_{z0,2}\right)}{l} = \left(\dfrac{a_y}{g} - \varphi_r\right)\cdot F_{z1} = \left(\mu-\varphi_r\right)\cdot F_{z1} \\ \left(C^*_{eff,2} + \dfrac{C_{M\alpha,2}}{l}\right)\cdot\alpha_2 + \dfrac{C_{M\alpha,1}}{l}\cdot\alpha_1 + F_{y0,2} - \dfrac{\left(M_{z0,1}+M_{z0,2}\right)}{l} = \left(\dfrac{a_y}{g} - \varphi_r\right)\cdot F_{z2} = \left(\mu-\varphi_r\right)\cdot F_{z2} \end{cases} \tag{A.75}$$

It is possible to solve the linear system depending on α_1 and α_2, which yields to the expressions of α_1 and α_2 as function of lateral acceleration $\mu = \dfrac{a_y}{g}$.

To do so, the second equation of the system in A.75 can be substituted with the sum of the two equations

$$\begin{cases} \left(C^*_{eff,1} - \dfrac{C_{M\alpha,1}}{l}\right)\cdot\alpha_1 - \dfrac{C_{M\alpha,2}}{l}\cdot\alpha_2 + F_{y0,1} + \dfrac{\left(M_{z0,1}+M_{z0,2}\right)}{l} = \left(\dfrac{a_y}{g} - \varphi_r\right)\cdot F_{z1} = \left(\mu-\varphi_r\right)\cdot F_{z1} \\ C^*_{eff,1}\cdot\alpha_1 + C^*_{eff,2}\cdot\alpha_2 + F_{y0,1} + F_{y0,2} = \left(\dfrac{a_y}{g} - \varphi_r\right)\cdot\left(F_{z1}+F_{z2}\right) = \left(\dfrac{a_y}{g} - \varphi_r\right)\cdot mg \end{cases}$$

The solutions are found by substitution, as follows:

$$\begin{cases} \left(C^*_{eff,1} - \dfrac{C_{M\alpha,1}}{l}\right)\cdot\alpha_1 - \dfrac{C_{M\alpha,2}}{l}\cdot\left(\dfrac{-C^*_{eff,1}\cdot\alpha_1 - F_{y0,1} - F_{y0,2} + \left(\dfrac{a_y}{g} - \varphi_r\right)\cdot mg}{C^*_{eff,2}}\right) \\ +F_{y0,1} + \dfrac{(M_{z0,1}+M_{z0,2})}{l} = \left(\dfrac{a_y}{g} - \varphi_r\right)\cdot F_{z1} \\ \alpha_2 = \dfrac{-C^*_{eff,1}\cdot\alpha_1 - F_{y0,1} - F_{y0,2} + \left(\dfrac{a_y}{g} - \varphi_r\right)\cdot mg}{C^*_{eff,2}} \\ \left(C^*_{eff,1} - \dfrac{C_{M\alpha,1}}{l} + \dfrac{C_{M\alpha,2}\cdot C^*_{eff,1}}{l\cdot C^*_{eff,2}}\right)\cdot\alpha_1 = \\ = \left(\dfrac{a_y}{g} - \varphi_r\right)\cdot\left(F_{z1} + \dfrac{C_{M\alpha,2}\cdot mg}{l\cdot C^*_{eff,2}}\right) - F_{y0,1} - \dfrac{(M_{z0,1}+M_{z0,2})}{l} - \dfrac{C_{M\alpha,2}\cdot(F_{y0,1}+F_{y0,2})}{l\cdot C^*_{eff,2}} \\ \alpha_2 = \dfrac{-C^*_{eff,1}\cdot\alpha_1 - F_{y0,1} - F_{y0,2} + \left(\dfrac{a_y}{g} - \varphi_r\right)\cdot mg}{C^*_{eff,2}} \end{cases} \tag{A.76}$$

$$\begin{cases} \alpha_1 = \left(\dfrac{lF_{z1}C^*_{eff,2} + C_{M\alpha,2}mg}{lC^*_{eff,1}C^*_{eff,2} - C_{M\alpha,1}C^*_{eff,2} + C_{M\alpha,2}C^*_{eff,1}} \right) \cdot (\mu - \varphi_r) + \\ \quad + \dfrac{-C^*_{eff,2}(M_{z0,1} + M_{z0,2} + lF_{y0,1} - C_{M\alpha,2}(F_{y0,1} + F_{y0,2})}{lC^*_{eff,1}C^*_{eff,2} - C_{M\alpha,1}C^*_{eff,2} + C_{M\alpha,2}C^*_{eff,1}} \\ \alpha_2 = \left(\dfrac{lF_{z2}C^*_{eff,1} - C_{M\alpha,1}mg}{lC^*_{eff,1}C^*_{eff,2} - C_{M\alpha,1}C^*_{eff,2} + C_{M\alpha,2}C^*_{eff,1}} \right) \cdot (\mu - \varphi_r) + \\ \quad + \dfrac{C^*_{eff,1}(M_{z0,1} + M_{z0,2} - lF_{y0,2}) + C_{M\alpha,1}(F_{y0,1} + F_{y0,2})}{lC^*_{eff,1}C^*_{eff,2} - C_{M\alpha,1}C^*_{eff,2} + C_{M\alpha,2}C^*_{eff,1}} \end{cases} \tag{A.77}$$

$$\begin{cases} \alpha_1 = K_{\alpha 1} \cdot \mu - K_{\alpha 1} \cdot \varphi_r + \alpha_{1,st} \\ \alpha_2 = K_{\alpha 2} \cdot \mu - K_{\alpha 2} \cdot \varphi_r + \alpha_{2,st} \end{cases} \tag{A.78}$$

A.7 Straight Motion Settings, Analytic Solutions' Derivation

In this section of the Appendix, the full derivation of the solutions presented in Chapter 7 is reported. The computations have been verified also using *Symbolic Math Toolbox*™ of *MATLAB*®.

The offsets of steering angle, steering torque, and vehicle attitude when driving straight ($a_y = 0$) are defined in Equations 7.1–7.3. They are rewritten using the coefficients reported below (introduced in Equations 7.11–7.14), to use more compact expressions

A: coefficients multiplying toe angles (front Ψ_{10}, rear Ψ_{20})

$$A_1 = \left(\frac{C_{F\alpha,10R} - C_{F\alpha,10L}}{C_{F\alpha,10R} + C_{F\alpha,10L}} \right) + \frac{F_{z1}}{2C_{F\alpha,10}} \cdot C_{el,\Psi,1} \cdot \varphi_r$$

$$A_2 = \left(\frac{C_{F\alpha,20R} - C_{F\alpha,20L}}{C_{F\alpha,20R} + C_{F\alpha,20L}} \right) + \frac{F_{z2}}{2C_{F\alpha,20}} \cdot C_{el,\Psi,2} \cdot \varphi_r$$

$$A_{1M} = -\left(\frac{C_{M\alpha,10R} - C_{M\alpha,10L}}{C_{M\alpha,10R} + C_{M\alpha,10L}} \right) - \frac{F_{z1}}{C_{M\alpha,10R} + C_{M\alpha,10L}} \cdot C_{M\alpha,el,\Psi,1} \cdot \varphi_r$$

B: coefficients multiplying camber (front γ_{10}, rear γ_{20})

$$B_1 = \left(\frac{C_{F\gamma,10R} - C_{F\gamma,10L}}{C_{F\alpha,10R} + C_{F\alpha,10L}} \right) + \frac{F_{z1}}{2C_{F\alpha,10}} \cdot C_{el,\gamma,1} \cdot \varphi_r$$

$$B_2 = \left(\frac{C_{F\gamma,20R} - C_{F\gamma,20L}}{C_{F\alpha,20R} + C_{F\alpha,20L}} \right) + \frac{F_{z2}}{2C_{F\alpha,20}} \cdot C_{el,\gamma,2} \cdot \varphi_r$$

$$B_{1M} = \left(\frac{C_{M\gamma,10R} - C_{M\gamma,10L}}{C_{M\alpha,10R} + C_{M\alpha,10L}} \right) - \frac{F_{z1}}{C_{M\alpha,10R} + C_{M\alpha,10L}} \cdot C_{M\alpha,el,\gamma,1} \cdot \varphi_r$$

C: coefficients related to tire characteristics' offsets ($F_{y0,i}$ and $M_{z0,i}$)

$$C_1 = \left(\frac{F_{y0,1R} + F_{y0,1L}}{C_{F\alpha,10R} + C_{F\alpha,10L}} \right)$$

$$C_2 = \left(\frac{F_{y0,2R} + F_{y0,2L}}{C_{F\alpha,20R} + C_{F\alpha,20L}} \right)$$

$$C_{1M} = \left(\frac{M_{z0,1R} + M_{z0,1L}}{C_{M\alpha,10R} + C_{M\alpha,10L}} \right)$$

D, E: coefficients multiplying road cross slope φ_r

$$D_1 = \frac{F_{z1}}{C^*_{eff,1,AL}}$$

$$D_2 = \frac{F_{z2}}{C^*_{eff,2,AL}}$$

$$E_{1M} = \frac{F_{z1}}{2C_{M\alpha,10}} \cdot \left(n - C_{M\alpha,roll,1} + C_{M\alpha,susp,1} - C_{M\alpha,steer,1} - C_{M\alpha,el,AL,1} + \frac{2C_{M\alpha,10}}{C^*_{eff,1,AL}} \right)$$

Thus, the offsets can be rewritten from Equations 7.1–7.3 according to Equations 7.8–7.10, reported below for convenience:

- Steering angle

$$\delta_{st} = -\Psi_{10} \cdot A_1 - \gamma_{10} \cdot B_1 - C_1 + \Psi_{20} \cdot A_2 + \gamma_{20} \cdot B_2 + C_2 + (-D_1 + D_2)\varphi_r$$

- Steering torque

$$M^*_{S,st} = -2C_{M\alpha,10} \cdot \left[\Psi_{10} \cdot (A_1 + A_{1M}) + \gamma_{10} \cdot (B_1 + B_{1M}) + C_1 + C_{1M} + E_{1M} \cdot \varphi_r \right]$$

- Vehicle attitude

$$\beta_{st} = \Psi_{20} \cdot A_2 + \gamma_{20} \cdot B_2 + C_2 + D_2\varphi_r$$

A.7.1 $M^*_{S,st} = 0, \beta_{st} = 0, \delta_{st} \neq 0$

$$\begin{cases} M^*_{S,st} = 0: & -2C_{M\alpha,10} \cdot [\Psi_{10} \cdot (A_1 + A_{1M}) + \gamma_{10} \cdot (B_1 + B_{1M}) + C_1 + C_{1M} + E_{1M} \cdot \varphi_r] = 0 \\ \beta_{st} = 0: & \Psi_{20} \cdot A_2 + \gamma_{20} \cdot B_2 + C_2 + D_2\varphi_r = 0 \end{cases}$$

$$\begin{cases} \Psi_{10} \cdot A_1 + \gamma_{10} \cdot B_1 + C_1 = -\Psi_{10} \cdot A_{1M} - \gamma_{10} \cdot B_{1M} - C_{1M} - E_{1M} \cdot \varphi_r \\ \Psi_{20} \cdot A_2 + \gamma_{20} \cdot B_2 + C_2 + D_2\varphi_r = 0 \end{cases} \quad (A.79)$$

$$\begin{cases} \Psi_{10} \cdot A_1 + \gamma_{10} \cdot B_1 + C_1 + (D_1 \cdot \varphi_r) = -\Psi_{10} \cdot A_{1M} - \gamma_{10} \cdot B_{1M} - C_{1M} - E_{1M} \cdot \varphi_r + (D_1 \cdot \varphi_r) \\ \Psi_{20} \cdot A_2 + \gamma_{20} \cdot B_2 + C_2 + D_2\varphi_r = 0 \end{cases}$$

$$\begin{aligned} \Rightarrow \delta_{st} &= [-\Psi_{10} \cdot A_1 - \gamma_{10} \cdot B_1 - C_1 + \Psi_{20} \cdot A_2 + \gamma_{20} \cdot B_2 + C_2] + (-D_1 + D_2)\varphi_r = \\ &= -\Psi_{10} \cdot A_1 - \gamma_{10} \cdot B_1 - C_1 - D_1\varphi_r = \\ &= \Psi_{10} \cdot A_{1M} + \gamma_{10} \cdot B_{1M} + C_{1M} + (E_{1M} - D_1) \cdot \varphi_r \end{aligned} \quad (A.80)$$

A.7.2 $M^*_{S,st} = 0,\ \delta_{st} = 0,\ \beta_{st} \neq 0$

$$\begin{cases} M^*_{S,st} = 0: & -2C_{M\alpha,10} \cdot [\Psi_{10} \cdot (A_1 + A_{1M}) + \gamma_{10} \cdot (B_1 + B_{1M}) + C_1 + C_{1M} + E_{1M} \cdot \varphi_r] = 0 \\ \delta_{st} = 0: & [-\Psi_{10} \cdot A_1 - \gamma_{10} \cdot B_1 - C_1 + \Psi_{20} \cdot A_2 + \gamma_{20} \cdot B_2 + C_2] + (-D_1 + D_2)\varphi_r = 0 \end{cases}$$

$$\begin{cases} \Psi_{10} \cdot A_1 + \gamma_{10} \cdot B_1 + C_1 = -\Psi_{10} \cdot A_{1M} - \gamma_{10} \cdot B_{1M} - C_{1M} - E_{1M} \cdot \varphi_r \\ \Psi_{10} \cdot A_1 + \gamma_{10} \cdot B_1 + C_1 + D_1\varphi_r = \Psi_{20} \cdot A_2\gamma_{20} \cdot B_2 + C_2 + D_2\varphi_r \end{cases}$$

$$\begin{cases} \Psi_{10} \cdot A_1 + \gamma_{10} \cdot B_1 + C_1 + (D_1 \cdot \varphi_r) = -\Psi_{10} \cdot A_{1M} - \gamma_{10} \cdot B_{1M} - C_{1M} - E_{1M} \cdot \varphi_r + (D_1 \cdot \varphi_r) \\ \Psi_{10} \cdot A_1 + \gamma_{10} \cdot B_1 + C_1 + D_1\varphi_r = \Psi_{20} \cdot A_2\gamma_{20} \cdot B_2 + C_2 + D_2\varphi_r \end{cases} \tag{A.81}$$

$$\begin{aligned} \Rightarrow \beta_s t &= \Psi_{20} \cdot A_2 + \gamma_{20} \cdot B_2 + C_2 + D_2\varphi_r = \\ &= -\Psi_{10} \cdot A_{1M} - \gamma_{10} \cdot B_{1M} - C_{1M} - \left(E_{1M} - D_1\right) \cdot \varphi_r \end{aligned} \tag{A.82}$$

A.7.3 $\delta_{st} = 0,\ \beta_{st} = 0,\ M^*_{S,st} \neq 0$

$$\begin{aligned} &\begin{cases} \beta_{st} = 0: & \Psi_{20} \cdot A_2 + \gamma_{20} \cdot B_2 + C_2 + D_2\varphi_r = 0 \\ \delta_{st} = 0: & \left[-\Psi_{10} \cdot A_1 - \gamma_{10} \cdot B_1 - C_1 + \Psi_{20} \cdot A_2 + \gamma_{20} \cdot B_2 + C_2\right] + \left(-D_1 + D_2\right)\varphi_r = 0 \end{cases} \\ &\begin{cases} \Psi_{20} \cdot A_2 + \gamma_{20} \cdot B_2 + C_2 + D_2\varphi_r = 0 \\ \Psi_{20} \cdot A_2 + \gamma_{20} \cdot B_2 + C_2 + D_2\varphi_r = \Psi_{10} \cdot A_1 + \gamma_{10} \cdot B_1 + C_1 + D_1\varphi_r \end{cases} \end{aligned} \tag{A.83}$$

$$\begin{aligned} \Rightarrow M^*_{S,st} &= -2C_{M\alpha,10} \cdot \left[\Psi_{10} \cdot \left(A_1 + A_{1M}\right) + \gamma_{10} \cdot \left(B_1 + B_{1M}\right) + C_1 + C_{1M} + E_{1M} \cdot \varphi_r\right] = \\ &= -2C_{M\alpha,10} \cdot \left[\Psi_{10} \cdot \left(A_1 + A_{1M}\right) + \gamma_{10} \cdot \left(B_1 + B_{1M}\right) + C_1 + C_{1M} + \left(E_{1M} + D_1 - D_1\right) \cdot \varphi_r\right] = \\ &= -2C_{M\alpha,10} \cdot \left[\Psi_{10} \cdot A_{1M} + \gamma_{10} \cdot B_{1M} + C_{1M} + \left(E_{1M} - D_1\right) \cdot \varphi_r\right] \end{aligned} \tag{A.84}$$

A.7.4 $\delta_{st} = 0,\ \beta_{st} = 0,\ M^*_{S,st} = 0$, Necessary Condition

With the *necessary condition* expressed by Equation 7.21, namely

$$\Psi_{10} \cdot A_{1M} + \gamma_{10} \cdot B_{1M} = -C_{1M} - \left(E_{1M} - D_1\right)\varphi_r$$

the steering torque offset becomes

$$\begin{aligned} M^*_{S,st} &= -2C_{M\alpha,10} \cdot \left[\Psi_{10} \cdot \left(A_1 + A_{1M}\right) + \gamma_{10} \cdot \left(B_1 + B_{1M}\right) + C_1 + C_{1M} + E_{1M} \cdot \varphi_r\right] = \\ &= -2C_{M\alpha,10} \cdot \left[\Psi_{10} \cdot \left(A_1 + A_{1M}\right) + \gamma_{10} \cdot \left(B_1 + B_{1M}\right) + C_1 + C_{1M} + \left(E_{1M} + D_1 - D_1\right) \cdot \varphi_r\right] = \\ &= -2C_{M\alpha,10} \cdot \left[\Psi_{10} \cdot A_1 + \gamma_{10} \cdot B_1 + C_1 + D_1 \cdot \varphi_r\right] \end{aligned} \tag{A.85}$$

Therefore

$$\begin{aligned} &\begin{cases} M^*_{S,st} = 0: & -2C_{M\alpha,10} \cdot \left[\Psi_{10} \cdot A_1 + \gamma_{10} \cdot B_1 + C_1 + D_1 \cdot \varphi_r\right] = 0 \\ \beta_{st} = 0: & \Psi_{20} \cdot A_2 + \gamma_{20} \cdot B_2 + C_2 + D_2\varphi_r = 0 \end{cases} \\ &\begin{cases} \Psi_{10} \cdot A_1 + \gamma_{10} \cdot B_1 + C_1 + D_1\varphi_r = 0 \\ \Psi_{20} \cdot A_2 + \gamma_{20} \cdot B_2 + C_2 + D_2\varphi_r = 0 \end{cases} \end{aligned} \tag{A.86}$$

$$\begin{aligned}\Rightarrow \delta_{st} &= \left[-\Psi_{10}\cdot A_1 - \gamma_{10}\cdot B_1 - C_1 + \Psi_{20}\cdot A_2 + \gamma_{20}\cdot B_2 + C_2\right] + \left(-D_1 + D_2\right)\varphi_r = \\ &= -\Psi_{10}\cdot A_1 - \gamma_{10}\cdot B_1 - C_1 - D_1\varphi_r = \\ &= 0 \end{aligned} \tag{A.87}$$

A.7.4.1 **Matrix Form of the System of Linear Equations** As an alternative, the same result can be obtained by expressing the linear system of equations in its corresponding matrix form. Since the condition in which the three offsets are simultaneously null is pursued, the linear system reads

$$\begin{cases} \beta_{st} = 0: & \Psi_{20}\cdot A_2 + \gamma_{20}\cdot B_2 + C_2 + D_2\varphi_r = 0 \\ \delta_{st} = 0: & \left[-\Psi_{10}\cdot A_1 - \gamma_{10}\cdot B_1 - C_1 + \Psi_{20}\cdot A_2 + \gamma_{20}\cdot B_2 + C_2\right] + \left(-D_1 + D_2\right)\varphi_r = 0 \\ M^*_{S,st} = 0: & -2C_{M\alpha,10}\cdot\left[\Psi_{10}\cdot\left(A_1 + A_{1M}\right) + \gamma_{10}\cdot\left(B_1 + B_{1M}\right) + C_1 + C_{1M} + E_{1M}\cdot\varphi_r\right] = 0 \end{cases} \tag{A.88}$$

By clustering the four unknowns (namely the wheel alignment angles Ψ_{10}, Ψ_{20}, γ_{10}, and γ_{20}) into a vector, the corresponding matrix form is

$$\begin{bmatrix} 0 & A_2 & 0 & B_2 \\ -A_1 & A_2 & -B_1 & B_2 \\ A_1 + A_{1M} & 0 & B_1 + B_{1M} & 0 \end{bmatrix} \begin{Bmatrix} \Psi_{10} \\ \Psi_{20} \\ \gamma_{10} \\ \gamma_{20} \end{Bmatrix} = \begin{Bmatrix} -C_2 - D_2\varphi_r \\ -\left(C_2 - C_1\right) - \left(D_2 - D_1\right)\varphi_r \\ -\left(C_1 + C_{1M}\right) - E_{1M}\varphi_r \end{Bmatrix} \tag{A.89}$$

Solving the matrix form yields the same solution already described.

A.7.5 Derivation of the Effective Axle Aligning Moment Expression with Wheel Alignment Angles

In this section, the derivation of Equation 7.9 is reported. Among the previously introduced coefficients, the ones pertaining to Equation 7.9 are reported here for convenience:

$$\begin{aligned} A_1 &= \left(\frac{C_{F\alpha,10R} - C_{F\alpha,10L}}{C_{F\alpha,10R} + C_{F\alpha,10L}}\right) + \frac{F_{z1}}{2C_{F\alpha,10}}\cdot C_{el,\Psi,1}\cdot\varphi_r \\ B_1 &= \left(\frac{C_{F\gamma,10R} - C_{F\gamma,10L}}{C_{F\alpha,10R} + C_{F\alpha,10L}}\right) + \frac{F_{z1}}{2C_{F\alpha,10}}\cdot C_{el,\gamma,1}\cdot\varphi_r \\ C_1 &= \left(\frac{F_{y0,1R} + F_{y0,1L}}{C_{F\alpha,10R} + C_{F\alpha,10L}}\right) \\ A_{1M} &= -\left(\frac{C_{M\alpha,10R} - C_{M\alpha,10L}}{C_{M\alpha,10R} + C_{M\alpha,10L}}\right) - \frac{F_{z1}}{C_{M\alpha,10R} + C_{M\alpha,10L}}\cdot C_{M\alpha,el,\Psi,1}\cdot\varphi_r \\ B_{1M} &= \left(\frac{C_{M\gamma,10R} - C_{M\gamma,10L}}{C_{M\alpha,10R} + C_{M\alpha,10L}}\right) - \frac{F_{z1}}{C_{M\alpha,10R} + C_{M\alpha,10L}}\cdot C_{M\alpha,el,\gamma,1}\cdot\varphi_r \\ C_{1M} &= \left(\frac{M_{z0,1R} + M_{z0,1L}}{C_{M\alpha,10R} + C_{M\alpha,10L}}\right) \\ E_{1M} &= \frac{F_{z1}}{2C_{M\alpha,10}}\cdot\left(n - C_{M\alpha,roll,1} + C_{M\alpha,susp,1} - C_{M\alpha,steer,1} - C_{M\alpha,el,AL,1} + \frac{2C_{M\alpha,10}}{C^*_{eff,1,AL}}\right) \end{aligned}$$

where

$$\frac{1}{C^*_{eff,i,AL}} = \frac{C_{el,AL,i}}{2C_{F\alpha,i0}}$$

$$C_{el,AL,i} = 1 + C_{roll,i} - C_{susp,i} + C_{steer,i} + \\ + \sigma_i \frac{l}{l-\alpha_i} \cdot \left[\xi_{\alpha,i}\left(-\alpha_{ply,iL} + \alpha_{ply,iR}\right) + \xi_{\gamma,i}\left(-\gamma_{con,iL} + \gamma_{con,iR}\right) \right] \tag{A.90}$$

$$C_{el,\Psi,i} = \sigma_i \frac{l}{l-a_i} \cdot 2 \cdot \xi_{\alpha,i}$$

$$C_{el,\gamma,i} = \sigma_i \frac{l}{l-a_i} \cdot 2 \cdot \xi_{\gamma,i}$$

and

$$C_{M\alpha,\Delta F_z,i} = \sigma_i \frac{l}{l-a_i} [\xi_{M\alpha,i}(2\Psi_{i0} - \alpha_{ply,iL} + \alpha_{ply,iR}) + \\ + \xi_{M\gamma,i}(-2\gamma_{i0} + \gamma_{con,iL} - \gamma_{con,iR})] = \\ = C_{M\alpha,el,ALi} + C_{M\alpha,el,\Psi,i} \cdot \Psi_{i0} + C_{M\alpha,el,\gamma,i} \cdot \gamma_{i0}$$

$$C_{M\alpha,el,AL,i} = \sigma_i \frac{l}{l-a_i} [\xi_{M\alpha,i}(-\alpha_{ply,iL} + \alpha_{ply,iR}) + \xi_{M\gamma,i}(\gamma_{con,iL} - \gamma_{con,iR})] \tag{A.91}$$

$$C_{M\alpha,el,\Psi,i} = \sigma_i \frac{l}{l-a_i} \cdot 2 \cdot \xi_{M\alpha,i}$$

$$C_{M\alpha,el,\gamma,i} = -\sigma_i \frac{l}{l-a_i} \cdot 2 \cdot \xi_{M\gamma,i}$$

The expression in Equation 7.9 is derived starting from Equation 6.20, by introducing the effective axle aligning moment expressed in Equation 2.52

$$M^*_{S,st} \approx -\varphi_r \cdot F_{z1} \cdot \left(n + \frac{C_{M\alpha,1}}{C^*_{eff,1}} \right) - C_{M\alpha,1} \cdot \frac{F_{y0,1}}{C^*_{eff,1}} - M_{z0,1} = \\ = -\varphi_r \cdot F_{z1} \cdot \left[n - (C_{M\alpha,roll,1} - C_{M\alpha,susp,1} + C_{M\alpha,steer,1} + C_{M\alpha,\Delta F_z,1}) + \frac{2C_{M\alpha,10}}{C^*_{eff,1}} \right] - \\ -[M_{z0,1L} + M_{z0,1R} - \Psi_{10} \cdot (C_{M\alpha,10R} - C_{M\alpha,10L}) + \gamma_{10} \cdot (C_{M\gamma,10R} - C_{M\gamma,10L})] - \\ -2C_{M\alpha,10} \cdot \frac{F_{y0,1}}{C^*_{eff,1}} = \\ = -\varphi_r \cdot F_{z1} \cdot \left[n - C_{M\alpha,roll,1} - C_{M\alpha,susp,1} + C_{M\alpha,steer,1} + C_{M\alpha,\Delta F_z,1}) + \frac{2C_{M\alpha,10}}{C^*_{eff,1}} \right] - \\ -(M_{z0,1L} + M_{z0,1R}) - \Psi_{10} \cdot (C_{M\alpha,10R} - C_{M\alpha,10L}) + \gamma_{10} \cdot (C_{M\gamma,10R} - C_{M\gamma,10L}) - \\ -2C_{M\alpha,10} \cdot \left[\Psi_{10} \cdot \left(\frac{C_{F\alpha,10R} - C_{F\alpha,10L}}{C_{F\alpha,10R} + C_{F\alpha,10L}} \right) + \gamma_{10} \cdot \left(\frac{C_{F\gamma,10R} - C_{F\gamma,10L}}{C_{F\alpha,10R} + C_{F\alpha,10L}} \right) + \right. \\ \left. + \left(\frac{F_{y0,1R} + F_{y0,1L}}{C_{F\alpha,10R} + C_{F\alpha,10L}} \right) \right] \tag{A.92}$$

Rewriting

$$C_{M\alpha,\Delta F_z,1} = C_{M\alpha,el,AL,1} + C_{M\alpha,el,\Psi,1}\cdot\Psi_{10} + C_{M\alpha,el,\gamma,1}\cdot\gamma_{10} \tag{A.93}$$

$$\begin{aligned}\frac{2C_{M\alpha,10}}{C^*_{eff,1}} &= 2C_{M\alpha,10}\cdot\left(\frac{C_{el,AL,1}}{2C_{F\alpha,10}} + \frac{C_{el,\Psi,1}\cdot\Psi_{10}}{2C_{F\alpha,10}} + \frac{C_{el,\gamma,1}\cdot\gamma_{10}}{2C_{F\alpha,10}}\right)\\ &= 2C_{M\alpha,10}\cdot\left(C^*_{eff,1,AL} + \frac{C_{el,\Psi,1}\cdot\Psi_{10}}{2C_{F\alpha,10}} + \frac{C_{el,\gamma,1}\cdot\gamma_{10}}{2C_{F\alpha,10}}\right)\end{aligned} \tag{A.94}$$

it is possible to introduce the coefficients A_1, A_{1M}, B_1, B_{1M}, and E_{1M}

$$\begin{aligned} M^*_{S,st} \approx & -\varphi_r\cdot F_{z1}\cdot\Big[n-(C_{M\alpha,roll,1} - C_{M\alpha,susp,1} + C_{M\alpha,steer,1} +\\ & + C_{M\alpha,el,AL,1} + C_{M\alpha,el,\Psi,1}\cdot\Psi_{10} + C_{M\alpha,el,\gamma,1}\cdot\gamma_{10}) +\\ & + 2C_{M\alpha,10}\cdot\left(\frac{C_{el,AL,1}}{2C_{F\alpha,10}} + \frac{C_{el,\Psi,1}\cdot\Psi_{10}}{2C_{F\alpha,10}} + \frac{C_{el,\gamma,1}\cdot\gamma_{10}}{2C_{F\alpha,10}}\right)\Big] -\\ & -2C_{M\alpha,10}\cdot\left[\Psi_{10}\cdot\left(\frac{C_{F\alpha,10R}-C_{F\alpha,10L}}{C_{F\alpha,10R}+C_{F\alpha,10L}} - \frac{C_{M\alpha,10R}-C_{M\alpha,10L}}{C_{M\alpha,10R}+C_{M\alpha,10L}}\right) + \right.\\ & + \gamma_{10}\cdot\left(\frac{C_{F\gamma,10R}-C_{F\gamma,10L}}{C_{F\alpha,10R}+C_{F\alpha,10L}} + \frac{C_{M\gamma,10R}-C_{M\gamma,10L}}{C_{M\alpha,10R}+C_{M\alpha,10L}}\right) +\\ & \left. + \left(\frac{F_{y0,1R}+F_{y0,1L}}{C_{F\alpha,10R}+C_{F\alpha,10L}}\right) + \left(\frac{M_{z0,1R}+M_{z0,1L}}{C_{M\alpha,10R}+C_{M\alpha,10L}}\right)\right] =\\ = & -\varphi_r\cdot F_{z1}\cdot\left[n-(C_{M\alpha,roll,1} - C_{M\alpha,susp,1} + C_{M\alpha,steer,1} + C_{M\alpha,el,AL,1}) + \frac{2C_{M\alpha,10}}{C^*_{eff,1,AL}}\right] -\\ & -2C_{M\alpha,10}\cdot\left[\Psi_{10}\cdot\left(\frac{C_{F\alpha,10R}-C_{F\alpha,10L}}{C_{F\alpha,10R}+C_{F\alpha,10L}} + \frac{F_{z1}}{2C_{F\alpha,10}}\cdot C_{el,\Psi,1}\cdot\varphi_r - \right.\right.\\ & \left. - \frac{C_{M\alpha,10R}-C_{M\alpha,10L}}{C_{M\alpha,10R}+C_{M\alpha,10L}} - \frac{F_{z1}}{2C_{M\alpha,10}}\cdot C_{M\alpha,el,\Psi,1}\cdot\varphi_r\right) +\\ & + \gamma_{10}\cdot\left(\frac{C_{F\gamma,10R}-C_{F\gamma,10L}}{C_{F\alpha,10R}+C_{F\alpha,10L}} + \frac{F_{z1}}{2C_{F\alpha,10}}\cdot C_{el,\gamma,1}\cdot\varphi_r + \right.\\ & \left. + \frac{C_{M\gamma,10R}-C_{M\gamma,10L}}{C_{M\alpha,10R}+C_{M\alpha,10L}} - \frac{F_{z1}}{2C_{M\alpha,10}}\cdot C_{M\alpha,el,\gamma,1}\cdot\varphi_r\right) +\\ & \left. + \left(\frac{F_{y0,1R}+F_{y0,1L}}{C_{F\alpha,10R}+C_{F\alpha,10L}}\right) + \left(\frac{M_{z0,1R}+M_{z0,1L}}{C_{M\alpha,10R}+C_{M\alpha,10L}}\right)\right] =\\ = & -2C_{M\alpha,10}\cdot[\Psi_{10}\cdot(A_1 + A_{1M}) + \gamma_{10}\cdot(B_1 + B_{1M}) + C_1 + C_{1M} + E_{1M}\cdot\varphi_r] \end{aligned} \tag{A.95}$$

It is highlighted that

$$-2C_{M\alpha,10}\cdot(C_1 + C_{1M}) = -2C_{M\alpha,10}\left(\frac{F_{y0,1R}+F_{y0,1L}}{C_{F\alpha,10R}+C_{F\alpha,10L}}\right) - M_{z0,1R} - M_{z0,1L} \leftrightsquigarrow -2\cdot PRAT_1$$

A.8 Tire Parameters Estimation

A.8.1 Tire Contact Patch Dimension

In literature, there are different formulae that can be used to have an estimation for tire footprint dimensions. In [71], a formula for tire contact width w_{tire} is reported, depending on tire normal load F_z, stiffness k_t, nominal width S_N, and outer radius r_o

$$w_{tire} \approx S_N \cdot \left[q_{rb2} \frac{F_z}{r_o \cdot k_t} + q_{rb1} \left(\frac{F_z}{r_o \cdot k_t} \right)^{\frac{1}{3}} \right] \tag{A.96}$$

Nevertheless, the estimation of tire stiffness provided in [71] is not used in this dissertation, since it involves the use of a parameter value (k_t) that is not available for the tires analyzed in this book.

Therefore, the formula reported in [59] for the footprint width is here used (tire size 205/50 *R*17)

$$w_{tire} \approx \left(-0.004 \cdot AR + 1.03\right) \cdot S_N = 170.15 \text{ mm} \tag{A.97}$$

A.8.2 Tire Stiffness

The formula reported in [59] (from the work developed in [60]) for the estimation of tire stiffness has provided satisfactory results based on experiments performed on 1000 tires of different manufacturers having 144 different sizes, namely

- rim codes: from 13 to 24
- section widths: from 175 to 305 mm.

Therefore, it has been used in this book to estimate the tire stiffness, starting from information on inflation pressure p_i and tire dimensions (footprint width w_{tire} and outer diameter D_{out})

$$\begin{aligned} k_t\left[\text{kg/mm}\right] &= 0.00028 \cdot p_i\left[kPa\right]\sqrt{w_{tire}\left[\text{mm}\right] \cdot D_{out}\left[\text{mm}\right]} + 3.45 \\ k_t\left[\text{N/m}\right] &= k_t\left[\text{kg/mm}\right] \cdot 9.81\frac{\text{N}}{\text{kg}} \cdot 1000\frac{\text{mm}}{\text{m}} \\ k_{t1} &= 259884 \text{ N/m} \\ k_{t2} &= 232759 \text{ N/m} \end{aligned} \tag{A.98}$$

A.8.3 Effective Rolling Radius

The effective rolling radius has a value between the loaded radius and the unloaded radius. It can be estimated from the vertical load on the wheel F_z, the unloaded outer radius r_o, and the previously introduced tire stiffness k_t [71]

$$r_{eff} = r_\Omega - \frac{F_{z0}}{k_t}\left[F_{reff} \cdot \frac{F_z}{F_{z0}} + D_{reff} \cdot arctan\left(B_{reff} \frac{F_z}{F_{z0}} \right) \right] = 310.5 \text{ mm} \tag{A.99}$$

where r_Ω is the free tire radius of the rotating tire that can be estimated according to [71] as

$$r_\Omega = r_o \cdot \left[q_{reo} + q_{V1} \left(\frac{r_o \cdot \Omega}{V_0} \right)^2 \right] \approx r_o \cdot q_{reo} = 317.6 \text{ mm} \tag{A.100}$$

The values for the coefficients F_{reff}, D_{reff}, B_{reff}, and q_{reo} are chosen as indicated in [71] (**Table A.1**).

TABLE A.1 Tire parameters and data.

Tire data				
205/50 *R*17				
Variable name	**Description**	**Formula**	**Value**	**Unit**
$C_{F\gamma}$	Force camber coefficient	$c_5 \cdot F_z$	F_z	N/rad
c_5			1	
$C_{M\gamma}$	Moment camber coefficient*	$c_6 \cdot \frac{(w_{tire}/2)^2}{r_{eff}} \cdot c_8 \cdot F_z$	$0.1049 \cdot F_z$	Nm/rad
c_6			0.3	
c_8			15	
w_{tire}	Footprint width	$(-0.004 \cdot AR + 1.03) \cdot S_N$	170.15	mm
$p_{i,1}$	Front tire inflation pressure		250	kPa
$p_{i,2}$	Rear tire inflation pressure		220	kPa
AR	Aspect ratio		50	%
S_N	Nominal section width		205	mm
H	Section height	$S_N \cdot AR/100$	102.5	mm
D_{rim}	Rim diameter		431.8	mm
D_{out}	Wheel outer diameter	$2H + D_{rim}$	636.8	mm
k_{t1}	Front tire stiffness	$0.00028 \cdot p_{i,1}\sqrt{w_{tire} \cdot D_{out}} + 3.45$	259884	N/m
k_{t2}	Rear tire stiffness	$0.00028 \cdot p_{i,2}\sqrt{w_{tire} \cdot D_{out}} + 3.45$	232759	N/m
r_Ω	Free tire radius	$r_0 \cdot q_{reo}$	317.6	mm
r_{eff}	Effective rolling radius	$r_\Omega - \frac{F_{z0}}{k_t}\left(F_{reff} \cdot \frac{F_z}{F_{z0}} + D_{reff} \cdot a\tan\left(B_{reff}\frac{F_z}{F_{s0}}\right)\right)$	310.5	mm
r_o	Unloaded tire radius	$D_{out}/2$	318.4	mm
q_{reo}			0.9974	
F_{z0}	Nominal load		4000	N
F_{reff}			0.07394	
D_{reff}			0.25826	
B_{reff}			8.386	

* In order to guarantee that the same parameters are used for the analytical computations and the comparative simulations, the aligning moment camber coefficient $C_{M\gamma,iOL/R}$ values for the different sets of tires of *MSC Adams*™ full-vehicle model have been evaluated with *MSC Adams™/Tire*. These values, obtained by the *PAC2002 Tire Data and Fitting Tool (TDFT)*, have been used for computations of the analytical expressions and the single-track model. The values pertaining to road cross slope φ_r = 0.02 rad are reported in Table A.2.

TABLE A.2 $C_{My,iL/R}$ values for the different sets of tires, extracted from *MSC Adams*™/*Tire* and used for the simulations and analytical computations. Values pertaining to road cross slope $\varphi_r = 0.02$ rad.

[Nm/rad]	Normal	Mod-A	Mod-B	Buffed
$C_{My,1L}$	-23.0	-19.6	-10.6	138.9
$C_{My,1R}$	-23.9	-16.0	-12.3	140.7
$C_{My,2L}$	-40.6	-37.7	-34.4	58.5
$C_{My,2R}$	-40.4	-37.0	-35.2	62.0

A.9 Vehicle Parameters and Data

The data that were not fixed for the considered vehicle have been estimated with realistic values according to [72].

A.9.1 Inertias and Center of Gravity

An investigation on mass properties and correlations with vehicle parameters is reported in [61]. It resulted in formulae to predict moments of inertia and center of gravity position by means of easily accessible data. The interpolating formulae were derived and tested by means of measurements performed with the *InTenso+* test rig at Politecnico di Milano on 24 different car models between 2000 and 2016.

One of the formulae reported in [61] links the center of gravity height to the roof height of the vehicle. Even if the mean error is of the order of 70 mm, it is considered sufficiently accurate for this dissertation

$$h_G = h_{roof} \cdot 0.38 = 0.4944 \text{ m} \tag{A.101}$$

The interpolating formulae for the moments of inertia prediction are stated to provide more accurate results, and correlate mass, length and width of the vehicle to moments of inertia (data are summarized in Tables A.3 and A.4)

$$\begin{aligned} I_z &= C_z \cdot m_{tot}^{q_z} \cdot Length^{r_z} \cdot Width^{s_z} = 2208.1 \text{ kg m}^2 \\ C_z &= 0.47 \\ q_z &= 0.85 \\ r_z &= 1.39 \\ s_z &= 0.46 \end{aligned} \tag{A.102}$$

$$\begin{aligned} I_x &= C_x \cdot m_{tot}^{q_x} \cdot Length^{r_x} \cdot Width^{s_x} = 534.8 \text{ kg m}^2 \\ C_x &= 1.32 \\ q_x &= 0.71 \\ r_x &= 0.25 \\ s_x &= 0.78 \end{aligned} \tag{A.103}$$

TABLE A.3 Vehicle data.

Vehicle data			
Sport compact coupé			
Variable name	**Description**	**Value**	**Unit**
m_{tot}	Total mass	1488	kg
I_z	Yaw moment of inertia	2208.1	kg m^2
I_x	Roll moment of inertia	534.9	kg m^2
m_{t1}	Unsprung mass front (single wheel)	40	kg
m_{t2}	Unsprung mass rear (single wheel)	35	kg
l	Wheelbase	2.550	m
a_1	Distance CoG-Front axle	0.978	m
a_2	Distance CoG-Rear axle	1.572	m
h_G	CoG height	0.494	m
h_1	Front roll center	0.0588	m
h_2	Rear roll center	0.0796	m
$2s_1$	Front track width	1.450	m
$2s_2$	Rear track width	1.470	m
h_{roof}	Roof height	1.301	m
Length	Vehicle length	4.133	m
Width	Vehicle width	1.810	m
k_1	Front suspension spring stiffness	20000	N/m
k_2	Rear suspension spring stiffness	24000	N/m
c_1	Front suspension damping (compression)	3800	Ns/m
c_2	Rear suspension damping (compression)	1300	Ns/m

TABLE A.4 Elastokinematic parameters.

Elastokinematic data			
Variable name	**Description**	**Value**	**Unit**
$k_{\varphi 1}$	Front rolling stiffness	40544.2	Nm/rad
$k_{\varphi 2}$	Rear rolling stiffness	41851.7	Nm/rad
h'	Distance roll axis-CoG	0.427	m
ε_1	Roll-steering coefficient (front)	−0.0245	
ε_2	Roll-steering coefficient (rear)	−0.0909	
τ_1	Roll-camber coefficient (front)	0.8912	
τ_2	Roll-camber coefficient (rear)	0.9137	
n	Caster offset	0.02	m
$c_{sf,i}$	Suspension compliance coefficient	$-1.25 \cdot 10^{-6}$	rad/N
$c_{\psi 1}$	Steer compliance	25000	Nm/rad
ξ_γ	Camber coefficient linearization	1	1/rad

index

www.ingramcontent.com/pod-product-compliance
Lightning Source LLC
LaVergne TN
LVHW070852210926
848482LV00016B/916